AF333766

Deformation Structures and Processes within the Continental Crust

Geological Society books refereeing procedures

The Society makes every effort to ensure that the scientific and production quality of its books matches that of its journals. Since 1997, all book proposals have been refereed by specialist reviewers as well as by the Society's Books Editorial Committee. If the referees identify weaknesses in the proposal, these must be addressed before the proposal is accepted.

Once the book is accepted, the Society Book Editors ensure that the volume editors follow strict guidelines on refereeing and quality control. We insist that individual papers can only be accepted after satisfactory review by two independent referees. The questions on the review forms are similar to those for *Journal of the Geological Society*. The referees' forms and comments must be available to the Society's Book Editors on request.

Although many of the books result from meetings, the editors are expected to commission papers that were not presented at the meeting to ensure that the book provides a balanced coverage of the subject. Being accepted for presentation at the meeting does not guarantee inclusion in the book.

More information about submitting a proposal and producing a book for the Society can be found on its website: www.geolsoc.org.uk.

It is recommended that reference to all or part of this book should be made in one of the following ways:

LLANA-FÚNEZ, S., MARCOS, A. & BASTIDA, F. (eds) 2014. *Deformation Structures and Processes within the Continental Crust*. Geological Society, London, Special Publications, **394**.

BORTHWICK, V. E., PIAZOLO, S., EVANS, L., GRIERA, A. & BONS, P. D. 2014. What happens to deformed rocks after deformation? A refined model for recovery based on numerical stimulations. *In:* LLANA-FÚNEZ, S., MARCOS, A. & BASTIDA, F. (eds) *Deformation Structures and Processes within the Continental Crust*. Geological Society, London, Special Publications, **394**, 215–234. First published online December 9, 2013, http://dx.doi.org/10.1144/SP394.11

GEOLOGICAL SOCIETY SPECIAL PUBLICATION NO. 394

Deformation Structures and Processes within the Continental Crust

EDITED BY

S. LLANA-FÚNEZ, A. MARCOS and F. BASTIDA
Universidad de Oviedo, Spain

2014
Published by
The Geological Society
London

THE GEOLOGICAL SOCIETY

The Geological Society of London (GSL) was founded in 1807. It is the oldest national geological society in the world and the largest in Europe. It was incorporated under Royal Charter in 1825 and is Registered Charity 210161.

The Society is the UK national learned and professional society for geology with a worldwide Fellowship (FGS) of over 10 000. The Society has the power to confer Chartered status on suitably qualified Fellows, and about 2000 of the Fellowship carry the title (CGeol). Chartered Geologists may also obtain the equivalent European title, European Geologist (EurGeol). One fifth of the Society's fellowship resides outside the UK. To find out more about the Society, log on to www.geolsoc.org.uk.

The Geological Society Publishing House (Bath, UK) produces the Society's international journals and books, and acts as European distributor for selected publications of the American Association of Petroleum Geologists (AAPG), the Indonesian Petroleum Association (IPA), the Geological Society of America (GSA), the Society for Sedimentary Geology (SEPM) and the Geologists' Association (GA). Joint marketing agreements ensure that GSL Fellows may purchase these societies' publications at a discount. The Society's online bookshop (accessible from www.geolsoc. org.uk) offers secure book purchasing with your credit or debit card.

To find out about joining the Society and benefiting from substantial discounts on publications of GSL and other societies worldwide, consult www.geolsoc.org.uk, or contact the Fellowship Department at: The Geological Society, Burlington House, Piccadilly, London W1J 0BG: Tel. +44 (0)20 7434 9944; Fax +44 (0)20 7439 8975; E-mail: enquiries@geolsoc.org.uk.

For information about the Society's meetings, consult *Events* on www.geolsoc.org.uk. To find out more about the Society's Corporate Affiliates Scheme, write to enquiries@geolsoc.org.uk.

Published by The Geological Society from:
The Geological Society Publishing House, Unit 7, Brassmill Enterprise Centre, Brassmill Lane, Bath BA1 3JN, UK

The Lyell Collection: www.lyellcollection.org
Online bookshop: www.geolsoc.org.uk/bookshop
Orders: Tel. +44 (0)1225 445046, Fax +44 (0)1225 442836

The publishers make no representation, express or implied, with regard to the accuracy of the information contained in this book and cannot accept any legal responsibility for any errors or omissions that may be made.

British Library Cataloguing in Publication Data

A catalogue record for this book is available from the British Library.
ISBN 978-1-86239-627-2
ISSN 0305-8719

Distributors
For details of international agents and distributors see:
www.geolsoc.org.uk/agentsdistributors

Typeset by Techset Composition India (P) Ltd, Bangalore and Chennai, India
Printed by Berforts Information Press Ltd, Oxford, UK

Contents

Acknowledgements

The papers in this volume arise from a Deformation Mechanisms, Rheology and Tectonics meeting held at the University of Oviedo from the 31st August to the 2nd September. The conference was attended by 100 participants from 18 countries. We thank all participants at the conference for the lively discussion at the lecture room and in the fieldtrips before and after the meeting.

We would like to thank the Geological Society Publishing House for their help and advice during the creation of the volume.

We gratefully acknowledge the work and constructive criticism by our colleagues who helped with the reviewing of manuscripts submitted for this volume.

Juan Luis Alonso, University of Oviedo, Spain

Auke Barnhoorn, Technical University Delft, The Netherlands

Hans de Bresser, Utrecht University, The Netherlands

Antonio Casas Sainz, University of Zaragoza, Spain

Javier Escuder Viruete, Instituto Geológico y Minero de España, IGME, Spain

Manèl Fernández, Institute of Earth Sciences 'Jaume Almera', CSIC, Spain

Laura Giambiagi, CONICET, Argentina

Djordje Grujic, Dalhousie University, Canada

Peter Hudleston, University of Minnesota, USA

Robert A. Hunter, Cornell University, USA

David Iacopini, University of Aberdeen, UK

Andreas K. Kronenberg, Texas A&M University, USA

Gweltaz Maheo, University Claude Bernard Lyon, ENS Lyon, France

Santanu Misra, ETH-Zürich, Switzerland

Luiz F. G. Morales, GFZ Helmholtz-Zentrum Potsdam, Germany

Wolfgang Preiss, Department of Primary Industries and Regions South Australia (PIRSA), Australia

Nick Timms, Curtin University, Australia

Joseph C White, University of New Brunswick, Canada

and six anonymous reviewers.

Deformation structures and processes within the continental crust: an introduction

SERGIO LLANA-FÚNEZ*, ALBERTO MARCOS & FERNANDO BASTIDA

Departamento de Geología, Universidad de Oviedo, calle Arias de Velasco s/n, 33005 Oviedo, Spain

**Corresponding author (e-mail: slf@geol.uniovi.es)*

Abstract: The collection of research papers in this volume provides an overview of current trends in the study of deformation structures and deformation processes in rocks from the continental crust. The volume has been divided into three sections to reflect the current interest in structures and processes: shear zones and folds; the interactions between magmatism and structure; and the relationship between microstructure and rheology. The range of topics includes the geometry of natural structures from field observations, the detailed analysis of the microstructure of deformed rocks, the timing of deformation by various geochronological techniques, the modelling of deformation processes to constrain the development of structures in rocks, and the use of laboratory experimentation to reproduce deformation fabrics in rocks. The mixture of techniques and methods treated illustrate the broad and multidisciplinary approach to the study of rock deformation in the continental crust.

Tectonic forces acting upon rocks result in a very large range of deformation structures that vary according to the intrinsic mechanical properties of the rocks' mineral constituents, the anisotropy of the rock bodies, the existence of fluids inside the rocks, the physical conditions under which the deformation process takes place and the boundary conditions that constrain the development of the structures. From very early on in the history of structural geology, the determination of the exact geometry of deformation structure in rocks has remained at the core of the study of structures formed in nature (e.g. Leith 1913; Billings 1954). The significance of the geometry of structures became even more important when the first attempts to quantify strain and characterize its distribution throughout rocks were made by John Ramsay. The mathematical modelling of rock structures initiated by his classical book at the time of plate tectonics 'discovery' was groundbreaking (Ramsay 1967). Two types of structures can be identified and have been the main focus of the quantification of strain: shear zones and folds, the former representing the type of structure that localizes deformation, the latter a special type of distributed deformation within rock sequences under certain specific conditions (e.g. the presence of anisotropy and its orientation at low angles to maximum compressive stress). To understand why some rock sequences fold or shear under tectonic stress there was a need to understand the internal processes that occur in deforming rocks, particularly when trying to understand what process induces localization within a rock.

The internal architecture of rocks at microscopic scale and the fact that the microstructure is determinant in how deformation mechanisms operate, also at microscopic scale, have led to the development of new lines of study. These have considered a number of new processes that happen at the scale of grain boundaries or even within grains, which ultimately may control the mechanical behaviour of large volumes of rocks (e.g. Knipe & Rutter 1990; Gapais *et al.* 2005). For example, to understand mantle flow, the observations of seismic anisotropy recorded by indirect geophysical methods are complemented by the study of crystallographic preferred orientation of olivine and the processes that generate this fabric in mantle rocks exposed at the surface (e.g. Prior *et al.* 2011). However, there is another geodynamic setting where rock microstructures, the development of structures and the tectonic processes at larger scale are intermingled: orogenic belts. The intensity of deformation at various crustal depths and the large amount of accumulated tectonic displacement along discrete structures, some acting as plate boundaries, allow the exhumation of deep rocks, not necessarily showing the geological record of the developing orogen, but perhaps its inherited geological history. Nevertheless, orogenic belts are certainly natural laboratories, albeit complex, that enable the exploration of the relationship between deformation processes and tectonic structures at various scales across different depths within the crust.

Some simplified versions of rock structures can be reproduced in the laboratory by using cores of

From: LLANA-FÚNEZ, S., MARCOS, A. & BASTIDA, F. (eds) 2014. *Deformation Structures and Processes within the Continental Crust*. Geological Society, London, Special Publications, **394**, 1–6.
First published online March 20, 2014, updated March 21, 2014, http://dx.doi.org/10.1144/SP394.13

natural rocks, or synthetic rocks or rock analogues. Of course, this approach requires a priori knowledge of the deformation process that produces the structure in the first place and a basic knowledge of the expected response to tectonic stress under ambient deformation conditions (e.g. temperature, pressure, fluids). If experimental rock deformation covers mainly the microstructure development in deformed rocks, numerical modelling tests the deformation process by comparing the resulting structure with natural cases. Both experimental and numerical approaches can not only reproduce a known geometry but also give feedback on the deformation process that produces the structure. It is through experimental deformation and numerical modelling of structures that we may progress in the understanding of the kinematics and dynamics of the development of tectonic structures in rocks.

Structures: shear zones and folds

Ductile structures receive considerable attention in the literature, the continuity of strain in otherwise heterogeneously deformed rocks is being tackled by dividing the rock into smaller homogeneously deformed volumes. If the original fabric of the rock is preserved, part of the accumulated strain can be quantified by observing the linearity of features, changes of angles between linear elements or the measurement of ellipticity in originally spherical objects. For simplicity most classical techniques assume that volume changes during deformation are absent or minimal. **Lisle (2013)** in this volume presents new methods of quantifying finite strain in shear zones where there are volume changes during deformation. A result of his analysis, which is a study from Marloes in Wales, Great Britain, is that significant changes of volume during deformation in shear zones invalidate the common assumption that simple shear predominates in shear zones.

The distribution of strain remains at the core of the understanding of structures in deformed rocks. Unfortunately, the information acquired in the field is usually insufficient to know the strain pattern in folded layers. Accordingly, it is very difficult to understand the processes that operated during folding by the analysis of data derived only from field observations. In order to improve our understanding of folding of rocks, a complementary methodology is necessary based on experimental or theoretical models and subject to geometrical or mechanical assumptions. Kinematic models of folding involve the compliance of certain geometrical rules (e.g. preservation of orthogonal thickness or the area on the fold profile, existence of a neutral surface, folded surfaces without longitudinal strain,

etc.), from which it is possible to obtain the strain pattern on the fold profile. In mechanical models it is necessary to specify initial forces operating at the boundaries of the layers and their rheological behaviour during folding. Analytical or numerical methods can provide insights into the evolution of the folding layer. One limitation of these models is that they often apply a simple 2D approach (single layers or simple multilayers) to a very complex reality. The kinematic and mechanical models can actually be considered complementary. In the case of folds it is known that strain is heterogeneously distributed; this is inherent to the fact that folds require an anisotropic medium to develop. It is in the fine detail of how and where various strain geometries nucleate and develop that we can understand the evolution of folds. **Frehner & Exner (2013)** explore numerically the distribution of strain within buckle folds using the finite element method by considering the refraction patterns in the foliations within buckle folds. Some of their findings are compared to real folds in Palaeozoic rocks formed during the development of the Variscan Orogeny in NW Spain.

Progressive deformation during orogenesis can lead to the superposition of similar-looking structures, formed at different times but under the same ambient conditions. On the basis of the geometrical arrangements between bedding, cleavage and thrust faults, **Calvín-Ballester & Casas (2013)** separate two consecutive deformation episodes in rocks from the foreland fold-and-thrust belt of the Variscan Orogeny in the Iberian Range.

The contributions mentioned so far have dealt with structures that affect only small portions of the continental crust; in contrast, the next article deals with a fold that can be regarded as crustal-scale, given its kilometric size. There are countless numbers of similar structures but the outstanding characteristic of the example studied by **Weisheit *et al.* (2013)** is that the tectonic deformation of this structure occurred over several million years in two separate orogenic cycles during the Early and Late Palaeozoic. They demonstrate the long period of polyorogenic tectonic activity by using various dating techniques in rocks affected by the folding.

Magmatism and structure

Structures that localize deformation and that affect the whole thickness of the continental crust are very likely related to some form of magmatism, whether they act as conduits for magmas to ascend or whether the temperature anomaly associated with the intrusion of igneous rocks promotes mechanical weakening and localization of deformation.

The Carboneras Fault in Spain is a strike-slip fault with well-exposed fault rocks, mostly cataclasites and fault gauges typical of deformation conditions in the brittle regime. The quality of the structures associated with the faulting process was used by Rutter *et al.* (1986) as a type locality to define the various types of shear fractures associated with a fault, from Riedel shears to P foliation and Y shears. This work was contemporary with the classic work in the brittle regime by Chester & Logan (1986) in the Punchbowl Fault, related to the San Andreas Fault. The Carboneras Fault is a relatively recent structure and a limited amount of erosion allows the preservation of structures formed in the upper crust. It does have volcanism associated with its tectonic activity. In this volume **Rutter *et al.* (2013)** show that the structure has a crustal scale, and they relate volcanism to tectonic activity; in fact, Ar/Ar dating of the volcanic rocks allows them to constrain the duration of the tectonic activity. The authors also show a clear link between the surface geology and tectonic activity in the upper mantle and interpret the crustal-scale structure as a stretching transform fault system based on the existence of velocity discontinuities in the mantle underneath. In this respect, the fault is thought to channel the ascent of magmas.

The use of the timing of magmatism as a way to constrain tectonic activity is also the subject of the next article in the volume, by **Rodríguez-Méndez *et al.* (2013)**, in which they have been able to constrain the final stages of the Permian magmatism in the Pyrenees using the SHRIMP technique applied to zircons from diabase dykes intruding Devonian rocks.

As mentioned earlier, crustal-scale structures often act as conduits for the ascent and emplacement of magmas. **Oriolo *et al.* (2013)** focus their attention on the development of cross-strike structures in the flat-slab segment of the Andean chain and put constraints on the tectonic evolution of this segment after the Miocene as a consequence of magmatism. Again, magmatism is used to constrain the timing of structure development.

During intrusion of a magmatic body the most significant effect on the host rock is an increase in temperature, a first-order physical parameter that strongly controls the mechanical behaviour of rocks. The duration of magmatic episodes at the scale of an orogen can last for tens of millions of years, albeit regionally the duration of magmatic events is down to millions of years (e.g. Rubatto *et al.* 2013). On an even more detailed scale, considering just one single intrusive event, the duration does not commonly exceed 1 Ma. **Caggianelli *et al.* (2013)** examine the thermal effect that one single magmatic intrusion has on the rheology of the crustal rocks that host the intrusion. They show through a thermo-rheological model that the brittle–ductile transition is affected by a single intrusion and that the rheological effect fades away 1 Ma after the intrusion occurred. For the crust to sustain mechanical weakening, therefore, requires the continuing ascent of magmatic bodies. These authors suggest, however, that in the regional context of tectonic extension this single intrusive event may be sufficient for the nucleation of a normal fault in the crust, and that faulting will further thermally weaken the crust by exhuming relatively hot crustal material during extension.

The determination of the temperature during rock deformation is fundamental to understanding tectonic processes since most deformation mechanisms and mineral reactions are strongly temperature dependent. A long-standing issue making this difficult is that the predominant rock-forming mineral quartz does not vary in composition with temperature. Recently, however, the content of titanium in quartz has been related to temperature through a number of calibrations in several rocks types (Wark & Watson 2006; Grujic *et al.* 2011; Thomas *et al.* 2010). **Morgan *et al.* (2013)** revisit the various calibrations in the thermal aureole of the Ballachulish Igneous Complex in Scotland. This particular intrusion and its metamorphic imprint on the host rock sequence is particularly well known and documented, and very suitable for a double check of the various calibrations of the titanium content of quartz. The work reported here by these authors shows clearly the disparities between the various methods. It also highlights the important effect of a heterogeneous distribution of titantium within quartz, which in some cases may hinder the precise determination of temperature during deformation unless it can be constrained by other means.

Microstructure and rheology

Besides environmental factors, such as temperature and pressure, the mechanical properties of a rock are determined in the first place by its composition and secondly by the microscopic arrangement of its constituents. One very well-known example of the importance of the microstructure in the mechanical behaviour of rocks is that of calcium carbonate rocks: a calcite marble and a micritic limestone nominally have the same composition (e.g. 99% calcite) but different grain size and a different arrangement of minor phases or impurities. In the case of a recrystallized marble, minor phases may be found in isolated grains within the calcite aggregate. In the case of a micritic limestone, Solnhofen limestone being a classic example, the minor phases are found along grain boundaries. As indicated in the literature, the presence of impurities along grain boundaries affects their mobility by pinning, requiring further temperature rises to match the migration rates or pure aggregates. The consequence of impurities is to produce an offsetting of the operation of intracrystalline plasticity to higher temperatures (e.g. Walker *et al.* 1990; Herwegh & Kunze 2002). Rocks with similar composition may have different mechanical behaviours due to the arrangement of the rocks' constituents, these being impurities along grain boundaries in this case. This particular property is exploited by **Llana-Fúnez & Rutter (2014)** to study the crystallographic preferred orientation (CPO) of calcite in experimentally deformed rock specimens. The experiments were performed at higher temperatures than would correspond to intracrystalline plasticity in a pure calcite rock, favouring an increase in ductility and in deformation rates but insufficient to produce a switch in deformation mechanism, due to the reduced mobility of grain boundaries as a consequence of the presence of impurities. The rock specimens, made from cylinders of Solnhofen limestone, were deformed in different experimental configurations to widen deformation geometries and corresponding CPO patterns. The orientation data were acquired using the electron back-scattered diffraction (EBSD) technique and the full crystallographic orientation allowed relating strain geometry and CPO pattern with the distribution in 3D of seismic velocities and propagation directions of P and S waves.

The possibility of relating full CPO to the microstructure is further explored by **Piazolo & Jaconelli (2013)** to define slip systems that are active during crystal plasticity of sillimanite. These authors go on to relate CPO to the rheology of various types of sillimanite aggregates on a high temperature Grt–Sil–Bt gneiss from Greenland. The determination of the orientation of the crystal with respect to the orientation of the grain boundaries is what the authors use to recognize the activity of certain slip systems with respect to others. For this, they also employed the EBSD technique, which is now becoming a basic tool for characterizing rock fabric in ductilely deformed rocks.

So far in this section, we have seen an experimental study into the development of microstructure in rocks and a second contribution mostly based on the fine tuning of an analytical technique; in both cases the aim has been to relate microstructure to deformation mechanism. The study of microstructure in deformed rocks is also feasible by numerical simulation. **Borthwick *et al.* (2013)** explore numerically the effects that recovery recrystallization has in rocks, at temperatures close to the deformation temperature, after the deformation process has stopped. They use the Elle-platform to analyse systematically the progress and consequences of recrystallization on the basis of varying dislocation type, mobility and size of interaction volume, showing that even at a temperature lower than that of peak deformation, the microstructure can be modified during recovery.

The current volume samples a range of approaches in the study of deformation mechanisms, rheology and tectonics in the continental crust from the field, to the lab and inside the computer. The selection of manuscripts includes studies that focus on describing particular types of structures and others that are process-oriented. The content of the volume highlights the multidisciplinary approach required to progress in the understanding of tectonic structures in rocks within the continental crust.

Most of the papers contained in this volume were presented at the *Deformation Mechanisms, Rheology and Tectonics* (DRT) meeting held in Oviedo (Spain) between 31 August and 2 September 2011. We would like to thank all contributors and participants at the meeting. Funding for this meeting was provided by research grants CGL2010-14890, CSD2006-00041, CGL2011-23628 and CGL2011-13171-E from the Spanish Ministry of Economy and Competitiness, and CNG11-29 and UNOV-2011-CONG-13 from the Asturias Regional Government and the University of Oviedo. Additional funding to run the meeting was provided by the sponsors Xstrata Zinc, Cajastur, Ciuden, Tectask, NPA-Fugro, Instituto Xeolóxico de Laxe and the Oviedo City Council. Special thanks go to the reviewers and authors of this volume, the staff at the Geological Society (A. Hills, T. Anderson and R. Kriefman) and the Society Book Editor at the Geological Society (R. Stephenson), for the support and help given during the production of this volume.

References

BILLINGS, M. P. 1954. *Structural Geology*. Prentice-Hall, Englewood Cliffs, New Jersey.

BORTHWICK, V. E., PIAZOLO, S., EVANS, L., GRIERA, A. & BONS, P. D. 2013. What happens to deformed rocks after deformation? A refined model for recovery based on numerical simulations. *In*: LLANA-FÚNEZ, S., MARCOS, A. & BASTIDA, F. (eds) *Deformation Structures and Processes within the Continental Crust*. Geological Society, London, Special Publications, **394**. First published online December 9, 2013, http://dx.doi.org/10.1144/SP394.11

CAGGIANELLI, A., RANALLI, G., LAVECCHIA, A., LIOTTA, D. & DINI, A. 2013. Post-emplacement thermorheological history of a granite intrusion and surrounding rocks: the Monte Capanne pluton, Elba Island, Italy. *In*: LLANA-FÚNEZ, S., MARCOS, A. & BASTIDA, F. (eds) *Deformation Structures and Processes within the Continental Crust*. Geological Society, London, Special Publications, **394**. First published online November 20, 2013, updated February 5, 2014, http://dx.doi.org/10.1144/SP394.1

CALVÍN-BALLESTER, P. & CASAS, A. 2013. Folded Variscan thrusts in the Herrera unit of the Iberian Range (NE Spain). *In*: LLANA-FÚNEZ, S., MARCOS, A. & BASTIDA, F. (eds) *Deformation Structures and Processes within the Continental Crust*. Geological Society, London, Special Publications, **394**. First published online November 21, 2013, http://dx.doi.org/10.1144/SP394.3

CHESTER, F. M. & LOGAN, J. M. 1986. Implications for mechanical properties of brittle faults from observations of the Punchbowl Fault Zone, California. *Pure and Applied Geophysics*, **124**, 79–106, http://dx.doi.org/10.1007/BF00875720

FREHNER, M. & EXNER, U. 2013. Strain and foliation refraction patterns around buckle folds. *In*: LLANA-FÚNEZ, S., MARCOS, A. & BASTIDA, F. (eds) *Deformation Structures and Processes within the Continental Crust*. Geological Society, London, Special Publications, **394**. First published online November 21, 2013, http://dx.doi.org/10.1144/SP394.4

GAPAIS, D., BRUN, J. P. & COBBOLD, P. R. (eds) 2005. *Deformation Mechanisms, Rheology and Tectonics: from Minerals to the Lithosphere*. Geological Society, London, Special Publications, **243**, http://dx.doi.org/10.1144/GSL.SP.2005.243.01.20

GRUJIC, D., STIPP, M. & WOODEN, J. L. 2011. Thermometry of quartz mylonites: importance of dynamic recrystallisation on Ti-in-quartz re-equilibration. *Geochemistry, Geophysics, Geosystems*, **12**, http://dx.doi.org/10.1029/2010GC003368

HERWEGH, M. & KUNZE, K. 2002. The influence of nano-scale second-phase particles on deformation of fine grained calcite mylonites. *Journal of Structural Geology*, **24**, 1463–1478, http://dx.doi.org/10.1016/S0191-8141(01)00144-4

KNIPE, R. J. & RUTTER, E. H. (eds) 1990. *Deformation Mechanisms, Rheology and Tectonics*. Geological Society, London, Special Publications, **54**, http://dx.doi.org/10.1144/GSL.SP.1990.054.01.48

LEITH, C. K. 1913. *Structural Geology*. Henry Hold & Co., New York.

LISLE, R. J. 2013. Strain analysis in dilatational shear zones, with examples from Marloes, SW Wales. *In*: LLANA-FÚNEZ, S., MARCOS, A. & BASTIDA, F. (eds) *Deformation Structures and Processes within the Continental Crust*. Geological Society, London, Special Publications, **394**. First published online November 22, 2013, http://dx.doi.org/10.1144/SP394.7

LLANA-FÚNEZ, S. & RUTTER, E. 2014. Effect of strain geometry on the petrophysical properties of plastically deformed aggregates: experiments on Solnhofen limestone. *In*: LLANA-FÚNEZ, S., MARCOS, A. & BASTIDA, F. (eds) *Deformation Structures and Processes within the Continental Crust*. Geological Society, London, Special Publications, **394**. First published online January 27, 2014, http://dx.doi.org/10.1144/SP394.12

MORGAN, D. J., JOLLANDS, M. C., LLOYD, G. E. & BANKS, D. A. 2013. Using titanium-in-quartz geothermometry, geospeedometry to recover temperatures in the aureole of the Ballachulish Igneous Complex, NW Scotland. *In*: LLANA-FÚNEZ, S., MARCOS, A. & BASTIDA, F. (eds) *Deformation Structures and Processes within the Continental Crust*. Geological Society, London,

Special Publications, **394**. First published online November 25, 2013, http://dx.doi.org/10.1144/SP394.8

ORIOLO, S., JAPAS, M. S., CRISTALLINI, E. O. & GIMÉNEZ, M. 2013. Cross-strike structures controlling magmatism emplacement in a flat-slab setting (Precordillera, Central Andes of Argentina). *In*: LLANA-FÚNEZ, S., MARCOS, A. & BASTIDA, F. (eds) *Deformation Structures and Processes within the Continental Crust*. Geological Society, London, Special Publications, **394**. First published online November 21, 2013, http://dx.doi.org/10.1144/SP394.6

PIAZOLO, S. & JACONELLI, P. 2013. Sillimanite deformation mechanisms within a Grt-Sil-Bt gneiss: effect of pre-deformation grain orientations and characteristics on mechanism, slip-system activation and rheology. *In*: LLANA-FÚNEZ, S., MARCOS, A. & BASTIDA, F. (eds) *Deformation Structures and Processes within the Continental Crust*. Geological Society, London, Special Publications, **394**. First published online November 22, 2013, http://dx.doi.org/10.1144/SP394.10

PRIOR, D. J., RUTTER, E. H. & TATHAM, D. J. (eds) 2011. *Deformation Mechanisms, Rheology and Tectonics: Microstructures, Mechanics and Anisotropy*. Geological Society, London, Special Publications, **360**, http://dx.doi.org/10.1144/SP360.1

RAMSAY, J. G. 1967. *Folding and Fracturing of Rocks*. McGraw-Hill, New York.

RODRÍGUEZ-MÉNDEZ, L., CUEVAS, J., ESTEBAN, J. J., TUBÍA, J. M., SERGEEV, S. & LARIONOV, A. 2013. Age of the magmatism related to the inverted Stephanian-Permian basin of the Sallent area (Pyrenees). *In*: LLANA-FÚNEZ, S., MARCOS, A. & BASTIDA, F. (eds) *Deformation Structures and Processes within the Continental Crust*. Geological Society, London, Special Publications, **394**. First published online November 21, 2013, http://dx.doi.org/10.1144/SP394.2

RUBATTO, D., CHAKRABORTY, S. & DASGUPTA, S. 2013. Timescales of crustal melting in the Higher Himalayan Crystallines (Sikkim, Eastern Himalaya) inferred from trace element-constrained monazite and zircon chronology. *Contributions to Mineralogy and Petrology*, **165**, 349–372, http://dx.doi.org/10.1007/s00410-012-0812-y

RUTTER, E. H., MADDOCK, R. H., HALL, S. H. & WHITE, S. H. 1986. Comparative microstructures of natural and experimentally produced clay-bearing fault gouges. *Pure and Applied Geophysics*, **124**, 3–30, http://dx.doi.org/10.1007/BF00875717

RUTTER, E. H., BURGESS, R. & FAULKNER, D. R. 2013. Constraints on the movement history of the Carboneras Fault Zone (SE Spain) from stratigraphy and $^{40}Ar-^{39}Ar$ dating of Neogene volcanic rocks. *In*: LLANA-FÚNEZ, S., MARCOS, A. & BASTIDA, F. (eds) *Deformation Structures and Processes within the Continental Crust*. Geological Society, London, Special Publications, **394**. First published online November 20, 2013, http://dx.doi.org/10.1144/SP394.5

THOMAS, J. B., WATSON, B. E., SPEAR, F. S., SHEMELLA, P. T., NAYAK, S. K. & LANZIROTTI, A. 2010. TitaniQ under pressure: the effect of pressure and temperature on the solubility of Ti in quartz. *Contributions to*

Mineralogy and Petrology, **160**, 743–759, http://dx. doi.org/10.1007/s00410-010-0505-3

WALKER, A. N., RUTTER, E. H. & BRODIE, K. H. 1990. Experimental study of grain-size sensitive flow of synthetic, hot-pressed calcite rocks. *In*: KNIPE, R. J. & RUTTER, E. H. (eds) *Deformation Mechanisms, Rheology and Tectonics*. Geological Society, London, Special Publications, **54**, 259–284, http://dx.doi. org/10.1144/GSL.SP.1990.054.01.24

WARK, D. & WATSON, B. E. 2006. TitaniQ: a titanium-in-quart geothermometer. *Contributions to Mineralogy and Petrology*, **152**, 743–754, http://dx.doi.org/10. 1007/s00410-006-0132-1

WEISHEIT, A., BONS, P. D., DANISÍK, M. & ELBURG, M. A. 2013. Crustal-scale folding: Palaeozoic deformation of the Mt Painter Inlier, South Australia. *In*: LLANA-FÚNEZ, S., MARCOS, A. & BASTIDA, F. (eds) *Deformation Structures and Processes within the Continental Crust*. Geological Society, London, Special Publications, **394**. First published online November 22, 2013, http://dx.doi.org/10. 1144/SP394.9

Strain analysis in dilatational shear zones, with examples from Marloes, SW Wales

RICHARD J. LISLE

School of Earth and Ocean Sciences, Cardiff University, Cardiff CF10 3AT, UK
(e-mail: lisle@cardiff.ac.uk)

Abstract: New and existing methods for the analysis of finite strains in shear zones involving volume changes are reviewed. By assuming that the wall rocks are undeformed, the position gradients tensor can be determined from data derived from deformed passive markers with or without knowledge of the orientation of cleavage within the shear zone. The new methods are both algebraic and graphical and include those based on off-axis Mohr circles. Application of these methods to deformed sandstones at Marloes Sands reveals important volume changes that would invalidate an approach that assumes simple shear.

Shear zones are parallel-sided zones of high strain enclosed within undeformed or weakly deformed wall rocks. They are known to be important geological structures for accommodating crustal deformation on a wide range of scales. Although such zones have been recorded as far back as the nineteenth century (e.g. Teall 1885), these structures have been intensively studied since the publication of the landmark paper by Ramsay & Graham (1970).

The presence of strongly curving tectonic foliation and the deflection of marker layers across shear zones are often suggestive of kinematics involving boundary-parallel shearing. However, in spite of their name, it has been shown theoretically that deformation within shear zones is not necessarily limited to boundary-parallel simple shear (Fig. 1). Specifically, Ramsay & Graham (1970) show that the boundary conditions that characterize shear zones with undeformed walls can be satisfied by an anisotropic volume-change deformation, simple shear, or some combination of these. Any volume increase or decrease would lead to a widening or narrowing of the shear zone respectively and the potential volumetric changes have important implications in relation to mineralization sited within shear zones (e.g. Robert & Brown 1986; Hodgson 1989). Examples of analyses of dilatational shear zones are Hancock (1972), Ramsay (1980), Ramsay & Huber (1987), Mohanty & Ramsay (1994), Schwerdtner (1982), and Srivastava *et al.* (1995).

More general strain models for shear zones such as that involving a component of pure shear in combination with simple shear have been considered (e.g. Sanderson 1982; Coward & Potts 1983; Tikoff & Fossen 1993), but these models either lead to high strain/discontinuities along the boundaries of the shear zone or require straining of the wall rocks. These features are not observed in the examples dealt with in this paper.

It is claimed that shear-zone deformation is better visualized by the technique referred to as factorization. This involves considering the finite strain to be the combined product of two or more ideal deformation types, for example simple shear, pure shear and isotropic volume change (Tikoff & Fossen 1993). Factorization is a way of describing the deformation in terms of the relative contributions from these ideal types. However, to be worthwhile this factorization procedure requires information regarding the relative timing of these contributions, since the final outcome in terms of finite strain depends on the order of superimposition. Since data of this kind are rarely available, factorization requires assumptions to be made regarding the strain history. Therefore factorization leads often to arbitrary results that depend on the kinematic assumptions made. For this reason, no attempt is made to factorize the deformations discussed in this paper but instead discussion is limited to the states of finite strain.

The aims of this paper are to present new techniques for the analysis of shear zones including those that involve volume change, and to illustrate their use by analysing strain in some Variscan shear zones in some classic outcrops in SW Wales.

Finite strain and boundary conditions

A state of two-dimensional finite strain (Fig. 2a) can be described by equations expressing the final coordinates of points (x', y') as a function of their initial coordinates (x, y):

$$\begin{aligned} x' &= ax + by \\ y' &= cx + dy. \end{aligned} \tag{1}$$

The coefficients in these equations (a, b, c, d) define a tensor that transforms undeformed vectors

From: Llana-Fúnez, S., Marcos, A. & Bastida, F. (eds) 2014. *Deformation Structures and Processes within the Continental Crust*. Geological Society, London, Special Publications, **394**, 7–20.
First published online November 22, 2013, http://dx.doi.org/10.1144/SP394.7

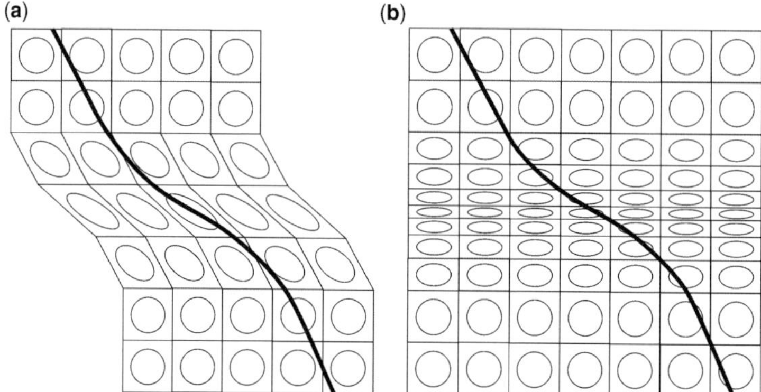

Fig. 1. Shear zones with distorted passive marker giving real and apparent lateral displacement: (**a**) simple shear zone; (**b**) dilatational shear zone.

into their deformed counterparts, using coordinates referred to the undeformed state:

$$\begin{bmatrix} a & b \\ c & d \end{bmatrix}. \tag{2}$$

This tensor has been given various names but here we refer to it as the position gradients tensor following Segel (1977) and Means (1983). If the finite strain is heterogeneous, some of the elements (a, b, c, d) will vary in value across the shear zone.

The changed position of a given point as a result of deformation can be described by displacement components, u and v (Fig. 2), where

$$u = x' - x = (a - 1)x + by \tag{3a}$$
$$v = y' - y = cx + (d - 1)y. \tag{3b}$$

The components u and v are the components of displacement vectors connecting the undeformed and deformed positions of points (Fig. 2c). These vectors have angles of slope β given by

$$\tan \beta = \frac{v}{u} = \frac{cx + (d - 1)y}{(a - 1)x + by}. \tag{4}$$

From Figure 2b it can be seen that the shear strain of lines parallel to the shear-zone boundary is

$$\gamma_x = u/y' = by/y' = by/(cx + dy). \tag{5}$$

The fractional volume change or dilatation, Δ, is related to the determinant of the position gradients tensor (Ramsay & Huber 1983, p. 287) according to

$$1 + \Delta = ad - bc. \tag{6}$$

The position gradients tensor shown in equation (2) is completely general. However, by taking into account the assumed boundary conditions of the shear-zone deformation, we find that the elements a and c of the tensor take on special values:

(a) The first boundary condition is that there is no wall-rock strain and no longitudinal strain of lines parallel to the x axis, that is, the shear-zone boundaries. This implies that $\partial u/\partial x$ equals zero. According to equation (3a), it follows that $a = 1$.

(b) There is no displacement in the y direction when $y = 0$, that is, $v = 0$ when $y = 0$. From equation (3b), it therefore follows that $c = 0$.

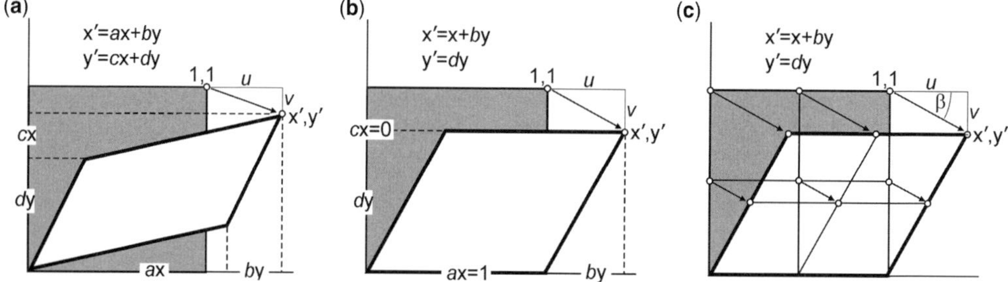

Fig. 2. Displacements and their relationship to elements a, b, c, d of the position gradients tensor: (**a**) general case; (**b**) shear zone with zero extension in the shear direction; (**c**) displacement vectors for the shear zone.

Since $a = 1$ and $c = 0$, the shear zones considered here have the following special properties:

(a) Since equation (4) simplifies to

$$\tan \beta = b/(d - 1) \qquad (7)$$

the directions of displacement vectors are independent of their positions, that is, all displacement vectors are parallel to each other (Fig. 2c).

(b) The shear strain equation (5) becomes

$$\gamma_x = b/d. \qquad (8)$$

(c) Equation (6) becomes

$$1 + \Delta = d \qquad (9)$$

that is the volume change, that produces a widening or narrowing of the shear zone, depends only on one element of the tensor.

(d) Therefore, the finite strain depends only on two quantities; the values of b and d.

The estimation of b and d from field data represents the aim of the techniques described in this paper. These estimates, together with the assumed values of a and c, allow the determination of standard finite strain parameters (e.g. strain ratio, ellipse orientation) to be derived, as explained in Ramsay & Huber (1983, appendix B).

Passive lines as shear strain markers

Equation (8) shows that knowledge of the shear strain parallel to the shear-zone boundary, γ_x, would help constrain the values of elements b and d. The shear strain γ_x has no simple meaning in relation to any possible component of simple shear to the deformation but is solely a measure of the angular rotation of a passive line originally perpendicular to the shear-zone boundary.

Single passive markers of general orientation do not provide sufficient information for the calculation of γ_x. The offset of the position of a vein on one boundary compared to the position of its straight line projection from the opposing wall can be referred to as the apparent displacement, u^* (Fig. 3). This distance, divided by the width of the shear zone, gives the apparent shear strain, γ^*.

It can be shown that the apparent shear strain, γ^*, depends on the difference in orientation of the marker inside and outside the shear zone, α and α' respectively and is given by

$$\gamma^* = \cot \alpha' - \cot \alpha. \qquad (10)$$

The relationship between the apparent shear strain, γ^*, and the true shear strain, γ_x, is clear from the following equations (derived in Appendix 1):

$$\gamma_x = \cot \alpha' - \cot \alpha/d, \text{ or} \qquad (11)$$

$$\gamma_x = \gamma^* + \cot \alpha (1 - 1/d). \qquad (12)$$

Thus the apparent shear strain only equals the true shear strain, γ_x, when $\alpha = 90°$ or when there is no volume change, that is, when $d = 1 + \Delta = 1$. Equation (12) indicates that a zero value of apparent shear strain shown by a particular passive marker (that does change direction on crossing the shear zone) does not imply zero shear strain, γ_x. It simply means that the finite displacement vectors in the shear zone are orientated parallel to the marker (Appendix A). Therefore, deductions regarding the age of marker layers such as dykes relative to deformation require caution.

Rearranging equation (11) gives

$$\cot \alpha' = \cot \alpha/d + \gamma_x, \qquad (13)$$

which describes the rotation of passive lines for the family of deformations considered here, and is therefore the equivalent of the well-known Wettstein's equation used for coaxial deformations (Ramsay & Huber 1983, p. 294).

It is apparent from equation (13), which involves the two unknowns d and γ_x, that a single deformed marker provides insufficient information for determining the true shear strain in the direction parallel to the shear-zone boundary, γ_x. Additional data are

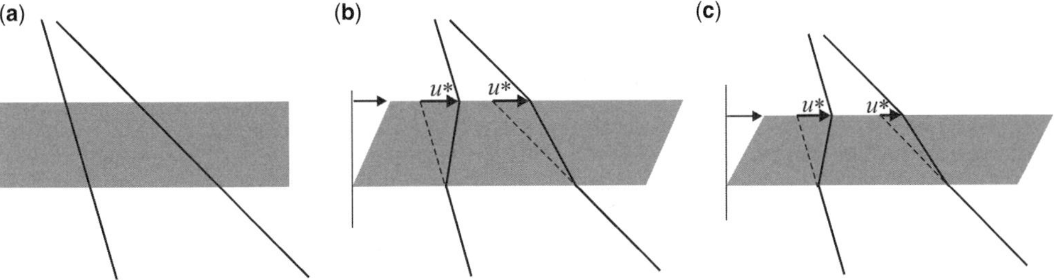

Fig. 3. Apparent shear strain: (**a**) undeformed state, two passive markers with different orientations; (**b**) simple shear deformation, displacements calculated from passive markers are correct; (**c**) dilatational shear zone, markers display inconsistent displacement estimates.

required to permit a complete analysis of the state of strain. Such information could be obtained from a second passive marker that is not parallel to the first. Alternatively, the rocks concerned may exhibit a foliation that provides evidence of the orientation of the maximum extension direction of the strain ellipse. Methods of analysing the strain from these two different datasets are given below.

Strain analysis using data consisting of two passive lines

Algebraic solution

The data consist of the inclination of two passive markers with respect to the shear-zone margin measured outside the shear zone, α_1 and α_2. In addition, γ_1^* and γ_2^*, the corresponding apparent shear strains for the two markers are required. The latter can be obtained by taking direct measurements of the offsets u_1^* and u_2^* (Fig. 3c) and dividing by the shear-zone width, or by calculation using angles α_1, α_2, and their deformed equivalents α_1', α_2' and equation (10).

Equation (12) can be written for each deformed marker.

$$\gamma_x = \gamma_1^* + \cot\alpha_1(1 - 1/d) \quad (14a)$$

$$\gamma_x = \gamma_2^* + \cot\alpha_2(1 - 1/d) \quad (14b)$$

Combining these two equations gives

$$\gamma_1^* + \cot\alpha_1(1 - 1/d) = \gamma_2^* + \cot\alpha_2(1 - 1/d)$$

and rearranging gives an equation for determining the dilatation:

$$d = 1 + \Delta$$
$$= (\cot\alpha_2 - \cot\alpha_1)/(\gamma_2^* - \gamma_1^* + \cot\alpha_2 - \cot\alpha_1).$$
$$(15)$$

After d is determined, it can be used in equation (14a) to solve for γ_x. Using equation (8), we can now determine b, the last remaining factor of the position gradients tensor, that is, $b = \gamma_x\, d$.

Ramsay (1980) presents another algebraic solution for the same problem.

To compute the strain ellipse parameters from the determined tensor elements a, b, c and d, we use the relations presented by Ramsay & Huber (1983, pp. 286–287). Since $a = 1$ and $c = 0$, these take simplified forms. The inclination of the maximum extension direction with respect to the shear-zone boundary, θ', is then given by

$$\tan 2\theta' = 2bd/(1 + b^2 - d^2) \quad (16)$$

and the principal stretches, s_1 and s_2, by

$$s_1, s_2 = \sqrt{\frac{p \pm \sqrt{p^2 - q}}{2}} \quad (17)$$

where $p = 1 + b^2 + d^2$ and $q = 4d^2$.

Mohr circle solution

Numerous constructions based on Mohr's circle have been proposed for geological strain analysis, for example, Brace (1961), Ramsay (1967), Means (1976), Allison (1984), Passchier (1990), Treagus (1990), Lisle (1992), Simpson & De Paor (1993), Ebner & Grasemann (2006) and Ragan (2009). Different variants of the Mohr construction have been devised specifically to deal with different strain tensors and various types of strain markers. For the solution of the current problem, which involves data consisting of the rotation angles of passive lines, a construction based on the Mohr circle for the reciprocal position gradients tensor (Means 1982, 1983) is useful.

This Mohr circle has the following properties. The circle passes through points each of which represents a deformed line of unique orientation. All lines are plotted as points in polar coordinates (r, θ) where r is the inverse stretch ($1/s$, where the stretch s is the ratio of deformed to original length) and θ equals ($\alpha' - \alpha$), that is, the angle of rotation of the line. In the present case where $a = 1$, and $c = 0$, the Mohr circle passes through points with Cartesian coordinates $(1, 0)$ and $(1/d, -b/d)$. These represent lines parallel to the x and y coordinate axes respectively, and these points plot diametrically opposite each other on the Mohr circle.

The angle between any pair of lines in the rock in the deformed state equals half the angle subtended at the Mohr circle's centre by the arc defined by the points representing the two lines. Therefore, if the points on the circle representing a pair of lines are each joined to an arbitrary third point on the circumference of the circle, the angle between the resulting lines on the Mohr diagram equals the real angle between those lines. Furthermore, if a certain position of the third point is chosen, the lines formed by joining the third point to points on the Mohr circle are parallel to the real lines represented by the points. The third point chosen this special way is called the pole of the Mohr circle.

The stages of the Mohr construction are:

(a) Plot the axes of the Mohr diagram with origin labelled O and an arbitrary point P on the positive part of x axis (Fig. 4a).

(b) Draw lines OA$_1$ and OA$_2$, through the origin O and inclined to the x axis at angles of $(\alpha_1' - \alpha_1)$ and $(\alpha_2' - \alpha_2)$ respectively (Fig. 4a).

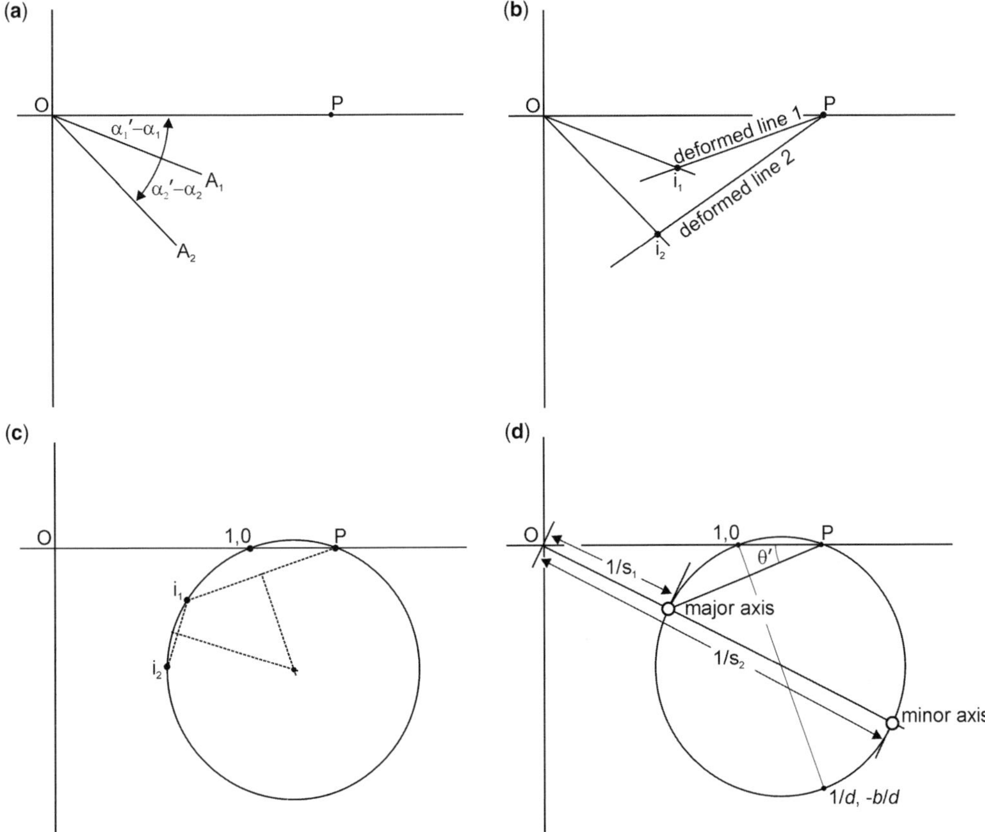

Fig. 4. A Mohr circle construction for strain analysis using two passive markers of different orientations. See text for explanation.

(c) The point P is assigned the status of the pole of the Mohr circle. Since the pole and the point (1, 0) both define a horizontal line, this implies that the line represented by point (1, 0), that is, the shear-zone boundary, is physically horizontal. Therefore we rotate the image of the shear zone so that boundary is parallel to the x axis of the Mohr diagram.

(d) Through the point P, draw lines parallel to the deformed orientations of the two markers in the rotated image. These will have inclination α_1' and α_2' with respect to the x axis, and intersect lines OA_1 and OA_2 at intersection points at i_1 and i_2 respectively (Fig. 4b).

(e) Construct the circle passing through the three points P, i_1 and i_2. This is the Mohr circle for the reciprocal positions gradients tensor (Fig. 4c). Besides at P, the circle usually cuts the x axis at a second point. This is point (0, 1). It defines the scale of the Mohr diagram.

(f) The lines of maximum and minimum elongation, s_1 and s_2 respectively, correspond to points on the circle closest to and most distant from the origin respectively. Their distances are equal to $1/s_1$ and $1/s_2$ respectively. The long axis orientation of the strain ellipse is found by joining the pole to the point representing the line of maximum elongation (Fig. 4d). The value of d, the proportional area change, is found from the x value of the point diametrically opposite to the point (1, 0).

Graphical solution

The most direct solution for the two-passive-line problem is a graphical one; it is also the simplest conceptually. This involves geometrically constructing material points from the intersection of the two passive markers present in the two undeformed walls of a shear zone. The offset of these points can be used as a marker of the displacement across the shear zone. This method is

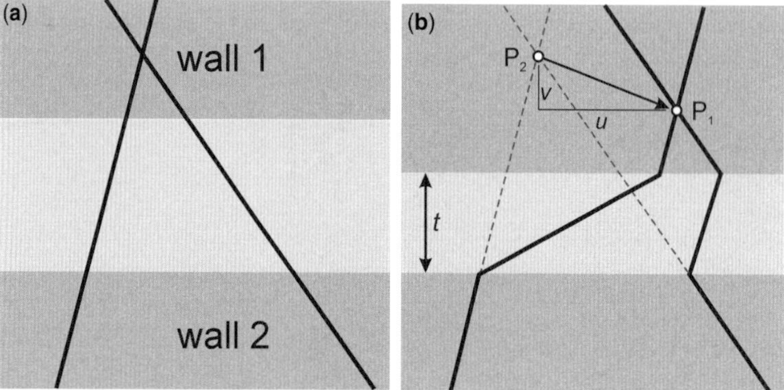

Fig. 5. (**a**) Undeformed state, two passive markers present in both walls of a future shear zone. (**b**) Deformed state, displacement is measured from the intersection points of the two markers in the two walls of the zone.

conceptually similar to techniques used for large-scale deformation in orogenic belts (Ramsay 1969) and for dyke dilation (Bussell 1989).

The construction involves the following steps:

(a) For both walls of the shear zone, labelled wall 1 and wall 2 in Figure 5a, determine the point of intersection of the two passive marker lines. These points, labelled P_1 and P_2 in Figure 5b, are material points associated with wall 1 and wall 2 respectively.

(b) The arrow pointing from P_2 to P_1 represents the displacement vector of the wall 1 relative to wall 2. This vector indicates the shear-zone normal and shear-zone parallel components of displacements u and v (Fig. 5b).

Measurements of u, v, as well as the zone thickness t, allow the shear strain in the direction of shear-zone boundary to be calculated as well as the volume change, that is

$$\gamma_x = u/t \tag{18}$$
$$d = v/t. \tag{19}$$

In addition, from equation (8)

$$b = uv/t^2.$$

Fig. 6. Conjugate shear zones at Marloes Sands, SW Wales. The shear zones on the lower edge of the photo are *c.* 15 cm wide.

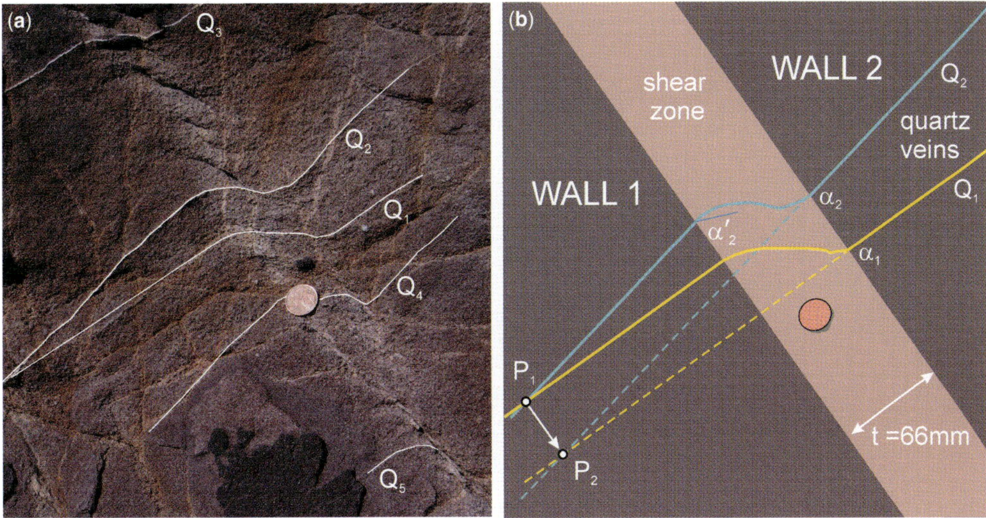

Fig. 7. (**a**) Shear zone at Marloes Sands where pre-shear-zone quartz veins (Q_1 to Q_5) serve as strain markers. (**b**) Graphical method. P_1, P_2 are points constructed by intersection of veins in wall 1 and wall 2, respectively, in order to determine the displacement vector (arrow).

The usual strain ellipse parameters s_1, s_2 and θ' are now found from equations (16) and (17).

In heterogeneous shear zones, the method could be applied to analyse the variation of strain and dilation across the zone. This would involve dividing the shear zone into a number of parallel strips, and using the method above on each pair of adjacent strips.

Examples

The methods above were applied to the well-known shear zones at Marloes Sands (SW Wales, Ordnance Survey map reference SM775077) described by Knipe & White (1979). The host rocks are coarse sandstones and conglomerates of Devonian age. The shear zones are perpendicular to bedding and the lack of stepping of the bedding plane indicates that the displacement vectors lie within the bedding plane. Therefore the bedding surfaces afford a true cross section of the shear zones (Fig. 6).

The shear zone in Figure 7 deforms thin quartz veins which serve as potential passive strain markers. A set of such veins seen in Figure 7 (veins Q_2, Q_3 and Q_4) make an angle of 100° with the shear-zone boundary outside the shear zone. They possess a curved form within the shear zone indicating heterogeneity of the internal strain, but make an average angle of 66° with the shear-zone boundary. One vein (Q_1 in Fig. 7) has an oblique trend and has an external angle of 89° and an average internal angle of 59°. The calculated apparent shear strains are 0.592 and 0.593 for vein Q_1 and vein Q_2 respectively.

For analysing the strain algebraically, the relevant data are: $\alpha_1 = 89°$, $\gamma_1^* = 0.592$ from vein Q_1 and $\alpha_2 = 100°$, $\gamma_2^* = 0.593$ from vein Q_2. Use of equation (15) gives a value of d of 1.005 corresponding to a negligible volume change, whilst equation (14) delivers a shear strain value γ_x of 0.592, and a b value of 0.595. Using equations (16) and (17) these tensor elements can be used to

Table 1. *Results of strain analyses of the shear zone in Figure 7a using three different methods*

	b	d	γ_x	s_1	s_2	Rs	θ'
Algebraic method	0.60	1.01	0.59	1.32	0.75	1.80	37°
Mohr circle method	0.66	1.13	0.58	1.44	0.78	1.85	41°
Graphical method	0.59	1.00	0.59	1.34	0.75	1.79	37°

Note: b and d are elements of the position gradients tensor; γ_x, shear strain in the direction of the shear-zone boundary; s_1 and s_2, principal stretches; Rs, strain ratio; θ', inclination of the s_1 direction to the shear strain boundary.

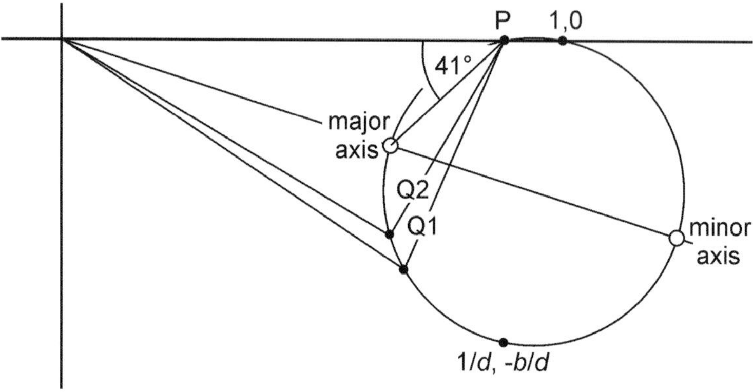

Fig. 8. Strain analysis of the shear zone in Figure 7a using the Mohr circle construction for the reciprocal position gradients tensor, illustrated in Figure 4. The similar orientations of veins Q1 and Q2 mean that the drawn circle is not tightly constrained and therefore neither are the estimates of the tensor elements b and d.

derive an average strain ellipse with $s_1 = 1.32$, $s_2 = 0.75$ and orientation $\theta' = 37°$ (Table 1).

The solutions obtained by the Mohr circle method and shown in Figure 8 are $d = 1.13$ (13% volume increase), $b = 0.66$ and average strain ellipse with $s_1 = 1.44$, $s_2 = 0.78$ and orientation $\theta' = 41°$ (Table 1).

The graphical method (Fig. 7b) gives hanging wall and footwall reference points with displacement components $u = 39$ mm, $v = 0$ mm. These displacements for a shear zone of width 66 mm produces a dextral shear strain of 0.59 and zero volume change, that is, $d = 1$. The strain ellipse parameters are given in Table 1.

The comparison of results produced by the three methods (Table 1) shows some variation, especially for d, the volume-change parameter. The graphical method (Fig. 7b) suggests how the quality of these sample data could be responsible for this variation. The two passive markers have similar

Fig. 9. Shear zone at Marloes Sands. The displacements associated with the shear zone caused a deflection of a set of older quartz veins. The strain concentration within the zone has also resulted in the formation of a foliation (pressure-solution cleavage) developed only in the shear zone. The shear zone is approximately 20 cm wide.

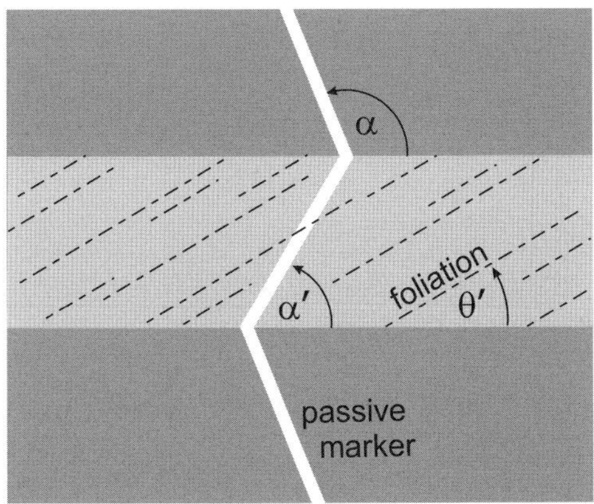

Fig. 10. An alternative dataset for strain estimation in shear zones: θ', the angle of inclination of the cleavage with respect to the shear-zone boundary; α, α' are angular inclinations of a single passive marker measured externally and internally, respectively.

orientations that produce instability of the intersection points used for determination of the displacement. The fact that the markers make a high angle with the shear-zone boundary further means that the instability affects the shear-zone normal components more than the shear-zone parallel components. Therefore any non-linearity of the markers or measurement errors will cause spurious estimates in the dilatational aspect of the strain. This

problem influences the results of all three methods, though the advantage of the graphical method is that it gives a visual appreciation of the suitability of the data being used.

Strain analysis from one passive line marker and the cleavage trace

This section treats the case of shear zones containing two different types of strain marker: a single passive marker and a foliation developed within the shear zone (Fig. 9). The necessary data (Fig. 10) consist of the external and internal angles of the passive marker, α and α' respectively, as well as the angle of inclination of the foliation with respect to the shear-zone boundary, θ'. Ramsay (1980) and Vitale & Mazzoli (2010) have previously tackled this problem by assuming that the internal foliation is a structure related to the strain within the shear zone and its trace is aligned parallel to the long axis of the finite strain ellipse.

Algebraic solution

Referring to Figure A1b and equation (1) the following relationships between the data (apparent shear strain, γ^*, α and θ') the unknowns b and d can be derived:

$$\gamma^* = (\cot\alpha + b - d\cot\alpha)/d \qquad (20)$$

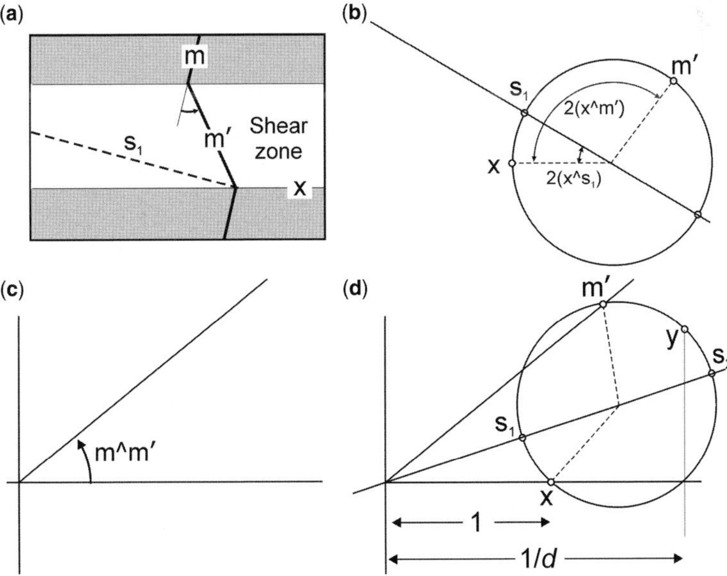

Fig. 11. The Mohr circle construction for strain determination from a deformed passive line and the cleavage orientation. See text for details.

Fig. 12. Shear zones at Marloes Sands with internal cleavage. A quartz vein serves as a passive strain marker. The coin has diameter 22 mm.

$$b = d\gamma^* + d\cot\alpha - \cot\alpha. \tag{21}$$

Assuming the cleavage trace to be parallel to the strain ellipse long axis,

$$\tan 2\theta' = 2bd/(1 + b^2 - d^2). \tag{22}$$

Using equation (21), we substitute for b to give

$$\tan 2\theta' = \frac{2(d\gamma^* + d\cot\alpha - \cot\alpha)d}{1 + (d\gamma^* + d\cot\alpha - \cot\alpha)^2 - d^2}. \tag{23}$$

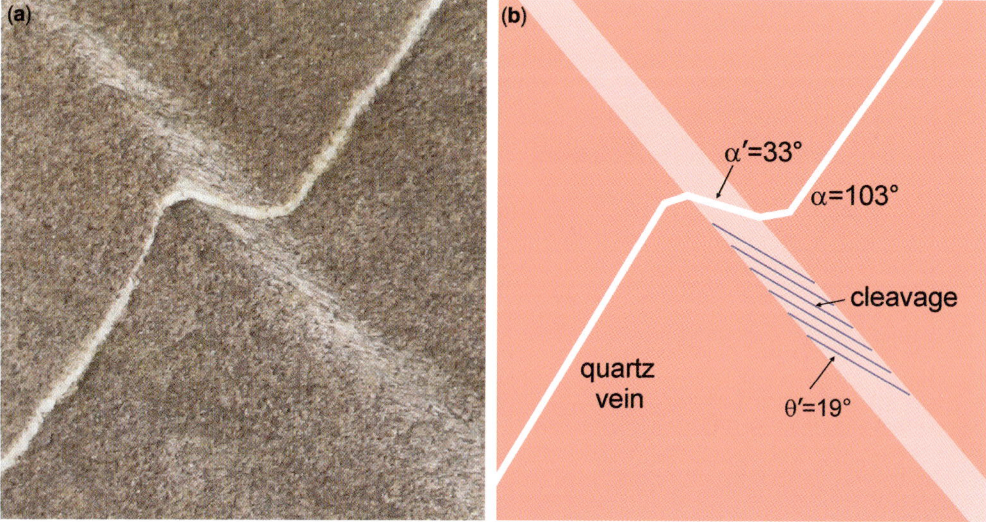

Fig. 13. (**a**) Close-up of Figure 12; (**b**) angular measurements used as data.

Table 2. *Results of strain analysis of shear zones in Figure 12*

	b	d	γ_x	s_1	s_2	Rs
Algebraic method	1.25	0.66	1.89	1.69	0.39	4.30
Mohr circle method	1.28	0.68	1.88	1.67	0.40	4.18

Expanding and rearranging produces a quadratic of the form:

$$pd^2 + qd + r = 0. \qquad (24)$$

where

$$p = \tan 2\theta'(\gamma^{*2} + \cot^2 \alpha + 2\gamma^* \cot \alpha - 1) - 2\gamma^* - 2\cot \alpha$$

$$q = 2\cot \alpha (1 - \cot \alpha \tan 2\theta' - \gamma^* \tan 2\theta')$$

$$r = \tan 2\theta'(1 + \cot^2 \alpha).$$

The roots of this quadratic (equation 24) give two potential values of $d = 1 + \Delta$ and the use of these values in equation (24) delivers two corresponding values of the element b. One of these solutions will be geologically viable. Principal strains are obtained from equation (17).

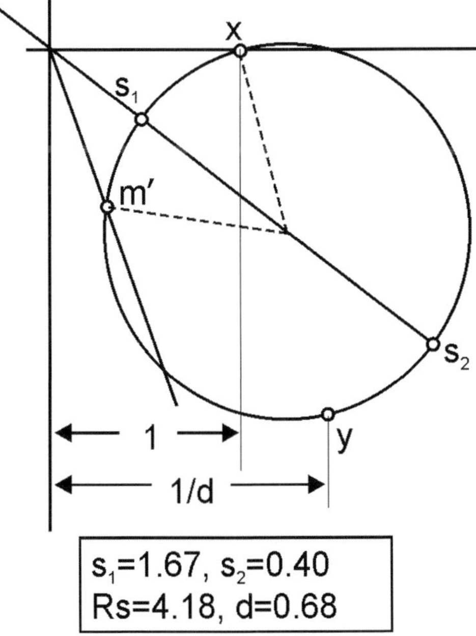

Fig. 14. Strain analysis of the shear zone in Figure 13 using the Mohr circle method.

Mohr construction

Since the orientation of the cleavage is measured in the deformed state, a convenient description of the deformation is the reciprocal position gradients tensor as described in the section 'Mohr circle solution'. The steps in the construction are:

(a) Record the angles of the cleavage, deformed marker (m') and undeformed marker (m) with the shear-zone boundary, the latter being used as the direction of the x axis of the Mohr diagram (Fig. 11a).

(b) Draw a circle on a transparent overlay, plot an arbitrary point x to represent the direction of the shear zone, and show the positions of the cleavage (s_1) and deformed marker m' as points on the circle at their doubled angles with x (Fig. 11b). Draw a long line by extending the radius through s_1.

(c) Draw the axes of the Mohr diagram and the rotation line: a line passing through the origin and inclined to the x axis at an angle equal to the rotation angle and rotation sense between m and m' (Fig. 11c).

(d) Overlay the Mohr circle of Figure 11b on the Mohr diagram (Fig. 11d). Rotate and move the overlay until x plots on the x axis, until m' lies on the rotation line and until the line passing through s_1 passes through the origin.

(e) When this is achieved, the coordinates of point x are (1, 0) and hence the scale of the diagram is fixed.

(f) The distances of the points labelled s_1, s_2 from the origin are equal to $1/s_1$, $1/s_2$ respectively. The principal stretches and their ratio are therefore determined.

(g) The point on the circle diametrically opposite to point x represents the y axis and has coordinates $(-b/d, 1/d)$. Hence d, the proportional volume change, can be found. Alternatively, d equals the product $s_1 s_2$.

(h) The fact that the centre of the Mohr circle plots above the x axis in Figure 11d indicates a sinistral shear sense.

Examples

The shear zones used are again from Marloes Sands (Fig. 12). They are thin (around 10 mm) and unlike many at this locality (cf. Fig. 6), tension gashes are notably absent. Pre-shearing quartz veins serve as passive markers. Figure 13 shows the angular measurements used in the estimating of strain, and the fact that these were taken in the centre of the shear zone.

If we consider the deformation in this shear zone to be simple shear, the small angle of inclination (19°) of the cleavage with respect to the shear-zone boundary would suggest a shear strain value of 2.55 with a strain ratio of 8.33. The algebraic method produces strain estimates (Table 2) that the deformation deviates significantly from simple shear and indicates a 34% volume decrease within the shear zone. The strain ratio is about half of that expected for simple shear. The Mohr circle method (Fig. 14) produces very similar results (Table 2).

Conclusions

This paper has presented a number of new techniques for strain analysis in shear zones and the treatment of datasets of various types. If two non-parallel passive markers are present, the graphical method is a direct visual method for estimating components of displacement parallel and perpendicular to the shear-zone boundary. This simple method has the advantage that it helps the user evaluate the quality of the data being used. The ideal markers are mutually perpendicular, whilst pairs of near-parallel markers place loose constraints on the strain components. In the latter case, markers at a high angle to the shear zone boundary lead to poor estimates of the normal components of displacements; markers at a low angle are not reliable for estimating the magnitudes of the shear-zone parallel components of movement.

When strain in the shear zone produces a foliation, its orientation provides useful data to help constrain the strain using two new methods.

The analysis of shear zones from SW Wales has yielded finite strain estimates, some of which involve significant volume changes and therefore are incompatible with a simple-shear deformation. Although such strains can be factorized into simple-shear and pure-shear components, this exercise has little real relevance in the absence of information on the details of the strain history.

Thanks are due to David Iacopini and an anonymous reviewer, both of whom provided helpful comments, and to Fernando Bastida for handling the reviewing process.

Appendix A: Deformation of passive lines in non-isochoric shear zones

We consider a homogeneously strained strip inside a shear zone (Fig. A1). Initially, the strip has unit width (Fig. A1a), but broadens or narrows to a width d as a result of volume change (Fig. A1b). A passive marker initially inclined to the shear-zone boundary at an angle α initially (Fig. A1a) undergoes rotation to an angle α' (Fig. A1b). Point A located on the boundary of the strip and on the passive marker undergoes a displacement to a new position B (Fig. A1b). The displacement vector is inclined at angle β to the shear-zone boundary.

The apparent shear strain, γ^*, across the strip equals the apparent displacement component u^* (distance CD) divided by the width d, where

$$u^* = \text{EB} - \text{EC} = d \cot \alpha' - d \cot \alpha \qquad \text{(A1)}$$

and therefore

$$\gamma^* = \cot \alpha' - \cot \alpha. \qquad \text{(A2)}$$

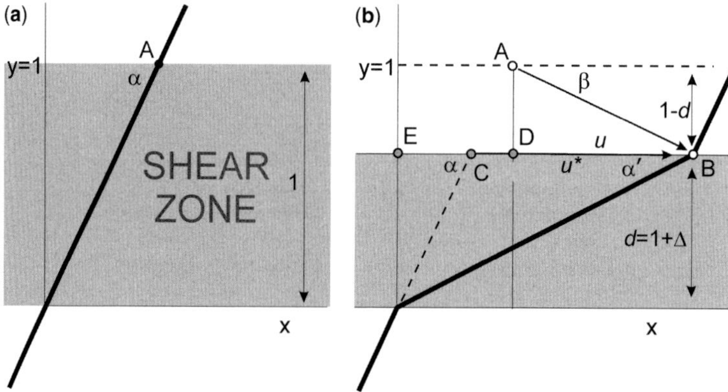

Fig. A1. True and apparent shear strain.

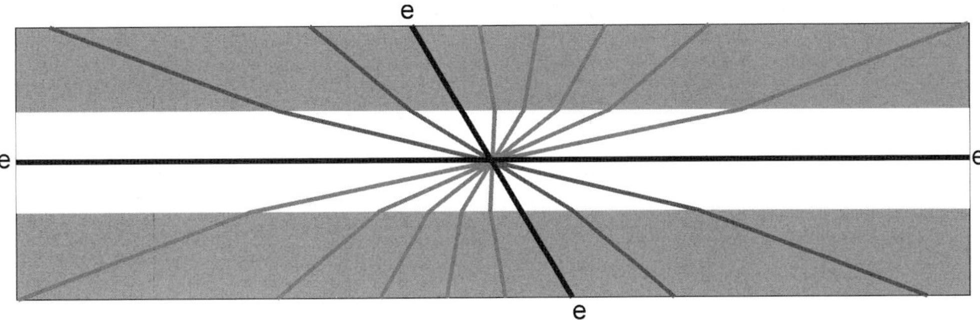

Fig. A2. Shear zone (white) with $d = 0.6$, $u = 0.2$ dextral. Red lines are passive markers with the 'wrong' sense of shear; green passive markers have sense of shear corresponding to the simple shear component. Blue lines are the eigenvectors of the deformation and define orientation fields of passive lines with different senses of shear rotation.

The true shear strain γ_x equals the true displacement u divided by the width d where $u = \mathrm{BD} = \mathrm{BC} - \mathrm{CD} = \mathrm{BC} - \cot \alpha \, (1 - d)$

Therefore,

$$\gamma_x = \gamma^* + \cot \alpha (1 - 1/d). \tag{A3}$$

If vector AB in Figure A1b is parallel to the marker in its undeformed orientation, α, the marker will remain straight so that $\alpha' = \alpha$. The apparent shear strain will be zero. For this special case, the marker is aligned with the displacement vector, that is, α is equal to α'. From Fig. A1b it can be seen that this places the following constraint on the ratio of shear strain to dilatation:

$$\cot \beta = \gamma_x / (1 - 1/d).$$

Markers will show different senses of apparent shear depending on whether α is greater or less than β. The special case of passive markers that have the same direction before and after deformation can be analysed using the concept of eigenvectors. The transformation matrix of the family of deformation types considered here has the form:

$$\begin{bmatrix} 1 & b \\ 0 & d \end{bmatrix}.$$

This matrix can be used to transform vectors from the undeformed to the deformed state.

$$\begin{bmatrix} 1 & b \\ 0 & d \end{bmatrix} \begin{bmatrix} x \\ y \end{bmatrix} = \begin{bmatrix} x' \\ y' \end{bmatrix}.$$

To determine the vectors that remain in the same direction, called the eigenvectors of the matrix, we first calculate the eigenvalues, λ, which produce zero value for the determinant

$$\begin{vmatrix} 1 - \lambda & b \\ 0 & d - \lambda \end{vmatrix}$$

so that $(1 - \lambda)(d - \lambda) = 0$. This is a quadratic whose roots give the values of the eigenvalues equal to d and 1 for the considered deformations.

If a given vector is an eigenvector, the following equation holds:

$$\begin{bmatrix} 1 - \lambda & b \\ 0 & d - \lambda \end{bmatrix} \begin{bmatrix} x \\ y \end{bmatrix} = 0$$

that is $(1 - \lambda)x + by = 0$

$$(d - \lambda)y = 0.$$

Substituting $\lambda = d$ and $\lambda = 1$ into the latter gives the two eigenvectors. The first is $\begin{bmatrix} b \\ 1 - d \end{bmatrix}$ and the second is parallel to the x axis.

These eigenvectors also define the range of orientations of passive lines showing different senses of rotation and thus different signs for the apparent shear strain, γ^* (Fig. A2). Simpson & De Paor (1993) discuss the eigenvectors of shear-zone deformations. On the Mohr diagram (e.g. Figs 4, 8 & 11) the points where the Mohr circle cuts the x-axis represent the eigenvectors.

References

ALLISON, I. 1984. The pole of the Mohr diagram. *Journal of Structural Geology*, **5**, 331–333.

BRACE, W. F. 1961. Mohr construction in the analysis of large geologic strain. *Bulletin of the Geological Society of America*, **72**, 1059–1079.

BUSSELL, M. A. 1989. A simple method for the determination of the dilation direction in intrusive sheets. *Journal of Structural Geology*, **11**, 679–687.

COWARD, M. P. & POTTS, G. J. 1983. Complex strain patterns developed at the frontal and lateral tips to shear zones and thrust zones. *Journal of Structural Geology*, **5**, 383–400.

EBNER, M. & GRASEMANN, B. 2006. Divergent and convergent non-isochoric deformation. *Journal of Structural Geology*, **28**, 1725–1733.

HANCOCK, P. L. 1972. The analysis of en echelon veins. *Geological Magazine*, **109**, 269–276.

HODGSON, C. J. 1989. The structure of shear-related, vein-type gold deposits: a review. *Ore Geology Reviews*, **4**, 231–273.

KNIPE, R. J. & WHITE, S. H. 1979. Deformation in low grade shear zones in the Old Red Sandstone, S.W. Wales. *Journal of Structural Geology*, **1**, 53–66.

LISLE, R. J. 1992. Strain analysis by simplified Mohr constructions. *Annales Tectonicae*, **5**, 102–117.

MEANS, W. D. 1976. *Stress and Strain*. Springer, New York.

MEANS, W. D. 1982. An unfamiliar Mohr circle construction for finite strain. *Tectonophysics*, **89**, T1–T6.

MEANS, W. D. 1983. Application of the Mohr-circle construction to problems of inhomogeneous deformation. *Journal of Structural Geology*, **5**, 279–286.

MOHANTY, S. & RAMSAY, J. G. 1994. Strain partitioning in ductile shear zones: an example from a Lower Pennine nappe of Switzerland. *Journal of Structural Geology*, **16**, 663–676.

PASSCHIER, C. W. 1990. Reconstruction of deformation and flow parameters from deformed vein sets. *Tectonophysics*, **180**, 185–199.

RAGAN, D. M. 2009. *Structural Geology: An Introduction to Geometrical Techniques*. 4th edn. Cambridge University Press, New York.

RAMSAY, J. G. 1967. *Folding and Fracturing of Rocks*. McGraw-Hill, New York.

RAMSAY, J. G. 1969. The measurement of strain and displacement in orogenic belts. *In*: KENT, P. E., SATTERTHWAITE, G. E. & SPENCER, A. M. (eds) *Time and Place in Orogeny*. Geological Society, London, Special Publications, **3**, 43–79.

RAMSAY, J. G. 1980. Shear zone geometry: a review. *Journal of Structural Geology*, **2**, 83–99.

RAMSAY, J. G. & GRAHAM, R. H. 1970. Strain variation in shear belts. *Canadian Journal of Earth Sciences*, **7**, 786–813.

RAMSAY, J. G. & HUBER, M. I. 1983. *The Techniques of Modern Structural Geology Vol 1: Strain Analysis*. Academic Press, London.

RAMSAY, J. G. & HUBER, M. I. 1987. *The Techniques of Modern Structural Geology Vol 2: Folds and Fractures*. Academic Press, London.

ROBERT, F. & BROWN, A. C. 1986. Archean gold-bearing quartz veins at the Sigma Mine, Abitibi greenstone belt, Québec. Part 2, Geologic relations and formation of the vein system. *Economic Geology*, **81**, 578–592.

SANDERSON, D. J. 1982. Models of strain variations in nappes and thrust sheets: a review. *Tectonophysics*, **88**, 201–233.

SCHWERDTNER, W. M. 1982. Calculation of volume change in ductile band structures. *Tectonophysics*, **4**, 57–62.

SEGEL, L. A. 1977. *Mathematics applied to Continuum Mechanics*. Macmillan, New York.

SIMPSON, C. & DE PAOR, D. G. 1993. Strain and kinematic analysis in general shear zones. *Journal of Structural Geology*, **15**, 1–20.

SRIVASTAVA, H. B., HUDLESTON, P. & EARLEY, D. 1995. Strain and possible volume loss in a high grade ductile shear zone. *Journal of Structural Geology*, **17**, 1217–1231.

TEALL, J. J. H. 1885. The metamorphosis of dolerite into hornblende schist. *Quarterly Journal of the Geological Society of London*, **41**, 133–145.

TIKOFF, B. & FOSSEN, H. 1993. Simultaneous pure and simple shear: the unifying deformation matrix. *Tectonophysics*, **217**, 267–283.

TREAGUS, S. H. 1990. The Mohr diagram for three-dimensional reciprocal stretch vs rotation. *Journal of Structural Geology*, **12**, 383–395.

VITALE, S. & MAZZOLI, S. 2010. Strain markers of heterogeneous ductile shear zones on the attitudes of planar markers. *Journal of Structural Geology*, **32**, 321–329.

Strain and foliation refraction patterns around buckle folds

MARCEL FREHNER[1]* & ULRIKE EXNER[2,3]

[1]*Geological Institute, ETH Zurich, Sonneggstrasse 5, 8092 Zurich, Switzerland*

[2]*Department for Geodynamics and Sedimentology, University of Vienna, Althanstrasse 14, 1090 Vienna, Austria*

[3]*Natural History Museum, Burgring 7, 1010 Vienna, Austria*

**Corresponding author (e-mail: marcel.frehner@erdw.ethz.ch)*

Abstract: Axial plane foliation associated with geological folds may exhibit a divergent or convergent fan. Commonly, the foliation is assumed to reflect the major principal finite strain orientation. Here, the strain orientation around numerically simulated single-layer buckle folds is analysed in detail. Four different strain measures are considered: (1) finite strain, (2) infinitesimal strain, (3) incremental strain (recording the strain history from a certain shortening value until the end), and (4) initially layer-perpendicular passive marker lines. In the matrix at the outer arc of the fold, all strain measures result in similar divergent fan patterns. Therefore, divergent foliation fans around natural folds cannot readily be associated with the finite strain orientation as they may reflect other strain measures. In the simulated folds, the convergent fans in the stronger layer show differences between the different strain measures, which are associated with a 90°-switch of the major principal strain from a layer-perpendicular to a layer-parallel orientation at the outer arc. A similar observation is made in one of three studied natural folds (near Ribadeo and Luarca, NW Spain). It is suggested that the convergent foliation fan pattern inside a fold is better suited for strain estimates than the divergent fan around a fold.

Rock deformation and foliation development are two intimately coupled processes. For example, the axial plane foliation in folded rocks has a very characteristic geometrical relationship to the folds, which can be used by field geologists to determine the geometry of unexposed parts of folds. Commonly, the foliation is refracted across boundaries between lithologies of different competence. The refraction angle has been related through analytical expressions to the viscosity ratio between the two lithologies (Treagus 1983, 1988, 1999; Talbot 1999; Mulchrone & Meere 2007), very similar to the refraction of optical light. Treagus & Sokoutis (1992) performed laboratory experiments in viscous materials to test the previously derived analytical expressions in simple-shear deformation. The findings of these studies have been applied, for example, to infer the relative strengths of metamorphic rocks in the Appalachians (Groome & Johnson 2006). To relate the analytical expressions and the laboratory experiments to the natural foliation refraction, all of these studies assume that the foliation orientation reflects the major principal finite strain orientation. This assumption stands since the early days of structural geology (Sharpe 1847; Sorby 1853; Haughton 1856; Cloos 1947; Ramsay 1967, pp. 180–182; Siddans 1972; Wood 1973, 1974; Tullis & Wood 1975; Price & Cosgrove 1990, pp. 450–453; Twiss & Moores 2007, pp.

400–405 & pp. 426–427), but it is, however, far from a proven fact.

For example, if the foliation develops late during the deformation history, it rather represents a cumulative strain (from onset of foliation development until the end of deformation) than the finite strain. Wood & Oertel (1980) compared the orientation of slaty cleavage with that of ellipsoidal strain markers in the Cambrian Slate Belt of Wales, assuming that the latter orientation represents the finite strain. They found a relationship but not an exact match between the two, suggesting that the cleavage is influenced by, but does not exactly follow, the orientation of the major principal finite strain. Oertel *et al.* (1989) compared the orientation of two different strain markers on the Appalachian Plateau, that is, the foliation orientation with the orientation of deformed crinoid columnals, in which the latter is assumed to represent the finite strain. Again, a relationship but not an exact match is found between the two. A review of the use and limitations of foliation orientation as an indicator for finite strain orientation is given by Oertel (1983).

If foliation refraction occurs around folded layers, more complex foliation refraction patterns can occur: the foliation fans (Ramsay 1967, pp. 403–405; Ramsay & Huber 1987, pp. 463–472). Two different types may occur: convergent and divergent foliation fans. The former occurs if the

From: LLANA-FÚNEZ, S., MARCOS, A. & BASTIDA, F. (eds) 2014. *Deformation Structures and Processes within the Continental Crust*. Geological Society, London, Special Publications, **394**, 21–37. First published online November 21, 2013, http://dx.doi.org/10.1144/SP394.4

foliation from the two limbs intersects in the inner arc of the fold; the latter occurs if the foliation from the two limbs intersects in the outer arc of the fold. Mechanically stronger layers tend to exhibit convergent fans while mechanically weaker layers tend to exhibit divergent fans. Such foliation fan patterns around folds are quite common features in outcrop-scale folds (e.g. Viola & Mancktelow 2005), but also occur on a smaller scale in thin sections (e.g. Aerden *et al.* 2010). Debacker *et al.* (2006) used foliation fan patterns to distinguish between folds amplifying prior to foliation development and folds amplifying contemporaneously with foliation development. Finally, Viola & Mancktelow (2005) demonstrated that foliation fans can even be folded themselves if the fold amplification is large enough.

Recently, Adamuszek *et al.* (2013) developed the Large Amplitude Folding theory, an analytical model to accurately calculate the geometry of high-amplitude folds. An extensive review of the information that may be gleaned from fold shapes can be found in Hudleston & Treagus (2010). However, there is no analytical method to calculate the strain inside and around a fold of large amplitude. A possible way to assess not only the finite strain orientation, but also the orientation of other strain measures is by conducting numerical experiments of buckle folding. Numerical methods allow for a direct calculation, visualization, and quantification of various strain measures inside and around a fold. Early finite-element (FE) simulations (Dieterich 1969; Dieterich & Carter 1969; Shimamoto & Hara 1976) calculated and visualized the orientation of the principal finite strain and of the principal stress to investigate the transition from inner-arc shortening to outer-arc extension in single-layer folds. The same phenomenon was investigated by Hudleston & Lan (1993), Hudleston & Lan (1995), and Lan & Hudleston (1995) by calculating the aspect ratios of the finite strain ellipses exactly on the fold axial plane (FAP) from their FE calculations. Recent studies that made extensive use of stress and strain orientation calculations in numerical FE simulations include Viola & Mancktelow (2005) studying foliation fans around folds, Frehner & Schmalholz (2006) studying the development of asymmetrical parasitic folds in multilayer systems, Reber *et al.* (2010) focusing on the stress orientation rotation on the limbs of amplifying folds and induced fracturing, and Frehner (2011) investigating the dynamic migration of the neutral lines through a single layer. All these studies use the finite strain, the infinitesimal strain, or the stress orientations. In contrast, the orientation and refraction patterns of foliation, which form during an arbitrary stage during folding, have not been investigated to date, and are the main subject of this study.

This study presents FE simulations of single-layer buckle folds for which various strain measures are calculated, including not only the finite strain and the infinitesimal strain, but also intermediate strain measures. The orientations of all of these strain measures inside the fold and in the surrounding matrix are visualized and quantified to describe the refraction pattern. These strain refraction patterns are compared to foliation refraction patterns in and around natural outcrop-scale folds in NW Spain.

Numerical method and model setup

For numerically simulating viscous buckle folding the FE method (e.g. Cuvelier *et al.* 1986; Hughes 2000; Zienkiewicz & Taylor 2000) is employed. The self-developed numerical code is implemented in MATLAB and simulates the two-dimensional (2D) incompressible plane-strain deformation of either linear (Newtonian) or power-law viscous isotropic materials in the absence of gravity. A detailed description of the numerical implementation is given in Frehner & Schmalholz (2006), Frehner (2011) and references therein. The code has successfully been benchmarked against the analytical fold growth-rate solution of both Fletcher (1977) for Newtonian rheology, and of Fletcher (1974) for power-law viscous rheology. Recent studies that also made use of this code investigated the development of asymmetrical parasitic folds in multilayers (Frehner & Schmalholz 2006), the dynamic behaviour of the neutral lines in amplifying single-layer folds (Frehner 2011), and the feasibility of dynamically unfolding mountain-scale folded cross-sections (Frehner *et al.* 2012).

In this study, upright symmetrical single-layer folds with dominant initial wavelength under horizontal shortening are studied. Out of all possible wavelengths, the dominant wavelength is the one with the highest initial growth rate and is therefore strongly related to the wavelength occurring in natural folds (Adamuszek *et al.* 2013). Both the folding layer and the surrounding matrix are isotropic and homogeneous. The initial setup for all simulations (Frehner 2011, fig. 3) comprises a rectangular box containing a horizontal layer embedded in a matrix of lower viscosity. The perfectly welded upper and lower layer interfaces exhibit a sinusoidal initial perturbation of dominant wavelength (Fletcher 1974, 1977) with an amplitude-to-layer thickness ratio of 0.01. The applied boundary conditions are: zero traction and zero boundary-perpendicular velocity (immobile free slip wall) at the lower and left boundaries; zero traction (free surface) at the top boundary; and zero traction with a prescribed horizontal velocity (moving free slip wall) at the right boundary. The prescribed horizontal velocity is adjusted during the simulations to maintain a constant background shortening strain rate. Both the upper and lower boundaries are set outside the zone of contact strain (Ramsay & Huber 1987, pp. 352–363) to avoid boundary effects.

Each time step of a FE simulation provides the velocity gradient tensor, $\dot{\mathbf{H}}$ with components $\dot{H}_{ij} = \partial v_i / \partial x_j$, at every integration point of the numerical domain in the Cartesian (x_1, x_2)-coordinate system, where v is velocity and x is the coordinate direction. The infinitesimal deformation gradient tensor at a given time step, m, is calculated as

$$\mathbf{D}_m = \left[\mathbf{I} + \Delta t \dot{\mathbf{H}}_m \right] \qquad (1)$$

where $\mathbf{I}$ is the 2×2 identity matrix, and Δt is the time increment. The infinitesimal deformation gradient tensor describes the deformation over one time increment. If the time increment chosen is small enough, it represents the instantaneous deformation. The finite deformation gradient tensor, $\mathbf{F}_m$ (Haupt 2002), at a given time step, m, is calculated using the Euler integration method as

$$\mathbf{F}_m = \prod_{k=1}^{m} \mathbf{D}_k \qquad (2)$$

where $\prod$ denotes the product. The finite deformation gradient tensor describes the entire deformation path from the beginning of the simulation ($k = 1$) to the current time step. Equation (2) can be modified to calculate the incremental deformation gradient tensor:

$$\mathbf{C}_{n-m} = \prod_{k=n}^{m} \mathbf{D}_k, \quad n < m. \qquad (3)$$

The incremental deformation gradient tensor records only the deformation path from an intermediate time step ($k = n$) to the current time step and is therefore an intermediate strain measure between the infinitesimal and finite strain. All three strain measures can be used to calculate a left Cauchy-Green tensor (Haupt 2002):

$$\mathbf{B}_m^D = \mathbf{D}_m \mathbf{D}_m^T, \quad \mathbf{B}_{n-m}^C = \mathbf{C}_{n-m} \mathbf{C}_{n-m}^T,$$
$$\mathbf{B}_m^F = \mathbf{F}_m \mathbf{F}_m^T. \qquad (4)$$

The eigenvectors and the square roots of the eigenvalues of these three left Cauchy-Green tensors correspond to the orientation and magnitudes of the infinitesimal ($\mathbf{B}_m^D$), incremental ($\mathbf{B}_{n-m}^C$), and finite ($\mathbf{B}_m^F$) strain ellipse axes (Frehner 2011) at a given time step m. The orientation of the long strain ellipse axes (major principal strains) will now be

visualized and quantified for various numerical simulations.

Numerical results

Figure 1 shows three subsequent shortening stages (top to bottom) of a single-layer folding simulation with a viscosity ratio of $R = 100$ between the folding layer and the surrounding matrix. For each stage the orientation and magnitude of the major principal strain is plotted as short lines for the three strain measures infinitesimal (left, Fig. 1a, e & j), incremental (centre), and finite strain (second from right, Fig. 1c, h & n). In addition, the orientation of initially vertical passive marker lines is plotted on the right-hand side (Fig. 1d, i & o). Naturally, the strain ellipses and their long axes are more elongated when more strain is recorded, which leads to a denser pattern for higher shortening values (top to bottom in Fig. 1) and for measures recording more strain (left to right in Fig. 1). Also, the lower-viscosity matrix experiences higher strain compared to the folding layer, which experiences more rotation. However, in the following the focus is on the orientation patterns of the major principal strains and not on their magnitudes.

Divergent fan pattern in the matrix

The major principal strain for all strain measures exhibits a very distinct divergent fan in the matrix at the outer arc of the fold close to the hinge. In many cases, these divergent fans even show horizontal orientations close to the layer interface. For a given shortening value (one row in Fig. 1), the patterns of the divergent fans are very similar for the infinitesimal and different incremental strain measures (e.g. Fig. 1a, b), but somewhat different for the finite strain (e.g. Fig. 1c). This may indicate that the divergent fan pattern is influenced by the pure-shear shortening prior to buckling initiation, which is recorded by the finite strain but not by the other strain measures. At low shortening values (Fig. 1a–c) the divergent fans occupy a large area around the fold. For increasing shortening values (Fig. 1j–n) the divergent fans are more restricted to the vicinity of the fold while further away the major principal strains are sub-parallel to the FAP and do not exhibit a fan. The orientation of the passive marker lines (Fig. 1d, i & o) hardly exhibits a convergent fan pattern in the matrix at the outer arc of the fold. If anything, the divergent fan weakly resembles the pattern of the finite strain. The fan patterns in the matrix at the inner arc of the fold are much less pronounced. At low shortening values ($s = 21.4\%$, Fig. 1a–c), a very slight divergent fan develops while at higher shortening

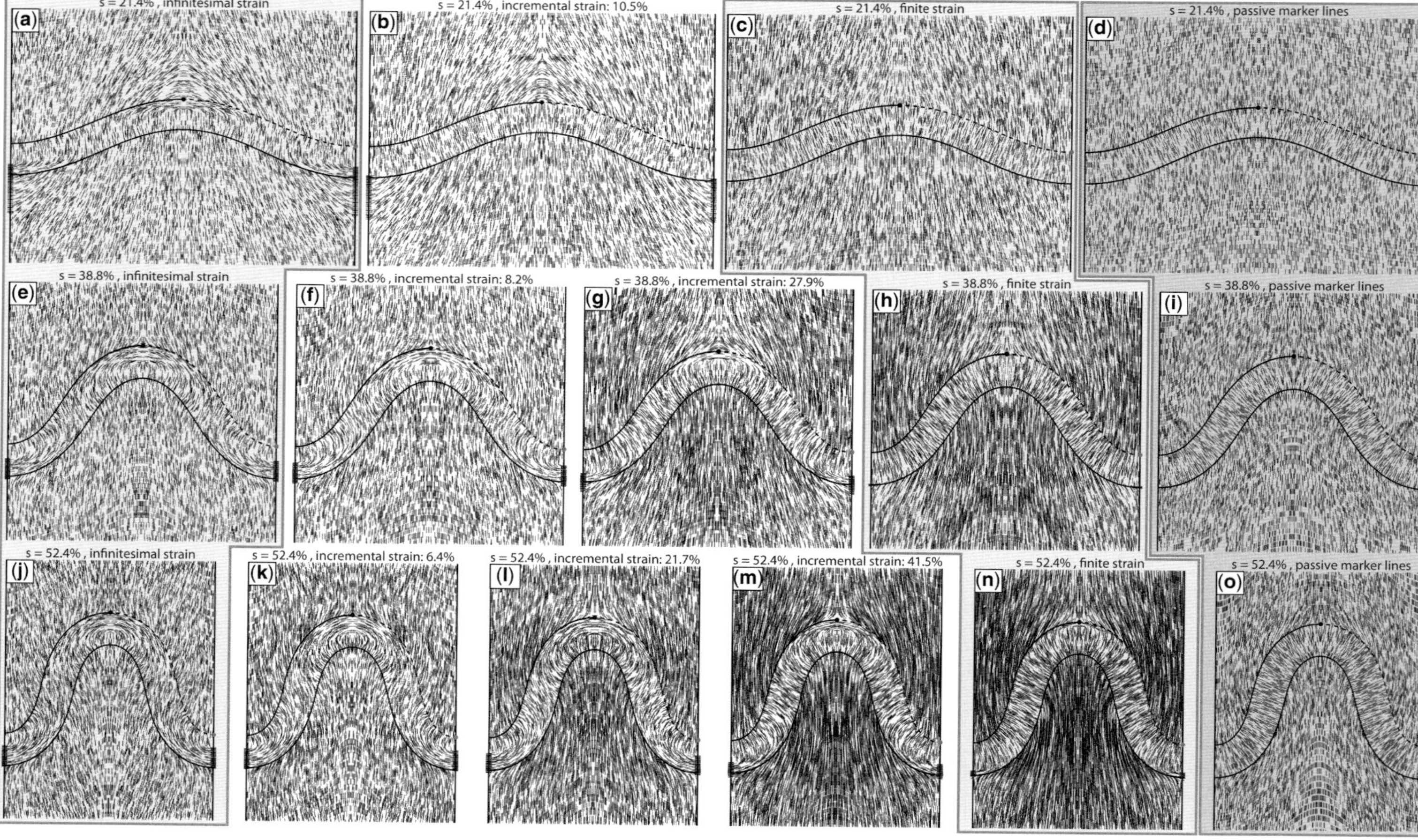

Fig. 1. Snapshots of a finite-element simulation of a single-layer buckle fold with a viscosity ratio of $R = 100$ between the folding layer and the surrounding matrix. Top to bottom represents increasing shortening, that is, $s = 21.4\%$, 38.8%, and 52.4%. Lines represent the orientation and magnitude (indicated by line length) of the major principal strain for the infinitesimal, finite, and incremental strain measures in snapshots on the left-hand side (**a, e, j**), in the second snapshots from the right (**c, h, n**), and the snapshots in-between (**b, f, g, k, l, m**), respectively. The amount of strain recorded by the incremental strain measure (equation (3)) is indicated above the respective snapshots. The snapshots on the right-hand side (**d, i, o**) show the orientation of initially vertical passive marker lines. The dashed fold interface connecting the antiform (black dot) with the synform (grey dot) is used in later figures.

values (Fig. 1e–h) a convergent fan develops that becomes more pronounced with increasing shortening (Fig. 1j–n).

Convergent fan pattern in the folding layer

The major principal strain for all strain measures exhibits a convergent fan at the inner arc of the higher-viscosity folding layer. Towards the outer arc of the layer the strain orientations are strongly modified and for most strain measures and shortening values the strain orientations switch from a convergent fan to a divergent fan pattern at the outer arc of the fold. This switch of strain orientation appears closer to the inner arc of the layer for strain measures recording smaller amounts of strain (e.g. infinitesimal compared to finite strain, Fig. 1j compared to Fig. 1n) and for higher shortening values (e.g. Fig. 1j compared to Fig. 1a). The switch from a convergent fan at the inner arc to a divergent fan at the outer arc can be associated with the migration of the neutral line from the outer towards the inner arc with increasing shortening (Frehner 2011). For the two end-member cases of either finite or infinitesimal strains the same switch of strain orientation was also observed by Dieterich (1969), Dieterich & Carter (1969), and Viola & Mancktelow (2005). The only case in Figure 1 where no such switch appears is for finite strain at a shortening of 21.4% (Fig. 1c). As a result of the switch of strain orientation in the hinge area, the major principal strain orientation patterns are somewhat, but not substantially, different for the different strain measures. The passive marker line orientation pattern in the folding layer (Fig. 1d, i & o) exhibits a convergent fan throughout the entire folding process and no switch of orientation to a divergent fan ever occurs. However, the convergent fan of the passive marker lines (Fig. 1d, i & o) resembles the part of the major principal finite strain orientation pattern (Fig. 1c, h & n) at the inner arc of the fold, where the latter is convergent.

Strain orientation quantification

For the simulation shown in Figure 1 the orientation of the fold interface and the major principal strains is plotted in Figure 2 in the same way as depicted in the figures for the natural folds (see 'Natural examples'). All orientations are measured from the black to the grey dot along the outer arc of the antiform exactly at the fold interface (dashed line in Fig. 1). Because the numerical folds are upright and symmetrical, the angle to the FAP (vertical axis in Fig. 2a, both axes in Fig. 2b) can be read as (i) limb dip for the fold interface data and (ii) deviation from vertical for the strain orientation data, respectively. Also, all FE simulations are 2D

and therefore only line orientations and not plane orientations can be calculated. However, to be consistent with the figures with the natural fold data, which are oriented in the 3D space, the term 'Dihedral angle' is used for the axis labels in Figure 2.

The strain orientations in the matrix are remarkably similar for all shortening values and for all strain measures. An exception is the finite strain at a shortening of 21.4% (Fig. 2a, b). In all other cases the major principal strains are perpendicular (−90°) to the FAP at the antiform hinge. Along the outer arc of the fold this value increases resulting in divergent fan patterns around the antiform (Fig. 1). At a shortening of 21.4% (Fig. 2a, b) the angles between the major principal strains in the matrix and the FAP are always negative and the divergent fans cover the entire fold from antiform to synform. Hence, the major principal strains also form a divergent fan in the inner arc of the adjacent synform (close to the grey dot in Fig. 1). In the higher shortening cases (Fig. 2c–f) the angles between the major principal strains in the matrix and the FAP become positive close to the adjacent synform indicating a convergent fan at the inner arc of the fold. However, these inner-arc convergent fans are much less pronounced than the outer-arc divergent fans. The transition from convergent to divergent fan (zero crossing) is on the antiform-side from the inflection point for all strain measures and closer to the antiform for strain measures recording smaller amounts of strain. The exceptional case of finite strain at a shortening of 21.4% (Fig. 2a, b) shows a negligible convergent fan at the inner arc of the synform and a divergent fan at the outer arc of the antiform, but without a switch of the major principal strain to −90°. In this case the transition from convergent to divergent fan is slightly on the synform-side from the inflection point. The passive marker lines in the matrix roughly follow the finite strain orientation, in particular close to the synform and for large shortening values. The passive markers do not rotate close to the fold hinge due to the traction-free boundary conditions and their orientation strongly deviates from that of the finite strain close to the antiform hinge. In the case of 21.4% shortening (Fig. 2a, b) the passive marker line orientation exhibits a convergent fan along the entire fold and resembles the major principal finite strain orientation only very weakly.

The strain orientations in the higher-viscosity folding layer (dashed lines in Fig. 2) share some common features for the different strain measures and for the different shortening values. With the exception of finite strain at 21.4% shortening, all major principal strains at the antiform hinge are perpendicular (−90°) to the FAP and parallel to the fold interface, which is represented by a 90° difference between the strain orientations and the

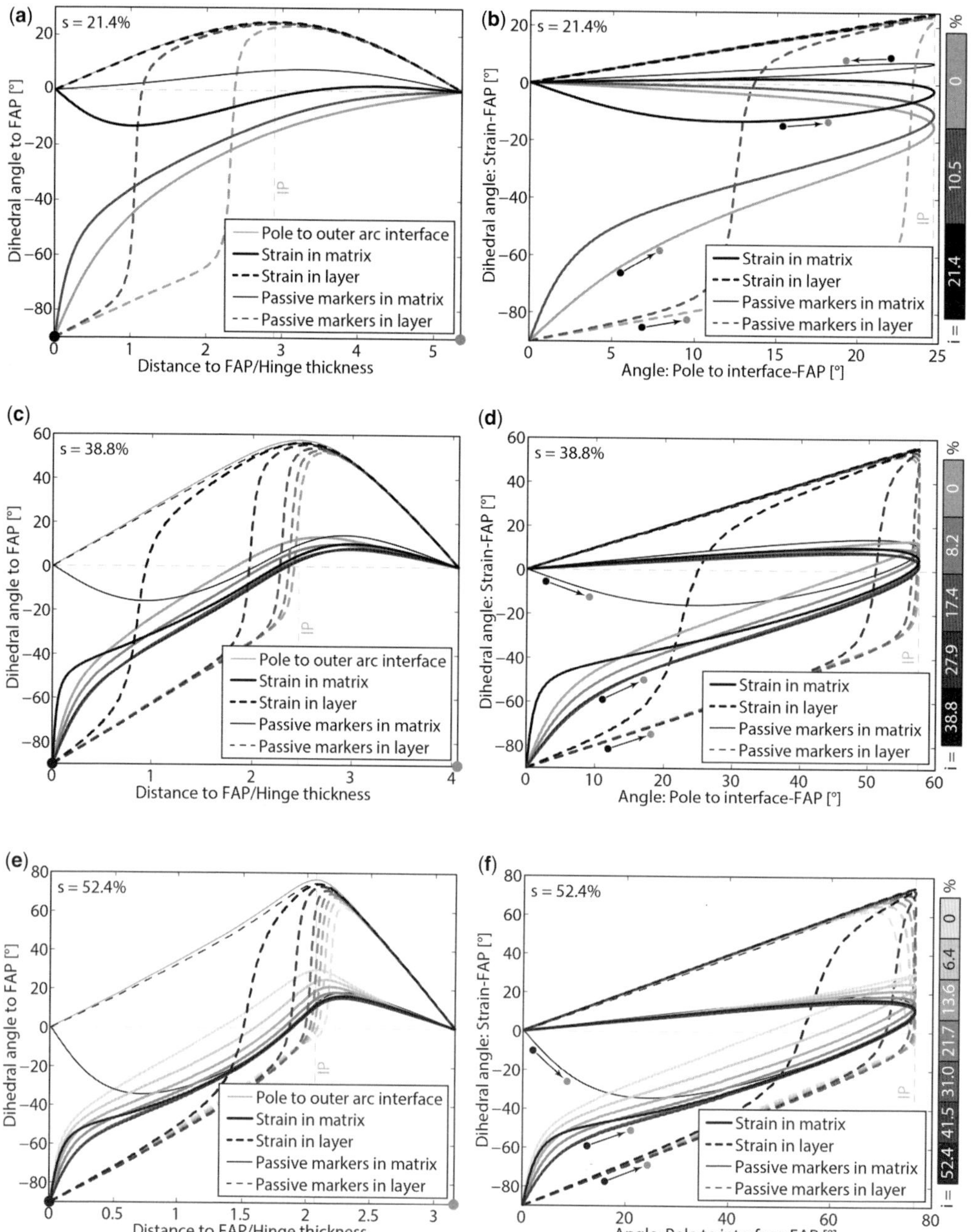

Fig. 2. Strain and fold interface orientation data for the same finite-element simulation as in Figure 1 (viscosity ratio $R = 100$) for the same three shortening values, s, as in Figure 1 (top to bottom). The data are taken from the outer arc of the fold along the dashed line from the black to the grey dot in Figure 1 (also indicated in (a), (c) and (e)). Different line grey levels represent different strain measures, whereas black represents finite strain, lightest grey infinitesimal strain, and intermediate grey levels different incremental strains, as indicated in the grey level-bar. Positive values for strain orientation indicate a convergent fan; negative values indicate a divergent fan. IP, inflection point. (a, c, e) Angle between major principal strains and the fold axial plane (FAP) plotted v. the normalized distance from the FAP. For the fold interface, the angle between the pole to the interface and the FAP is plotted. (b, d, f) Angle between major principal strains and the FAP plotted v. the angle between the pole to the fold interface and the FAP. Small arrows indicate the direction from the antiform towards the synform (black towards grey dot in Fig. 1).

fold interface in Figure 2a, c, and e. Along the fold, the major principal strains maintain their sub-parallel orientation to the fold interface before they switch to a layer-perpendicular orientation indicated by negligible difference between the strain orientations and the fold interface close to the synform. The 90°-switch from a layer-parallel to a layer-perpendicular strain orientation happens within a very short distance on the antiform-side from the inflection point and closer to the antiform for strain measures recording larger amounts of strain. At higher shortening (Fig. 2e) and for strain measures recording small amounts of strain the 90°-switch happens slightly on the synform-side from the inflection point. This 90°-switch of strain orientation can also be identified in Figure 2b, d, and f, where the major principal strain orientations plot with a slope of about one for a large part of the fold, which indicates either layer-parallel or layer-perpendicular orientations. The finite strain at 21.4% shortening (Fig. 2a, b) maintains a layer-perpendicular orientation along the entire fold. The

same is true for the passive marker lines in the folding layer (also Fig. 1d, i & o).

Varying viscosity ratios

FE simulations were also performed for viscosity ratios (R) of 25 and 40 between the folding layer and the surrounding matrix. For a shortening of 38.8% the fold interface, strain, and passive marker line orientation data are plotted in Figure 3 and some simulation snapshots are shown in Figure 4. The strain orientations in the matrix along the fold interface for a small viscosity ratio of $R = 25$ (solid lines in Fig. 3a, b) are different for the different strain measures, while the different strain measures are comparable for higher viscosity ratios, particularly on the synform-side from the inflection point ($R = 40$ and $R = 100$; Figs 3c, d & 2c, d). However, the absolute values of the dihedral angles are generally smaller for lower viscosity ratios, meaning that the fan patterns are less pronounced. At a viscosity ratio of 25 (Fig. 3a, b), the

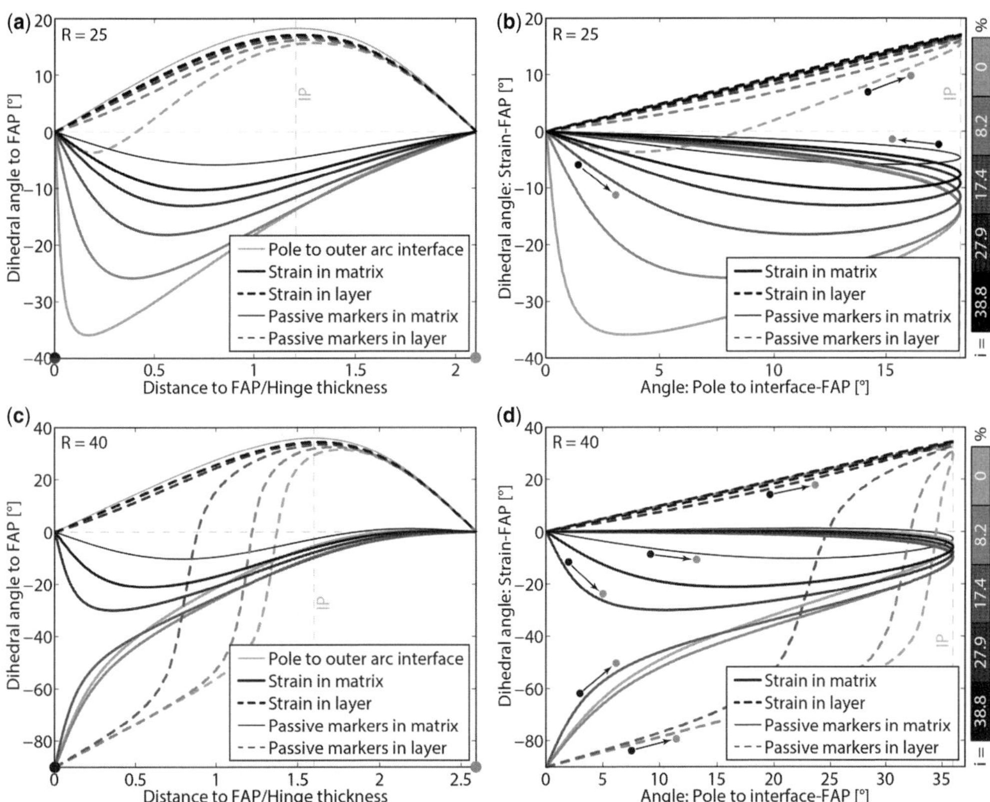

Fig. 3. Strain and fold interface orientation data for two finite-element simulations with viscosity ratio $R = 25$ (upper) and 40 (lower) for a shortening of $s = 38.8\%$. The left (**a, c**) and right (**b, d**) representations of the data are the same as in Figure 2.

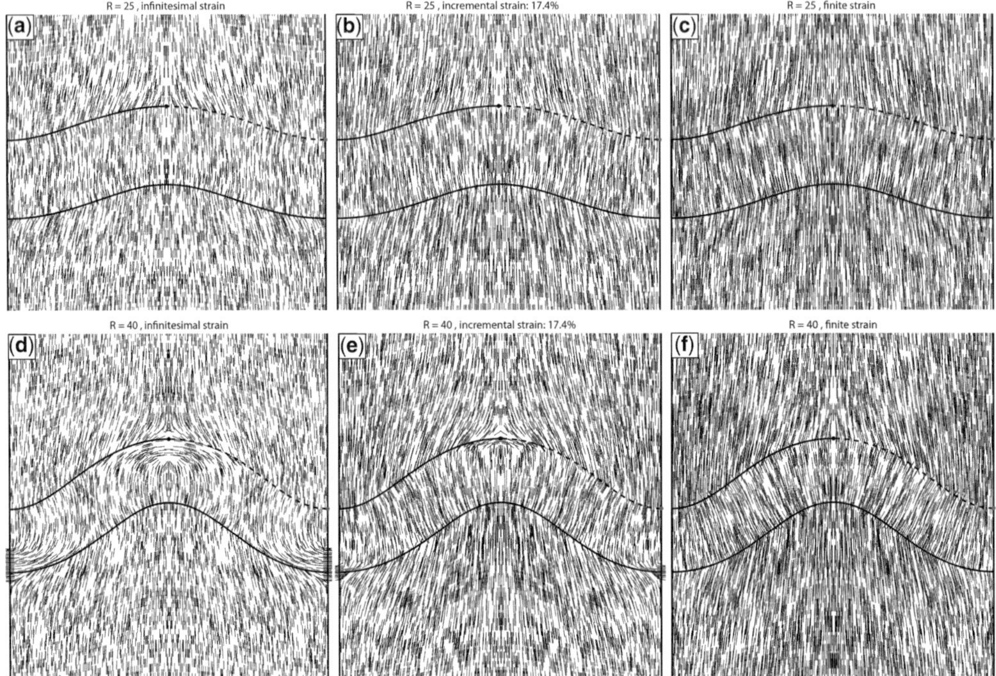

Fig. 4. Snapshots of two finite-element simulations of a single-layer buckle fold with viscosity ratios of $R = 25$ (top, **a–c**) and 40 (bottom, **d–f**) for a shortening of $s = 38.8\%$. Lines represent the orientation and magnitude (indicated by line length) of the major principal strain for the infinitesimal, incremental, and finite strain measures in snapshots on the left-hand side (a and d), in the centre (b and e), and on the right-hand side (c and f), respectively. The amount of strain recorded by the incremental strain measure (equation (3)) is 17.4%.

major principal strains in the matrix exhibit divergent fan patterns along the entire fold. For increasing viscosity ratios ($R = 40$ and $R = 100$; Fig. 2c, d) an increasing portion of the matrix at the inner arc of the synform exhibits convergent fans and Figure 3c and d mark about the transition, where the major principal strains are parallel to the FAP in the matrix at the inner arc of the synform (also Fig. 4d–f). For a low viscosity ratio of $R = 25$ the major principal strains do not switch to a layer-parallel orientation at the antiform hinge (0° dihedral angle at the antiform in Fig. 3a, b; also Fig. 4a, b); for the intermediate viscosity ratio of $R = 40$ only the strain measures recording a small amount of strain show such a switch (Figs 3c, d & 4d–f); and for a high viscosity ratio ($R = 100$) all strain measures show this switch (Figs 2b, c & 1e–h).

In the folding layer with higher viscosity the strain orientations are very similar for all strain measures in the case of a viscosity ratio of 25 (Fig. 3a, b). The major principal strain orientations are close to layer-perpendicular and exhibit convergent fans along the entire fold. Only the infinitesimal strain measure shows a slight divergent fan in a very restricted area around the hinge,

but exactly at the hinge the major principal infinitesimal strain returns to a parallel orientation to the FAP and does not show the 90°-switch as is the case for higher viscosity ratios (Figs 3c, d & 2c, d). The resulting convergent fan is negligible and not even detectable in Figure 4a. For an intermediate viscosity ratio of 40 the finite strain and the incremental strain measure recording the most strain maintain a layer-perpendicular orientation along the entire fold. The strain measures recording a smaller amount of strain switch by 90° to a layer-parallel orientation on the antiform-side from the inflection point. For a high viscosity ratio of 100 all strain measures show this 90°-switch (Fig. 1e–h).

Power-law viscous rheology

Various numerical FE simulations were performed using power-law viscous rheology with various combinations of the power-law exponent of the folding layer and the surrounding matrix. Generally, the observations made above for the Newtonian cases do not change for the power-law viscous cases. For example, Figure 5 shows the major principal strain orientations plotted against

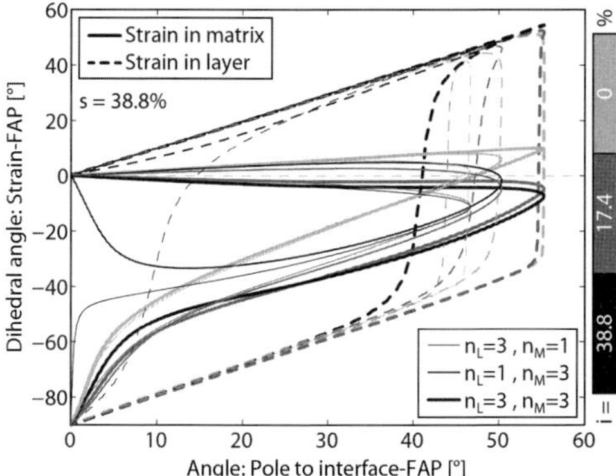

Fig. 5. Angle between major principal strains and the fold axial plane (FAP) plotted v. the angle between the pole to the fold interface and the FAP at a shortening of $s = 38.8\%$ for three different power-law viscous FE simulations with a reference viscosity ratio of $R = 40$ and a combination of power-law exponents of the layer (n_L) and the matrix (n_M) of $n_L = 3$, $n_M = 1$ (thin lines), $n_L = 1$, $n_M = 3$ (intermediate lines), and $n_L = 3$, $n_M = 3$ (thick lines). Different line grey levels represent different strain measures, whereas black represents finite strain, intermediate grey incremental strain, and lightest grey infinitesimal strain. The amount of strain recorded by the incremental strain measure (equation (3)) is 17.4%.

the fold interface orientations for FE simulations using a power-law exponent of 3 for the layer and/or the matrix, a reference viscosity ratio of 40, and a shortening of 38.8%. Three strain measures for each simulation are shown in Figure 5: infinitesimal, finite, and incremental strain, the latter recording the last 17.4% shortening. The different power-law viscous rheologies lead to different amplification rates of the dominant wavelength (Fletcher 1974) and therefore to different maximum limb dips at the same shortening value. Apart from this, the strain orientation patterns are similar for all power-law viscous cases and similar to the Newtonian case (Fig. 3d). These similarities include: (i) a 90°-switch of the major principal strains in the folding layer from a layer-perpendicular orientation close to the synform to a layer-parallel orientation close to the antiform, whereas this switch happens on the antiform-side from the inflection point and (ii) a smooth transition of the major principal strain orientations in the matrix from a divergent fan close to the antiform hinge to a weak or negligible convergent fan at the inner arc of the adjacent synform.

Natural examples

In the following section, the aim of analysing the foliation orientation in natural folds is not a quantitative one-to-one comparison between natural and numerical folds, but a qualitative one. The geometry of the selected natural fold examples is much more complex than the relatively simple symmetrical single-layer folds of the performed numerical study. Nevertheless, a qualitative comparison allows foliation-orientation features to be identified that can be reproduced by the numerical simulations. For a one-to-one reproduction of the natural fold geometries and the corresponding foliation orientations, inverse numerical methods may be applied (e.g. Lechmann *et al.* 2010; Frehner *et al.* 2012). However, for accurate modelling results using such methods, various physical parameters during fold formation have to be known quite precisely, such as pressure, temperature, or the rheological flow-law.

For the qualitative comparison of the numerical results with natural examples, three outcrop-scale folds with well-developed foliation refraction patterns were selected. The studied outcrops are located in the Western Asturian–Leonese Zone (WALZ) of the Variscan belt in NW Spain (Marcos 1973; Bastida & Pulgar 1978; Laner 2010; inset in Fig. 6). The WALZ comprises metasedimentary rocks of lower Palaeozoic age, unconformably overlying Precambrian turbidites (Bastida *et al.* 1986). The entire zone was affected by Variscan nappe stacking with top-to-the-east thrusting and co-eval folding of both the sedimentary cover and the underlying basement units. Although three deformation phases have been distinguished, the fold axes for these events are generally aligned parallel (Fernández *et al.* 2007, and references therein).

Folds at two sites on the Cantabrian coast were investigated in this study. Two fold examples have been selected from a site at Portizuelo, east of Luarca, Asturias; a third example is from a site at Benquerencia, west of the town of Ribadeo, Galicia. At Portizuelo, Lower Ordovician rocks are summarized as Luarca Slates, comprising quartzites, sandstones, and slates deposited in an extensional regime on a passive continental margin (Marcos 1973). The abundant intercalations of layers of different strength are deformed into upright to slightly east-verging folds with wavelengths of up to several tens of metres (related to D3 folding according to Bastida *et al.* 1986). At Benquerencia, Cambrian metasediments of the Cándana Group are deformed into recumbent folds in the normal limb of the Foz-Tapia Anticline, a major fold of the Mondoñedo nappe (Bastida *et al.* 1986). Folds of several hundred metres amplitude and accompanying asymmetrical second-order folds of several metres in size can be identified in the intercalated sandstones and shales, which deformed under greenschist facies metamorphic conditions (Fernández *et al.* 2007).

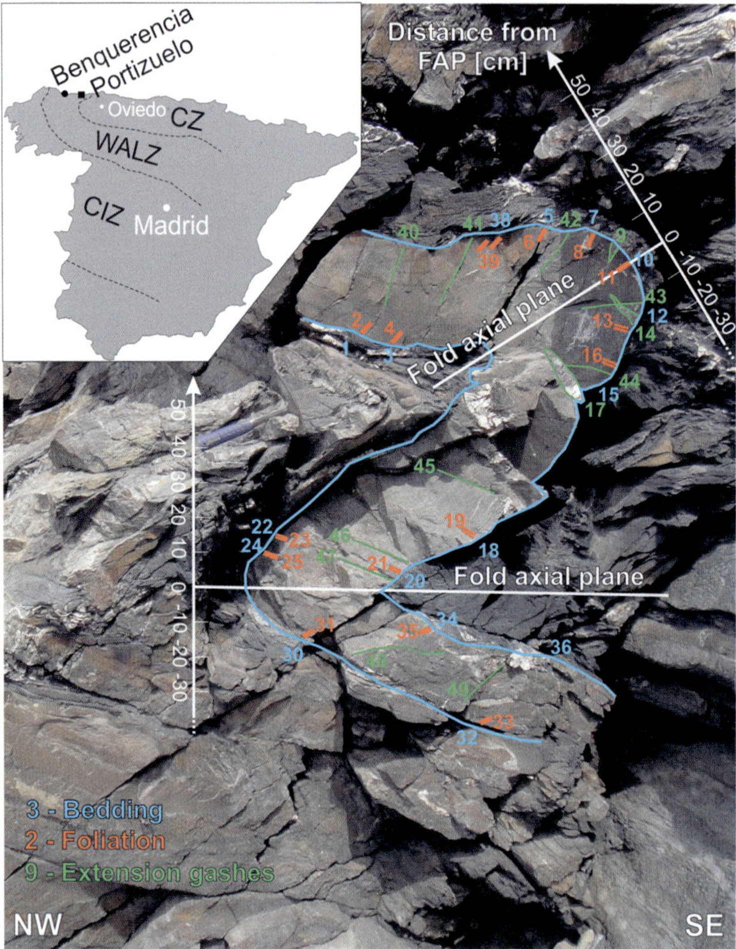

Fig. 6. Outcrop picture of a SE-verging synform/antiform fold train at Portizuelo, NW Spain. Orientations of bedding, foliation and extension gashes were measured at the positions indicated by numbers exclusively within the sandstone layer. The fold axial planes (FAP) of the upper and lower fold are not parallel and both are used in Figure 7. Inset in the upper-left corner shows the location of both investigated sites and their positions in the Hercynian orogenic units (CZ, Cantabrian Zone; WALZ, Western Asturian–Leonese Zone; CIZ, Central Iberian Zone) of the Iberian Peninsula (modified after Bastida *et al.* 1986).

Outcrop example 1 (Portizuelo)

The first outcrop example is located at the Portizuelo site (43°32′56″N, 6°30′36″W). A sandstone layer is folded into a SE-vergent synform/antiform train, embedded in an indistinct shale matrix (Fig. 6). In addition to the axial plane foliation, numerous extension gashes overprinting the sedimentary bedding in the sandstone layer were measured. Some broader extension gashes are clearly related to outer-arc extension (e.g. measurement numbers 9, 14, 17 in Fig. 6). In contrast, the more abundant thinner veins (measurement numbers 40–49 in Fig. 6) are not restricted to the outer-fold arcs, but are cross-cut by the broader veins. They may reflect an earlier deformation stage, possibly related to burial of the sediments, or movement and opening along an early refracted foliation surface (Ramsay 1967). The two folds shown in Figure 6 have two non-parallel FAPs (upper and lower). Therefore, both FAPs are considered in Figure 7, where the relationships between bedding, foliation, and extension gashes are plotted against the distance to the FAPs (Fig. 7a) and the orientation of the FAPs (Fig. 7a, b).

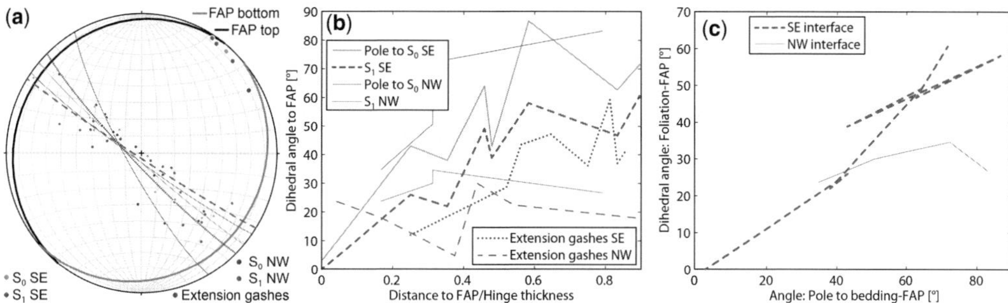

Fig. 7. Field measurements of the outcrop shown in Figure 6. (**a**) Lower-hemisphere equal area projection. Small dots represent poles to plane measurements. Coloured lines and large dots represent the best fitting great circle with its pole to the data of the same colour. FAP bottom and FAP top denote the fold axial plane of the bottom and top fold in Figure 6; S_0 SE and S_1 SE denote the bedding and foliation at the southeastern fold interface (right blue line in Fig. 6); S_0 NW and S_1 NW denote the bedding and foliation at the northwestern fold interface (left blue line in Fig. 6). (**b**) Dihedral angle between plane measurements and the FAP plotted v. the normalized distance of the measurements from the FAP. Measurements at the southeastern fold interface are plotted with respect to the upper FAP; measurements at the northwestern fold interface are plotted with respect to the lower FAP. Positive values for foliation (S_1) indicate a convergent foliation fan; negative values indicate a divergent fan. For the bedding (S_0), the angle between the pole to the bedding and the FAP is plotted instead of the dihedral angle. (**c**) Dihedral angle between the foliation and the FAP plotted v. the angle between the pole to the bedding and the FAP. In (b) and (c) measurements 9, 14, and 17 (Fig. 6) are omitted because they exhibit a different orientation trend from the other extension gashes.

Outcrop example 2 (Portizuelo)

The second studied fold crops out close to the first one at Portizuelo (43°32′49″N, 6°30′45″W). Several sandstone layers are folded into an upright, slightly SE-verging synform (Fig. 8). Although a low viscosity matrix is not exposed at this location, the sandstone layers show varying orientation of axial plane foliation in this fold. The more tightly folded inner sandstone layer (above the blue line in Fig. 8) shows a narrow convergent foliation fan (measurement points 19–24 in Fig. 8). Below the blue line, the outer sandstone layer exhibits a more widely spaced convergent foliation fan (measurement points 2, 4, 6, …, 18 in Fig. 8). Both of these foliation fans indicate a layer-parallel shortening direction. In addition, layer-perpendicular shortening structures can be identified close to the fold hinge at the outer arc of the outer layer. A quartz-filled extension vein is folded in layer-perpendicular direction and intersected and deformed by layer-parallel pressure-solution surfaces (inset in Fig. 8). At the outer arc of the fold these pressure-solution surfaces can be identified in the field but are hardly abundant enough to form a bedding-parallel foliation. Similar to Figure 7, the relation between the structural elements and their position along the fold is shown in Figure 9.

Outcrop example 3 (Benquerencia)

At the Benquerencia site, a third fold was examined for this study (43°34′6″N, 7°12′49″W, Fig. 10),

which formed at slightly higher metamorphic conditions compared to the first site, as indicated by the higher intensity of axial plane foliation. The relation between the structural elements and their position along the fold are shown in Figure 11. The angles between the foliation in the quartzite and in the shale measured at the identical position around the fold are rather small (Fig. 11), which suggests that the viscosity ratio between the layers is smaller during deformation compared to the first site.

Discussion

In this study, various strain measures (finite, incremental, and infinitesimal strain, and passive marker lines) are considered, which may represent the onset of foliation development at different stages during deformation. If so, finite strain would represent an early foliation in the folding history, which continues to be modified by the changing stress state during folding. Initially vertical passive marker lines represent a foliation development prior to buckling initiation, which is not modified any more during folding and only rotates passively. A foliation developing at a given time during folding does not record the early pure-shear shortening and thickening of the layer prior to buckling initiation and is represented by the incremental strain. In contrast, infinitesimal strain represents a foliation development during the very last stage of folding. Brittle fractures, for example, may develop during this stage.

Fig. 8. Upright, slightly SE-verging synform at Portizuelo developed in sandstone beds. Bedding and foliation were measured at the indicated positions. Inset: Zoom of one extension gash in the outermost sandstone layer slightly NW of the fold axial plane (FAP), which is folded in layer-perpendicular direction and partially truncated by layer-parallel pressure-solution surfaces. Traced in red are the few pressure-solution surfaces that can be identified as such in the field, but there may be more such surfaces that are not traced here.

In both numerical and natural folds the strain pattern depends on the geometry and amplitude of the initial perturbation on the layer interfaces (Mancktelow 2001). For example, a layer with large initial perturbation amplifies earlier than one with small initial perturbation and experiences less pure-shear shortening prior to buckling initiation. Therefore, the initial perturbation also influences the orientation pattern of the strain measures recording this early deformation stage (i.e. finite strain and incremental strain recording a large amount of strain).

In the following, the geometrical relationships for the various scenarios of foliation development inferred from the strain measure orientation patterns are discussed and qualitatively compared to the field observations. It is emphasized that a quantitative one-to-one comparison is not feasible because the numerical simulations do not cover all the natural complexities, such as multilayer folding (Fig. 8), non-parallel FAPs in adjacent folds (Fig. 6), or strong anisotropy (Fig. 10).

Comparison between numerical and natural folds

The data curves for the natural data are not as smooth as for the numerical data due to measurement inaccuracies and natural variations. In addition, measurements made on both sides of the FAP of a natural fold are combined in the data plots. If the folds are not perfectly symmetrical with respect to the FAP, this introduces some data variability. Nevertheless, some common features between natural and numerical data can still be recognized. For example, the foliation in the weaker shale around the fold in Figure 10 exhibits a clear divergent fan (negative values in Fig. 11b, c). With increasing distance from the FAP the absolute value of the dihedral angle between the foliation in the shale and the FAP first increases and then decreases again. The foliation in the stiffer sandstone exhibits a convergent fan (positive values in Fig. 11b, c) and the foliation orientation almost mirrors that in the shale. In the numerical

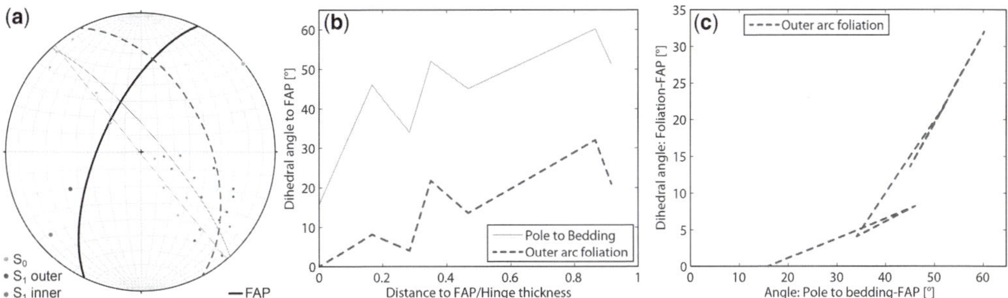

Fig. 9. Same as Figure 7 but for field measurements of the outcrop shown in Figure 8. (**a**) S_1 outer is the foliation at the outer arc of the fold; S_1 inner is the foliation at the inner arc of the fold. (**b, c**) Only the measurements at the outer arc of the fold are shown.

simulations, similar orientation patterns can be identified for strain measures recording a large amount of strain (e.g. finite strain) or for the passive marker lines in simulations with low viscosity ratios (Fig. 3). This suggests that in the case of the fold in Figure 10 the foliation started developing early during the folding history and that the mechanical contrast between the quartzite and the shale was not very large during folding. However, the natural data show a large angle between the

Fig. 10. NW-verging, recumbent fold at Benquerencia, developed in quartzitic and shale beds. The bedding interface and foliations in the shale and quartzite layer were measured at the indicated locations. Note the coin in the fold hinge at the inner-fold arc for scale.

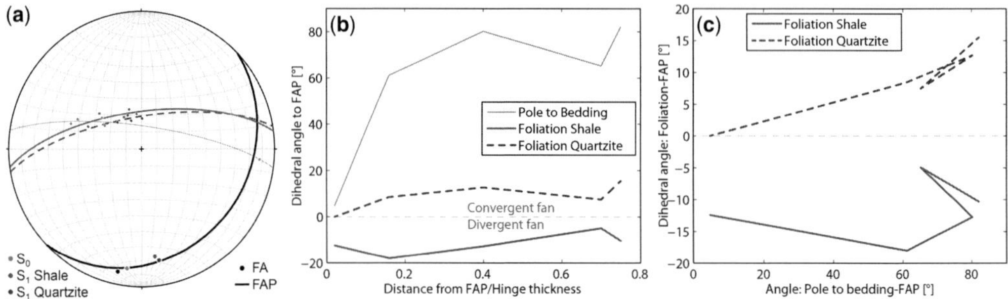

Fig. 11. Same as Figure 7 but for field measurements of the outcrop shown in Figure 10. (**a**) S_0 denotes bedding; S_1 Shale, the foliation in the shale; S_1 Quartzite, the foliation in the quartzite; FA, the measured fold axis; FAP, the measured fold axial plane. In (**b**) and (**c**) all the measured data are shown.

pole to bedding and the foliation in the quartzite (mean slope < 1 in Fig. 11c). This is different from the numerical results, for which the major principal strains in the layer are almost layer-perpendicular for strain measures recording large amounts of strain (slope approximately one in Fig. 3b, d).

In all natural cases the foliation in the stiffer folding layer follows the trend of the bedding orientation with various amounts of angular offset. In Figure 7b the angular offset is small and the foliation in the layer is almost perpendicular to the bedding (slope approximately one in Fig. 7c). The orientation of the extensional gashes in Figure 6 also follows the trend of the bedding, but with a larger angular offset than the foliation (Fig. 7b). In the other two natural cases the foliation orientation in the fold has a larger angular offset to the bedding, but still follows the trend of bedding orientation (Figs 9b, c & 11b, c). The numerical results show a similar behaviour but only for strain measures recording a large amount of strain (e.g. finite strain) or for the passive marker lines in the case of small viscosity ratios (Fig. 3). For higher viscosity ratios (Figs 2 & 3c, d) and for strain measures recording smaller amounts of strain, the major principal strains in the folding layer exhibit an almost sudden 90°-switch from a layer-perpendicular orientation close to the synform to a layer-parallel orientation close to the antiform. Such a switch is not observed in the natural folds. One possible interpretation is that all the examined natural folds are characterized by small mechanical contrasts to the surrounding matrix and/or that the foliation in the folds developed very early during the folding history. However, it might also be possible that such a 90°-switch of foliation orientation is not identifiable in the examined folds because the foliation is not developed strongly enough. Also, it has to be emphasized that a layer-parallel foliation is much more challenging to identify as such in the field because it is parallel to the sedimentary

bedding and might be misinterpreted as a sedimentary or diagenetic feature.

One case of layer-perpendicular shortening is present in the fold in Figure 8, at the outermost arc of the fold (inset in Fig. 8). A quartz-filled extensional vein is folded in layer-perpendicular direction and intersected by layer-parallel pressure-solution surfaces. Layer-perpendicular shortening has also been observed in other natural folds, but mainly on the fold limbs. For example, Viola & Mancktelow (2005) observe folding in layer-perpendicular direction of the earlier axial plane foliation on the limbs of metre-scale folds in Namibia. They also describe pressure solution accentuating the existing sedimentary bedding (called pseudobedding in Viola & Mancktelow 2005), which is a manifestation of layer-perpendicular shortening. In Figure 8 the extension gash is clearly older than both its small-scale folding and the pressure solution cutting it and possibly developed due to outer-arc layer-parallel tensional stresses. The younger folding structure of the extension gash and pressure-solution surfaces developed due to layer-perpendicular compression. Both structures indicate that the largest compressive stress acted in layer-perpendicular direction during their formation. In the ideal case of viscous rheology, this corresponds to a layer-parallel orientation of the major principal strain. In the numerical simulations, such orientations are observed for large viscosity ratios (Figs 2 & 3c, d) and for strain measures recording a small amount of strain, and can be explained by the migration of the neutral line from the outer towards the inner arc of the fold (Frehner 2011). Therefore, in contrast to the interpretation above, it may also be inferred that both the quartz vein and the pressure-solution surfaces developed late during the folding history and that the folding layer exhibits a large mechanical contrast to the surrounding matrix. The observed structures would also be promoted by large initial perturbations on the sedimentary interfaces, as they would lead to faster fold amplification. However, the initial perturbation is almost impossible to estimate in natural folds.

Foliation as strain indicator?

The amount of shortening and the viscosity ratio between layer and matrix can be estimated from the fold shape, using for example the Fold Geometry Toolbox (Adamuszek *et al.* 2011). However, the amount of strain recorded by the foliation is difficult to estimate. The numerical simulations demonstrate that the major principal strain orientations only have small differences for the different strain measures. It is expected that for most natural folds these differences are masked by the natural variations or the errors on measurements of the foliation orientation. In particular, the divergent fan in the matrix around the outer arc of the fold is not very sensitive to the amount of recorded strain for most of the fold (Figs 1–4) and is therefore not a good proxy for strain estimates; on the other hand, the convergent fan in the higher-viscosity layer depends more strongly on the recorded strain (Figs 1 & 4).

When measured exactly at the fold interface, this dependency is not so obvious for all shortening values and viscosity ratios (e.g. dashed lines in Fig. 2e or Fig. 3a, b), but the pattern inside the layer can vary strongly for the different strain measures (Figs 1 & 4). Therefore, it is essential to characterize the entire foliation pattern in natural folds, and not only measure the foliation along the fold interface. Close to the FAP, two perpendicular foliations may overprint each other, because the strain switches by 90° almost instantaneously when the neutral line passes a certain point on the FAP (Frehner 2011). However, this overprinting relationship can be challenging to recognize in natural folds because one of the foliations is bedding-parallel and possibly only weakly developed (e.g. only a few pressure-solution surfaces in the inset in Fig. 8). Other structures (e.g. folded and intersected vein in Fig. 8) can help the interpretation of the 90°-switch of foliation orientation. Especially in areas of polyphase folding or in the absence of additional structures this might be a challenging decision for field geologists. Because of these difficulties, the foliation fan pattern inside the folding layer is also not very well suited as a proxy for strain estimates.

major principal strain orientation around buckle folds. In particular the divergent fan pattern in the weak matrix at the outer arc of the folds is nearly insensitive to the amount of strain recorded by the strain measure. Therefore, the divergent foliation fan in the matrix at the outer arc of a natural fold does not necessarily represent the finite strain orientation, but may also represent infinitesimal or an incremental strain. The major principal strain orientation inside the higher-viscosity folding layer is different for the different strain measures, particularly towards the outer arc of the fold, where the strain switches by 90° from a layer-perpendicular to a layer-parallel orientation. Where exactly this 90°-switch happens depends on the external shortening and the amount of strain recorded by the strain measure. However, in the field it can be difficult to identify a layer-parallel foliation because it is parallel to sedimentary or diagenetic structures. Therefore, the foliation fan pattern inside the folding layer is also not very well suited as a proxy for strain estimates.

The natural foliation fan patterns exhibit some of the features observed in the numerical simulations. However, natural orientation variations and noisy data result in foliation orientation plots that are difficult to interpret. No final conclusion can be drawn in terms of the amount of strain that is recorded by the foliation fan patterns as different interpretations are possible. It is expected that a foliation developing early during folding records more strain and resembles the major principal finite strain orientation and that a foliation developing late during folding rather resembles the major principal infinitesimal strain. However, such a one-to-one correlation between natural and numerical data is not possible for the presented case study.

The organizers of the DRT Conference 2011, in particular S. Llana-Fúnez, are acknowledged for their effort in putting together this exciting conference in Oviedo, Spain. A field trip of the authors prior to, and many intense discussions during, the DRT Conference 2011 motivated this work. R. Laner is greatly acknowledged for sharing the scientific findings of his Diploma thesis, as well as for many useful tips on accessible roads and outcrops. This work was supported by the Austrian Science Fund project V151-N22. The referee P. Hudleston is thanked for his constructive criticism of the original manuscript.

Conclusions

The numerical FE simulations demonstrated that strain measures recording different amounts of strain result in similar refraction patterns to the

References

ADAMUSZEK, M., SCHMID, D. W. & DABROWSKI, M. 2011. Fold geometry toolbox – Automated determination of fold shape, shortening, and material properties. *Journal of Structural Geology*, **33**, 1406–1416, http://dx.doi.org/10.1016/j.jsg.2011.06.003

ADAMUSZEK, M., SCHMID, D. W. & DABROWSKI, M. 2013. Theoretical analysis of large amplitude folding of a single viscous layer. *Journal of Structural Geology*, **48**, 137–152, http://dx.doi.org/10.1016/j.jsg.2012.11.006

AERDEN, D. G. A. M., SAYAB, M. & BOUYBAOUENE, M. L. 2010. Conjugate-shear folding: a model for the relationships between foliations, folds and shear zones. *Journal of Structural Geology*, **32**, 1030–1045, http://dx.doi.org/10.1016/j.jsg.2010.06.010

BASTIDA, F. & PULGAR, J. A. 1978. La estructura del manto de Mondoñedo entre Burela y Tapia de Casariego (Costa Cantabrica, NW de España). *Trabajos de Geologia*, **10**, 75–159.

BASTIDA, F., MARTINEZ-CATALAN, J. R. & PULGAR, J. A. 1986. Structural, metamorphic and magmatic history of the Mondonedo nappe (Hercynian belt, NW Spain). *Journal of Structural Geology*, **8**, 415–430, http://dx.doi.org/10.1016/0191-8141(86)90060-X

CLOOS, E. 1947. Oolite deformation in South Mountain Fold Maryland. *Geological Society of America Bulletin*, **58**, 843–918, http://dx.doi.org/10.1130/0016-7606(1947)58[843:ODITSM]2.0.CO;2

CUVELIER, C., SEGAL, A. & VAN STEENHOVEN, A. A. 1986. *Finite Element Methods and the Navier-Stokes Equations*. D. Reidel Publishing Company, Dordrecht.

DEBACKER, T. N., VAN NOORDEN, M. & SINTUBIN, M. 2006. Distinguishing syn-cleavage folds from pre-cleavage folds to which cleavage is virtually axial planar: Examples from the Cambrian core of the Lower Palaeozoic Anglo-Brabant Deformation Belt (Belgium). *Journal of Structural Geology*, **28**, 1123–1138, http://dx.doi.org/10.1016/j.jsg.2006.03.027

DIETERICH, J. H. 1969. Origin of cleavage in folded rocks. *American Journal of Science*, **267**, 155–165.

DIETERICH, J. H. & CARTER, N. L. 1969. Stress-history of folding. *American Journal of Science*, **267**, 129–154.

FERNÁNDEZ, F. J., ALLER, J. & BASTIDA, F. 2007. Kinematics of a kilometric recumbent fold: The Courel syncline (Iberian massif, NW Spain). *Journal of Structural Geology*, **29**, 1650–1664, http://dx.doi.org/10.1016/j.jsg.2007.05.009

FLETCHER, R. C. 1974. Wavelength selection in the folding of a single layer with power-law rheology. *American Journal of Science*, **274**, 1029–1043.

FLETCHER, R. C. 1977. Folding of a single viscous layer: Exact infinitesimal-amplitude solution. *Tectonophysics*, **39**, 593–606, http://dx.doi.org/10.1016/0040-1951(77)90155-X

FREHNER, M. 2011. The neutral lines in buckle folds. *Journal of Structural Geology*, **33**, 1501–1508, http://dx.doi.org/10.1016/j.jsg.2011.07.005

FREHNER, M. & SCHMALHOLZ, S. M. 2006. Numerical simulations of parasitic folding in multilayers. *Journal of Structural Geology*, **28**, 1647–1657, http://dx.doi.org/10.1016/j.jsg.2006.05.008

FREHNER, M., REIF, D. & GRASEMANN, B. 2012. Mechanical versus kinematical shortening reconstructions of the Zagros High Folded Zone (Kurdistan region of Iraq). *Tectonics*, **31**, http://dx.doi.org/10.1029/2011TC003010

GROOME, W. G. & JOHNSON, S. E. 2006. Constraining the relative strengths of high-grade metamorphic rocks using foliation refraction angles: an example from the Northern New England Appalachians. *Journal of Structural Geology*, **28**, 1261–1276, http://dx.doi.org/10.1016/j.jsg.2006.03.023

HAUGHTON, S. 1856. On slaty cleavage and the distortion of fossils. *Philosophical Magazine*, **12**, 409–421.

HAUPT, P. 2002. *Continuum Mechanics and Theory of Materials*. Springer Verlag, Berlin.

HUDLESTON, P. J. & LAN, L. 1993. Information from fold shapes. *Journal of Structural Geology*, **15**, 253–264, http://dx.doi.org/10.1016/0191-8141(93)90124-S

HUDLESTON, P. J. & LAN, L. 1995. Rheological information from geological structures. *Pure and Applied Geophysics*, **145**, 605–620, http://dx.doi.org/10.1007/BF00879591

HUDLESTON, P. J. & TREAGUS, S. H. 2010. Information from folds: a review. *Journal of Structural Geology*, **32**, 2042–2071, http://dx.doi.org/10.1016/j.jsg.2010.08.011

HUGHES, T. J. R. 2000. *The Finite Element Method: Linear Static and Dynamic Finite Element Analysis*. Dover Publications, Mineola.

LAN, L. & HUDLESTON, P. J. 1995. The effects of rheology and the strain distribution in single layer buckle folds. *Journal of Structural Geology*, **17**, 727–738, http://dx.doi.org/10.1016/0191-8141(94)00095-H

LANER, R. 2010. *Fluid assisted cataclastic deformation in quartzitic rocks (Portizuelo Antiforme, Luarca, NW Spain)*. Master thesis, Faculty of Earth Sciences, Geography and Astronomy, University of Vienna (Austria), http://othes.univie.ac.at/10591/1/2010-07-14_9725921.pdf

LECHMANN, S. M., SCHMALHOLZ, S. M., BURG, J.-P. & MARQUES, F. O. 2010. Dynamic unfolding of multilayers: 2D numerical approach and application to turbidites in SW Portugal. *Tectonophysics*, **494**, 64–74, http://dx.doi.org/10.1016/j.tecto.2010.08.009

MANCKTELOW, N. S. 2001. Single layer folds developed from initial random perturbations: the effects of probability distribution, fractal dimension, phase and amplitude. *In*: KOYI, H. A. & MANCKTELOW, N. S. (eds) *Tectonic Modeling: A Volume in Honor of Hans Ramberg*. Geological Society of America Memoirs, Boulder, **193**, 69–87.

MARCOS, A. 1973. Las Series del Paleozoico Inferior y la estructura herciniana del occidente de Asturias (NW de España). *Trabajos de Geología*, **6**, 1–113.

MULCHRONE, K. F. & MEERE, P. A. 2007. Strain refraction, viscosity ratio and multi-layer deformation: a mechanical approach. *Journal of Structural Geology*, **29**, 453–466, http://dx.doi.org/10.1016/j.jsg.2006.10.004

OERTEL, G. 1983. The relationship of strain and preferred orientation of phyllosilicate grains in rocks – a review. *Tectonophysics*, **100**, 413–447, http://dx.doi.org/10.1016/0040-1951(83)90197-X

OERTEL, G., ENGELDER, T. & EVANS, K. 1989. A comparison of the strain of crinoid columnals with that of their enclosing silty and shaly matrix on the Appalachian Plateau, New York. *Journal of Structural Geology*, **11**, 975–993, http://dx.doi.org/10.1016/0191-8141(89)90048-5

PRICE, N. J. & COSGROVE, J. W. 1990. *Analysis of Geological Structures*. Cambridge University Press, Cambridge.

RAMSAY, J. G. 1967. *Folding and Fracturing of Rocks*. McGraw-Hill Book Company, New York.

RAMSAY, J. G. & HUBER, M. I. 1987. *The Techniques of Modern Structural Geology, Volume 2: Folds and Fractures*. Academic Press, London.

REBER, J. E., SCHMALHOLZ, S. M. & BURG, J.-P. 2010. Stress orientation and fracturing during three-dimensional buckling: Numerical simulation and application to chocolate-tablet structures in folded turbidites, SW Portugal. *Tectonophysics*, **493**, 187–195, http://dx.doi.org/10.1016/j.tecto.2010.07.016

SHARPE, D. 1847. On slaty cleavage. *Quarterly Journal of the Geological Society of London*, **3**, 74–105.

SHIMAMOTO, T. & HARA, I. 1976. Geometry and strain distribution of single-layer folds. *Tectonophysics*, **30**, 1–34, http://dx.doi.org/10.1016/0040-1951(76)90135-9

SIDDANS, A. W. B. 1972. Slaty cleavage: a review of research since 1815. *Earth-Science Reviews*, **8**, 205–232.

SORBY, H. C. 1853. On the origin of slaty cleavage. *Edinburgh New Philosophical Journal*, **55**, 137–148.

TALBOT, C. J. 1999. Can field data constrain rock viscosities? *Journal of Structural Geology*, **21**, 949–957, http://dx.doi.org/10.1016/S0191-8141(99)00037-1

TREAGUS, S. H. 1983. A theory of finite strain variation through contrasting layers, and its bearing on cleavage refraction. *Journal of Structural Geology*, **5**, 351–368, http://dx.doi.org/10.1016/0191-8141(83)90023-8

TREAGUS, S. H. 1988. Strain refraction in layered systems. *Journal of Structural Geology*, **10**, 517–527, http://dx.doi.org/10.1016/0191-8141(88)90038-7

TREAGUS, S. H. 1999. Are viscosity ratios of rocks measurable from cleavage refraction? *Journal of Structural Geology*, **21**, 895–901, http://dx.doi.org/10.1016/S0191-8141(99)00018-8

TREAGUS, S. H. & SOKOUTIS, D. 1992. Laboratory modeling of strain variations across rheological boundaries. *Journal of Structural Geology*, **14**, 405–424, http://dx.doi.org/10.1016/0191-8141(92)90102-3

TULLIS, T. E. & WOOD, D. S. 1975. Correlation of finite strain from both reduction bodies and preferred orientation of mica in slate from Wales. *Geological Society of America Bulletin*, **86**, 632–638, http://dx.doi.org/10.1130/0016-7606(1975)86<632:COFSFB>2.0.CO;2

TWISS, R. J. & MOORES, E. M. 2007. *Structural Geology* 2nd edn. H. W. Freeman & Company, New York.

VIOLA, G. & MANCKTELOW, N. S. 2005. From XY tracking to buckling: axial plane cleavage fanning and folding during progressive deformation. *Journal of Structural Geology*, **27**, 409–417, http://dx.doi.org/10.1016/j.jsg.2004.10.011

WOOD, D. S. 1973. Patterns and magnitudes of natural strain in rocks. *Philosophical Transactions of the Royal Society London A – Mathematical Physical and Engineering Sciences*, **274**, 373–382, http://dx.doi.org/10.1098/rsta.1973.0066

WOOD, D. S. 1974. Current views of the development of slaty cleavage. *Annual Reviews of Earth and Planetary Sciences*, **2**, 369–401.

WOOD, D. S. & OERTEL, G. 1980. Deformation in the Cambrian Slate Belt of Wales. *Journal of Geology*, **88**, 309–326.

ZIENKIEWICZ, O. C. & TAYLOR, R. L. 2000. *The Finite Element Method, Volume 1: The Basis*. Butterworth-Heinemann, Oxford.

Folded Variscan thrusts in the Herrera Unit of the Iberian Range (NE Spain)

PABLO CALVÍN-BALLESTER* & ANTONIO CASAS

Departamento de Ciencias de la Tierra, Universidad de Zaragoza, 50009 Zaragoza, Spain

**Corresponding author (e-mail: calvinballester@gmail.com)*

Abstract: The Variscan structure of the Herrera unit (Iberian Chain, NE Spain) is characterized by a system of NNW–SSE-striking, east-verging, foreland-dipping thrusts, generated in a thin-skinned context, whose formation was favoured by the presence of two main detachment levels (Precambrian and Silurian shales). During the formation of the thrust system (first phase of deformation, D_1), the thrust sheets were deformed internally, mainly by asymmetrical folds with axial surface cleavage and east-verging thrusts; the normal evolution of the thrust system and the emplacement of an underlying thrust sheet resulted in progressive tilting and stacking, thus generating a foreland-dipping thrust system. These structures were folded and tilted by the emplacement of the Datos thrust (second phase of deformation, D_2), generating an associated deformation characterized by subvertical folds with axial surface cleavage, with slight east vergence. These structures can obliterate the previous D_1 structures in the sector near the Datos thrust. D_1 and D_2 structures have not been observed in the same outcrops, but the relationship between bedding and cleavage as well as cleavage relationship with other structures allows discrimination between the two cleavage sets. Based on the features of Palaeozoic rocks of the Iberian Range and the Iberian Massif, we support the idea that the Herrera unit belongs to the Cantabrian Zone.

Several examples in the literature cover the issues of folded thrusts and duplex structures from different points of view: structural analysis, cross-section reconstruction, numerical modelling, analytical modelling, and so on. (Boyer & Elliot 1982; Banks & Warburton 1986; Butler 1987; Alonso 1987, 1989*a*; Muñoz 1992; Wilson & Schumaker 1992; Casas-Sainz & Cortés-Gracia 1996; Gutiérrez-Alonso 1996; Couzens-Schultz *et al.* 2003 and references therein). Couzens-Schultz *et al.* (2003) constructed several analogue models in order to find the parameters that control the duplex formation and geometry; these authors concluded that both the strength of the detachment and the shortening rate appear to play a significant role in duplex evolution.

The Iberian Variscan belt provides several examples of duplex structures and folded-imbricate thrust systems, mostly located in the fold-and-thrust belt of the Cantabrian Zone (CZ) and its contact with the slate belt (Western Asturian–Leonese Zone, WALZ) in its hinterland. Structures of this kind are not well known in areas separated from the main outcrops of Palaeozoic rocks in the Iberian Peninsula (Pyrenees, Catalan Coastal Ranges and Iberian Chain), probably due to the limited extension of the outcrops of Variscan rocks (see, for example, García-Sansegundo *et al.* 2011), complicating the interpretation of these structures, especially in the absence of a clear regional structural framework. Nevertheless, in the Variscan units of the Iberian Chain the folded-imbricate thrust

model may be of key importance to understanding and interpreting the overall structure. In the Herrera unit of the Iberian Range, the Palaeozoic series lies uncomformably on Precambrian shales (Álvaro *et al.* 2008), which are the regional detachment level for the Variscan belt in this area. The Palaeozoic series consists mainly of alternating shales and sandstones with interbedded carbonates, including a 1000-m thick shaly unit of Silurian age. The mechanical stratigraphy is suitable for the development of multiple detachment levels and the formation of duplex systems.

The interest for the study of this area is twofold. Firstly, the area presents opportunities for detailed structural studies: on the one hand, thrusts develop together with two cleavage systems that allow for detailed studies of kinematics of structures and the determination of thrust–cleavage relationships; on the other hand, an imbricate, foreland-dipping thrust system has been recognized in this sector, allowing insights into the geometry and kinematics in thick sedimentary sequences. Secondly, an improved characterization of the Variscan structure of the Iberian Chain may facilitate its correlation with the Iberian Massif. Geological mapping and structural analysis (mainly considering bedding and cleavage relationships), a study on the vergence of the structures, and specially the integration of all these data into geological cross-sections, have allowed us to reconstruct the structure of the Herrera unit and establish the existence of refolded

From: LLANA-FÚNEZ, S., MARCOS, A. & BASTIDA, F. (eds) 2014. *Deformation Structures and Processes within the Continental Crust.* Geological Society, London, Special Publications, **394**, 39–52.
First published online November 21, 2013, http://dx.doi.org/10.1144/SP394.3

thrusts and two main phases of deformation during the Variscan evolution of the area.

Geological setting

The Aragonese Branch of the Iberian Range (NE Spain) constitutes a NW–SE alignment of structures generated during the Alpine intraplate deformation of the Iberian Peninsula (Casas-Sainz *et al.* 2000). The core for this region is formed by Precambrian and Palaeozoic rocks, deformed during the Variscan Orogeny according to NW–SE and NNW–SSE structural trends.

The relationship between this region and other Palaeozoic outcrops (e.g. the Iberian Massif, western Spain) is difficult to establish because the areas in between are covered with Mesozoic and Tertiary rocks. Julivert & Martínez (1983), Tejero & Capote (1987) and Álvaro (1991) support the view that the entire Iberian Range is included in the

WALZ. Conversely, Gozalo & Liñán (1988), using tectonostratigraphic criteria, proposed that the division between the CZ and the WALZ is in the north of the Iberian Range, where it appears as the Datos or Jarque thrusts, these being first-order structures with a NW–SE trend and NE vergence (Fig. 1a), separating the Badules Unit (to the west) and the Herrera Unit (to the east).

The study area is located at the northern edge of the Iberian Range, within the Herrera unit (Fig. 1b). This unit constitutes the NE limb of a kilometre-scale Variscan major anticline, with a predominant NW–SE to NNW–SSE trend and NE vergence. This anticline is associated with the Datos thrust that cuts across its core and superimposes Cambrian–Ordovician rocks of the Badules Unit on to the Upper Cambrian–Devonian rocks of the Herrera unit. In the core of the anticline there are Upper Proterozoic–Cambrian slates, which constitute the lower regional detachment level. Classically, the Variscan structure of the Herrera

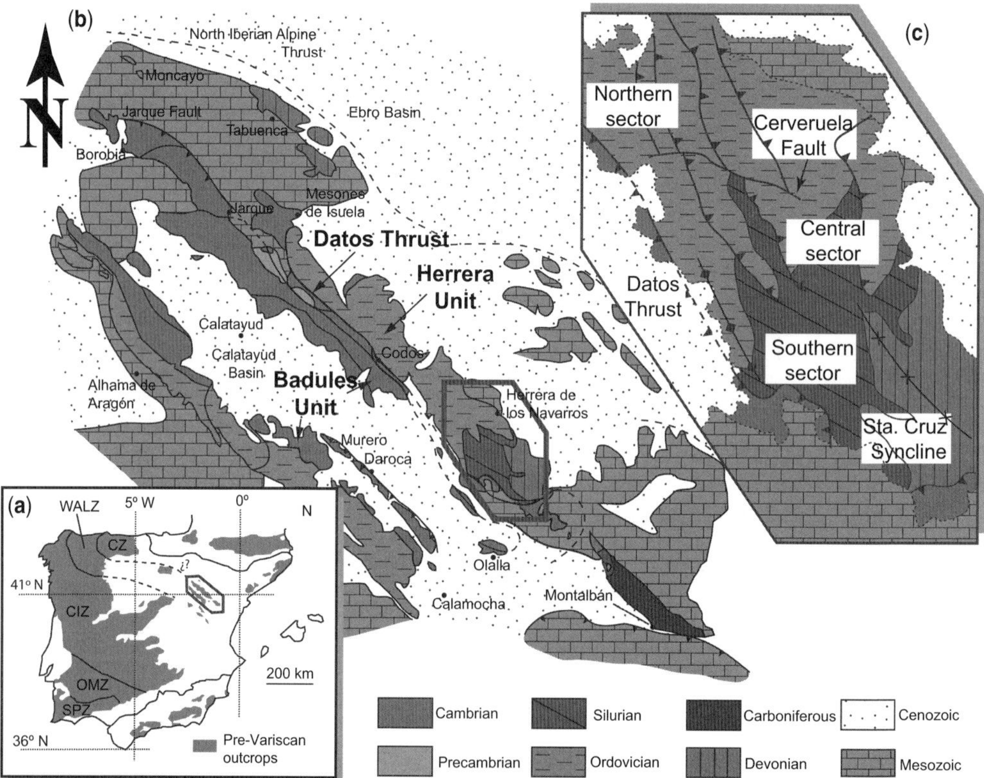

Fig. 1. (**a**) Pre-Variscan outcrops of the Iberian Peninsula showing their main tectonostratigraphic subdivisions: SPZ, South Portugese Zone; OMZ, Ossa-Morena Zone; CIZ, Central Iberian Zone; WALZ, West Asturian–Leonese Zone; CZ, Cantabrian Zone; red outlined area shown in (b). (**b**) Simplified geological map of the Aragonese Branch of the Iberian Chain; red outlined area shown in (c). (**c**) Structural sketch of the study area showing the main structures and sectors.

unit is considered to have arisen from a polyphase deformation (Tejero 1987; Tejero & Capote 1987), characterized by a set of kilometre-scale, east-verging folds later affected by an east-directed thrust system associated with the Datos thrust.

A more than 9000-m thick series characterizes the stratigraphic succession of the Herrera unit (Fig. 2). The lower level, Upper Cambrian–Silurian, is formed by alternating, hectometre-thick sandstone and shale packages, topped by a shale unit approximately 1000 m thick. The upper level, Upper Silurian–Devonian, is composed of shales, sandstones and carbonates.

All these pre-Variscan rocks are uncomformably covered by a Lower Permian volcanoclastic and detrital succession, and by Lower Triassic (Buntsandstein facies) conglomerates and sandstones (Lago *et al.* 2004), overlain by Mesozoic and Tertiary units. The Permian unit is restricted to the Fombuena graben, limited by north–south normal faults. All in all, the Permian to Tertiary sequences reach a maximum thickness of about 2000 m (Cortés-Gracia & Casas-Sainz 1996).

Description of the structure

Macrostructure

The macrostructure represented on the geological map (Fig. 3) shows three main sectors according to their cartographic and structural features (Fig. 1c), consistent with a NNW–SSE structural trend (Fig. 4). The northern sector is characterized by repetitions of the stratigraphic series by NNW–SSE-striking, east-dipping thrusts. In general, bedding shows steep dips to the east, and overturned, west dips in the western sector. The units involved in this system are Cambrian to Silurian in age. Between the localities of Aladrén and Cerveruela the series is apparently parallel to the thrust surfaces and only two Cambro-Ordovician units consisting of sandstone and shales are involved. Most thrusts of this system abut against an east–west fault (that can be interpreted as a tear fault) limiting this sector to the south.

South of the study area, the Cambrian units disappear and the Silurian shales crop out extensively. In this sector (central sector in Fig. 1c), a hanging-wall ramp, with younger rocks towards the south, can be observed. To the east, the tip line for each thrust is shifted northwards with respect to the thrust sheet below, except for the easternmost sheet, which reaches a more southerly position than the previous sheets and could cut the Devonian rocks. In the central sector the Silurian rocks are folded with a sub-horizontal envelope, allowing for extensive outcrops.

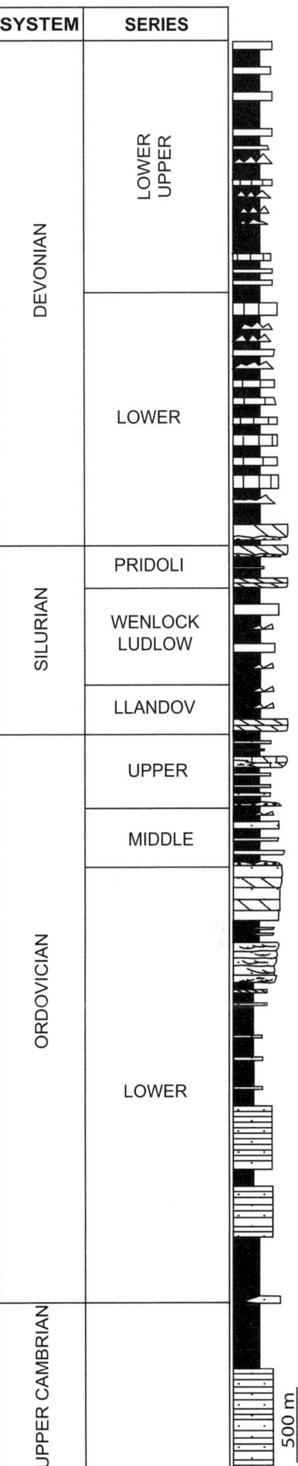

Fig. 2. Synthetic Palaeozoic stratigraphic profile of the Herrera unit.

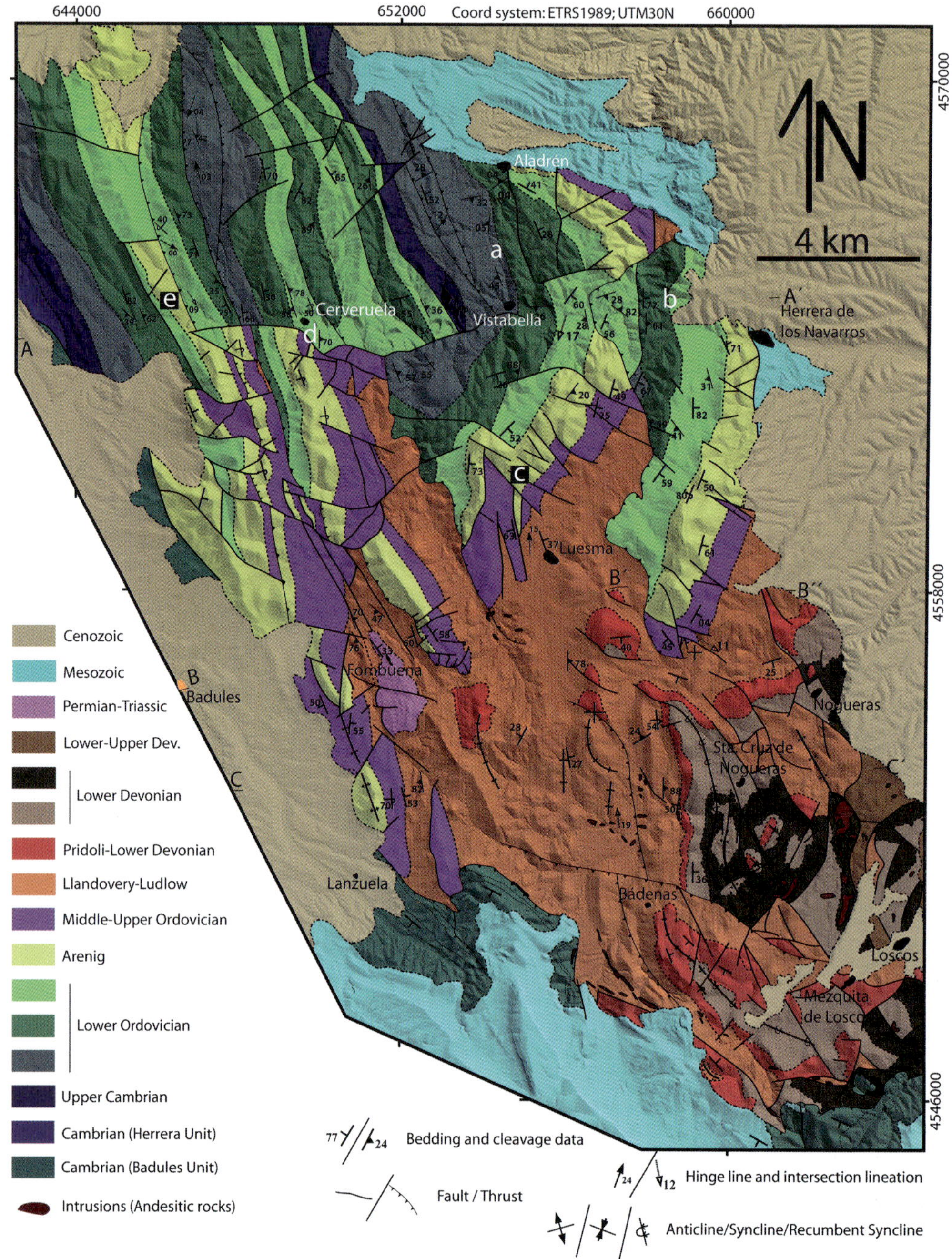

Fig. 3. Geological map of the study area showing the cross-section and location of the main outcrops, (**a**) to (**e**), appearing in Figure 7.

In the southern sector younger rocks (Silurian to Devonian) crop out extensively, affected by an east-verging thrust-and-fold system, with shallow dips to the west, favouring the presence of klippes with Devonian rocks in their hanging walls. The most representative structures are two synclines, trending NNW–SSE in the north and NW–SE in the south, becoming sub-parallel to the Datos

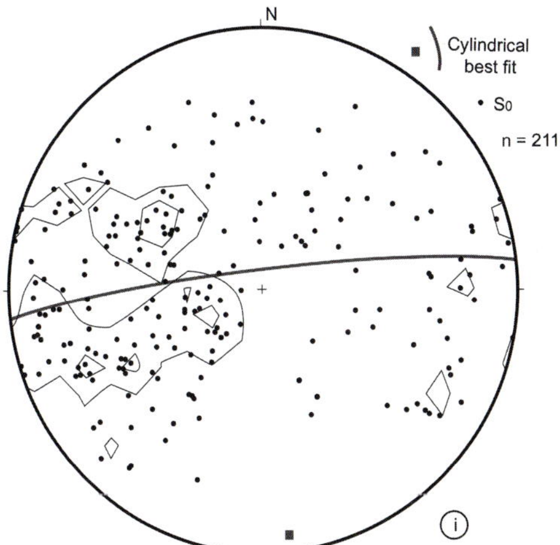

Fig. 4. Stereoplot (Schmidt net, lower hemisphere) of bedding planes recorded throughout the study area; these are consistent with NNW–SSE folding.

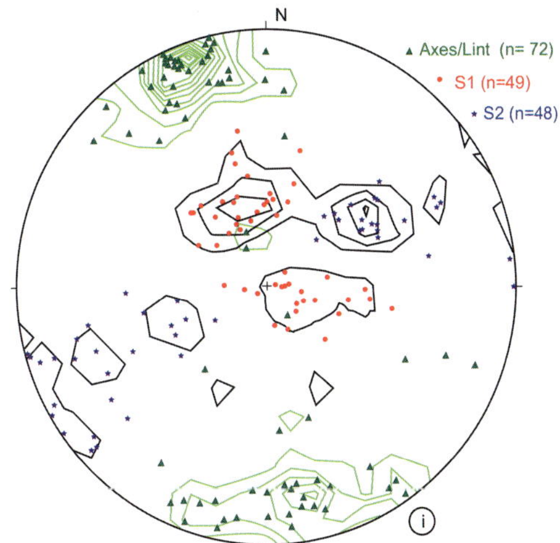

Fig. 5. Stereoplot (Schmidt net, lower hemisphere) of the orientation of the main structural features recognized in the Lower Structural Level (Upper Cambrian–Silurian).

thrust. In this sector klippes of Cambrian rocks associated with the Datos thrust (that delineates the study area to the west and south) can be observed. The Datos thrust strikes WNW–ESE in the south, dipping 30° to the south, progressively changing to a NNW–SSE strike and dipping steeply to the north.

The Palaeozoic rocks are surrounded by outcrops of Mesozoic rocks, folded in WNW–ESE direction, and unconformably covering the Datos thrust. To the south, they show similar dips to the Cambrian units. In the northern sector, Mesozoic units with NE dip lie unconformably on the Ordovician–Silurian rocks. Only at the northern and southern limits of the studied area do the Paleozoic rocks show a similar attitude to Mesozoic rocks.

Mesostructure

The most visible structures at outcrop scale are folds, minor thrusts and two sets of cleavage, related to thrusts and folds (Fig. 5). Cleavage surfaces show different attitudes depending on their relationships with tectonic structures, and can be ascribed to three main families.

The best developed set of structures in the study area is an east-verging system of thrusts and asymmetrical, cleavage-related folds. The folds (F_1) are asymmetrical (Z-shape) and have an eastward vergence (Fig. 6a), with shallow-dipping axial surfaces (15°E to 20°W), locally leading to formation of synformal anticlines (Fig. 6a–c). They can either be associated with thrusts or appear as decametric-scale trains of folds. The associated

foliation (S_1) is a slaty cleavage in shales and rough cleavage in sandstones, with shallow dips both to the east and west. A pervasive S_0/S_1 intersection lineation (L_{int-1}), parallel to F_1 axes (and also to axes of F_2 and L_{int-2}, see below) can be observed.

Throughout the study area, but mainly in the overturned limb of the large anticline related to the Datos thrust there are NNW–SSE-trending, upright-to-overturned, tight folds (F_2) with associated axial surface cleavage (S_2). These structures are developed particularly in shales, masking previous structures. S_2 is a slaty cleavage, dipping steeply (60–90°) both to the west (dominant) and east. The folds are asymmetrical (S-shape) and subvertical, with slight east vergence. Cleavage can obliterate the stratification, recognizable only from subtle changes in lithology (Fig. 6d). As it occurs with F_1, the fold axes and the intersection lineation (S_0–S_2) show NNW–SSE trends and a shallow plunge both to the north and south (Fig. 6).

Finally, back thrusts and brittle shear zones (Fig. 6e) with shallow dips (20–30°) deform the previously described structures. Locally, these structures can appear folded and tilted because of late deformation, probably of Alpine age.

Apart from indirect correlation according to orientation and relationship with other structures, it is difficult in the study area to define the relationships between S_1 and S_2, because they do not appear together in the same outcrops. S_1 can be seen in alternating sandstone–shale sequences, far from the Datos thrust, where S_2 is less visible. On the contrary, S_2 appears mainly in shaly series

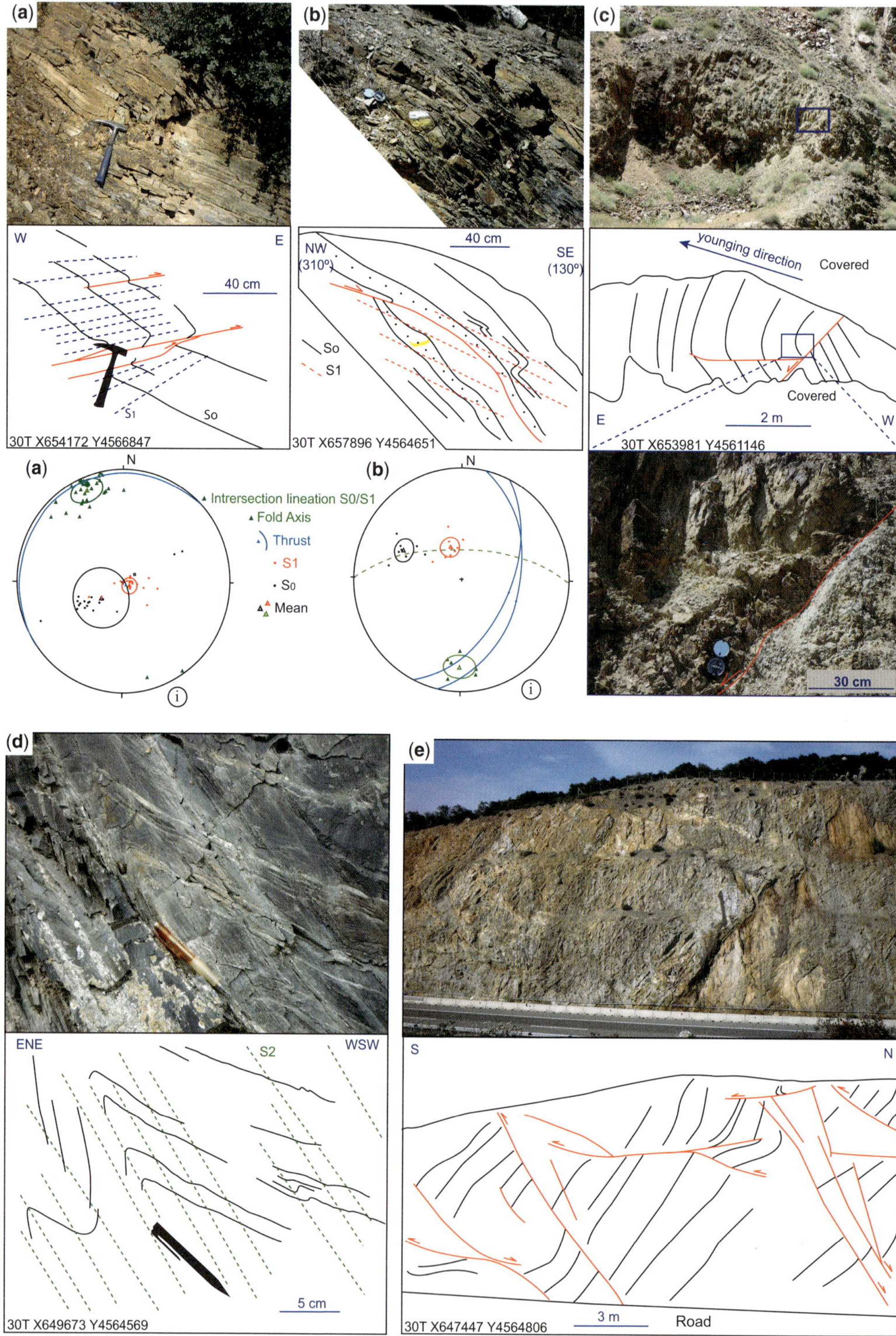

Fig. 6. Structural features of selected outcrops relevant to the interpretation proposed; UTM coordinates are shown, and their location is also shown in Figure 3. (**a**) NNW–SSW-striking east-verging, thrust-related fold, with axial surface

near the Datos thrust, probably obliterating previous structures.

Interpretation

The Herrera Unit can be interpreted as having a thin-skinned deformation style, in which the Precambrian–Cambrian shales would constitute the regional detachment level; these rocks can be observed in small outcrops along the trace of the Datos thrust (Fig. 1b), although they are absent in the study area. Other shaly levels within the Cambro-Silurian sequence may form minor detachment levels, strongly conditioning the structure. The most important comprises the Silurian shales, located approximately in the middle of the pretectonic sequence, accommodating the deformation in two ways: (i) as a partial roof décollement, especially in the northern sector, of the thrust detached in the Precambrian–Ordovician shales (Fig. 7a, b) and (ii) as the floor décollement for the thrusts affecting the Devonian rocks (Fig. 7c). For this reason, we divided the stratigraphic series into a Lower Structural Level (LSL), made up of Cambro-Silurian rocks, and an Upper Structural Level (USL), formed mainly of Devonian rocks. The Carboniferous, syn-tectonic sequence, with structural features similar to the USL, crops out only in the Montalbán anticline, south of the study area (Ferreiro et al. 1991).

The interpretation of the structure of the LSL is shown in cross-sections A–A' and B–B'–B'' (Fig. 4a, b). In the cross-sections, east-dipping thrusts with eastward transport direction are interpreted, in accordance with the minor folds and thrusts and the cross-cutting relationships (hanging-wall and footwall cut-offs). Beds show steeper dips than thrusts, indicating the existence of hanging-wall, low-angle ramps. The apparent parallelism between the thrusts and the beds in their foot-walls suggests the presence of flats except in the easternmost thrust, where a footwall ramp can be interpreted.

Towards the south, the thrusts cut progressively younger units in their hanging walls and cannot be followed to the south, where extensive outcrops of Silurian shales can be found. Our interpretation is that the upper part of the Silurian shales partly constituted a roof décollement, thus forming a duplex system. However, the geometry of the easternmost thrust, cutting across the Devonian rocks, indicates that the geometry did not exactly correspond to a duplex system, and that there was probably an early, pre-thrusting stage of folding involving the Silurian and Devonian rocks.

Comparing cross-sections A–A' and B–B'–B'' (Fig. 7) and considering the geological map (Fig. 3), a decrease in the number of thrusts to the south can be observed, coinciding with the location of the Cerveruela fault (Fig. 1c). This supports the hypothesis that the Cerveruela fault could be a tear fault, and the central thrust of the northern sector may be a complexity of the structure, forming a small duplex system favoured by décollements in the shale levels.

Towards the south the structure is simpler (Fig. 7c) and only a fold-and-thrust system with east vergence and shallow dips (consistent with the existence of klippes) can be observed.

The relationships between the northern-central sector and the southern sector are shown in Figure 8. The Herrera Thrust System in the north cuts progressively younger beds towards the south by means of oblique ramps, so that in the southern sector it is located above the present-day topographic surface, cutting a Silurian–Devonian fold-and-thrust system. At the same time, the ramps of each hanging wall represent the boundary of each thrust towards the foreland (to the east), and also probably represent the lateral termination of the thrusts.

Finally, the Datos thrust, with along-strike dip changes, and steepening in a northward direction, cuts from the west the system described above. The similar dip between the hanging wall of the Datos thrust and the overlying, unconformable Mesozoic beds in the southernmost sector of the studied area indicates a sub-horizontal original dip in this area, consistent with the existence of klippes of Cambrian rocks present in this sector. These changes of dip are consistent with a large NNW–SSE-trending anticline (Fig. 8): a plunge of nearly 30° towards the south of the whole structure allows one to observe a steep limb in the northern and central sectors (Fig. 7a, b), with a straight NNW–SSE trace (Fig. 3); in the southern sector, the hinge of this structure (Fig. 7c), with a shallow south-dipping and irregular WNW–ESE trace, can be observed (Fig. 3).

The original Variscan structure apparently does not show major changes due to the Alpine

---

Fig. 6. (*Continued*) cleavage (S$_1$) and stereoplot of the structures measured in the outcrop. (**b**) Early structure similar to (a), subsequently tilted to the east, showing a current dip towards the east and stereoplot of the structures measured in the outcrop. (**c**) Early thrust with associated drag fold, consistent with east-verging structures; as in (b), it has been subsequently tilted towards the east. (**d**) Overturned, east-verging folds, with axial surface cleavage. (**e**) Back thrust in Ordovician sandstones; note the brittle component of these structures that cut across beds formed during the main stages.

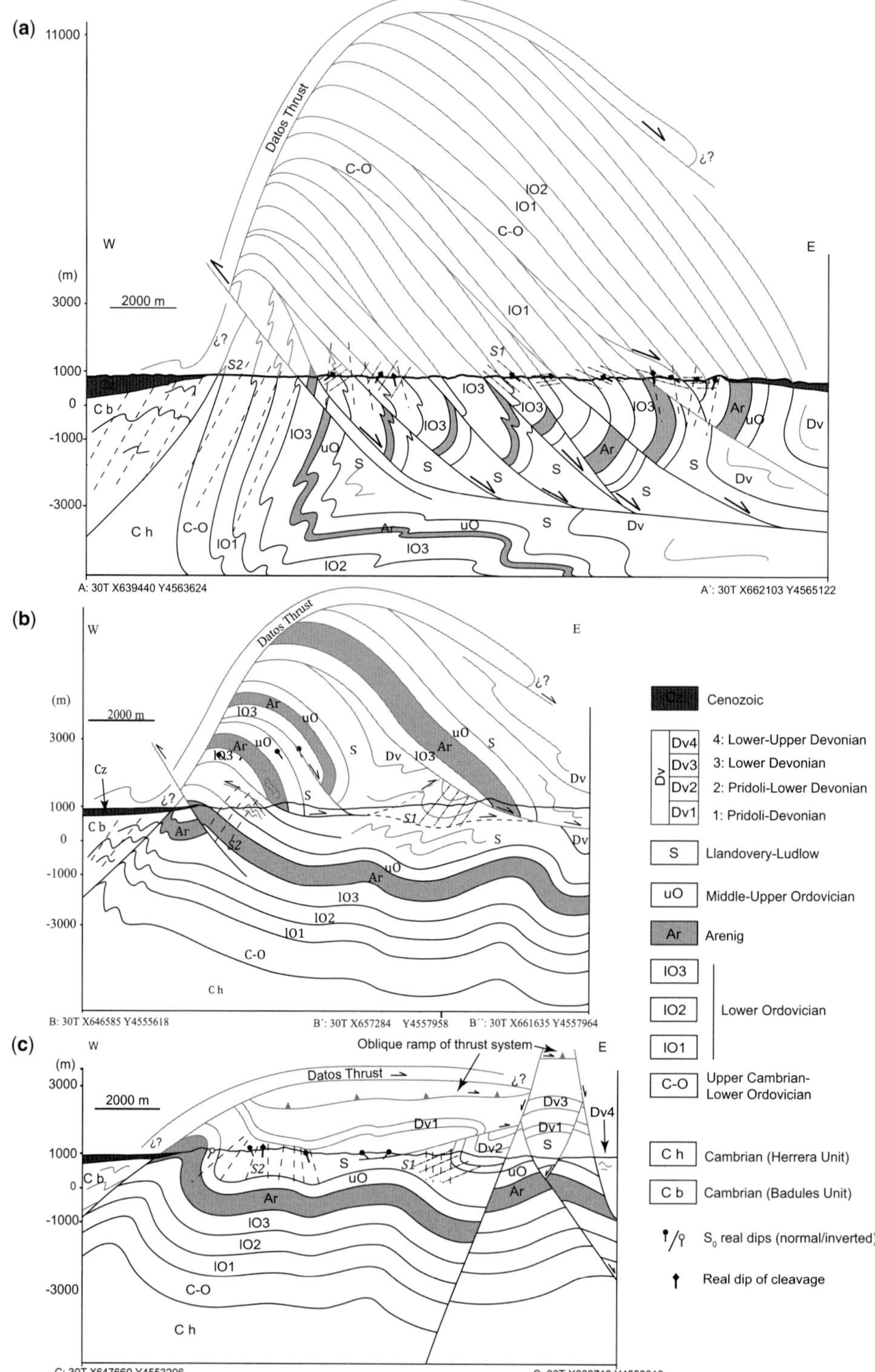

Fig. 7. Geological cross-sections showing the main structural features of the study area. Location of the cross-sections is shown in Figure 3. (**a**) Northern cross-section A−A′. (**b**) Central cross-section B−B′−B″. (**c**) Southern cross-section C−C′.

Fig. 8. 3D model shows the attitudes of the study area and the relationship between the northern and southern sector.

Orogeny. Only at the northern and southern limits of the studied area can Palaeozoic beds be observed that are folded according to the east–west trend, typical of Tertiary compressional deformation in this area (Cortés-Gracia & Casas-Sainz 1996). In the central sector, scattering of bedding poles from a cylindrical fit along a girdle defined by a NNW–SSE fold (Fig. 4) can be indicative of limited rotation and folding of Variscan structures in a later compressive stage, probably during the Tertiary compression.

Tectonic evolution

The structure of the Herrera Unit in the studied area can be interpreted as being the result of three stages of deformation, probably some of them overlapping in time in different areas; with each stage being considered as leading to the complete series of structures with consistent vergence and structural style. D_1 structures are related to the formation of the east-verging thrust system involving the Cambro-Ordovician series and folding observed in the Devonian rocks. Folds (F_1), and associated S_1 formed simultaneously during the emplacement of the successive sheets of this thrust system.

The emplacement of underlying thrust sheets (Fig. 9a, b) resulted in progressive tilting and stacking, thus generating a foreland-dipping thrust system. This kind of imbricate structure appears when the distance between ramps is lower than the displacement on each thrust (Butler 1987). The sequence of emplacement cannot be totally defined, because of the absence of chronological criteria. Butler (1987) suggests that the formation of this type of tectonic stacking develops more easily in a piggyback thrust sequence. According to observations, a piggyback thrust system is more compatible with the structure; the first thrust sheet emplaced may be the easternmost, favouring its cutting across Devonian folds. The later locations of the other sheets generate progressive tilting in the latter, with steeper dips to the foreland to the east.

D_2 structures are concentrated mainly around the Datos thrust. They show steep dips and are consistent with large-scale eastward-verging folds, in agreement with the major structure (Fig. 9c). The cross-cutting relations between S_0-S_2 are consistent with the large antiform associated with the

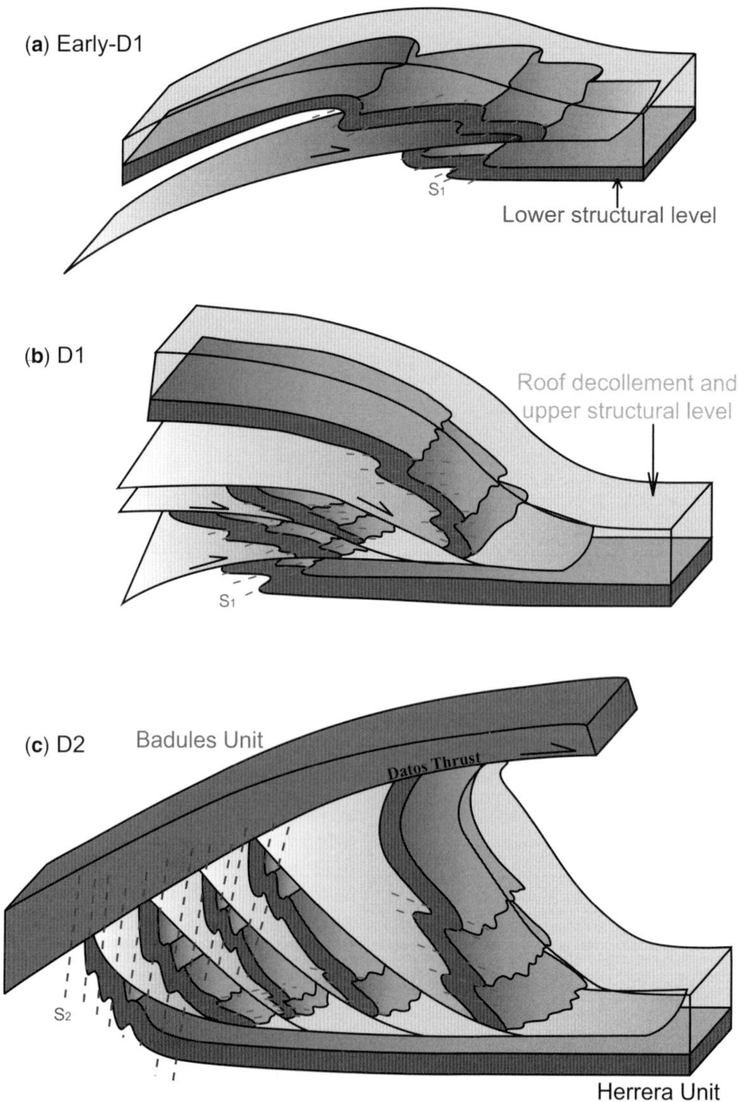

Fig. 9. Evolutionary model of the main structure north of the study area. (**a**) Formation of the first thrust and related structures (F_1 and S_1), east vergence. (**b**) Evolution of the duplex system, progressive tilting and stacking and formation of foreland-dipping duplex. (**c**) Formation of the out-of-sequence Datos thrust and folding with generation of axial surface cleavage (S_2).

Datos thrust, and the intensity of deformation decreases with the distance to this fault, showing the relationship between these structures and the formation of the Datos thrust. However, the Datos thrust is also folded for this stage, indicating an early origin of this structure within D_2.

Late structures (third phase of deformation, D_3) are scarce in the study area and do not seem related to any regional-scale tectonic process. Most of them are brittle structures, indicative of the transition to surface conditions, therefore suggesting that much of the tectonic stack was eroded. This could happen either during the later stages of the Variscan Orogeny (i.e. Permian) or during the Tertiary compressional stage, but to date there are no clear criteria to include them within a particular period.

Discussion

In summary, the structure of this sector of the Herrera Unit is characterized by a foreland-dipping imbricate thrust system affecting the Cambrian–Silurian series. The Silurian shales lie on a large hanging-wall flat that could possibly define a roof thrust for a foreland duplex system. Because of the erosion level, this hypothesis cannot be fully confirmed; none the less there are several lines of evidence that support it.

(1) Large outcrops of Silurian shales, with folds showing a sub-horizontal envelope, can be interpreted as a footwall flat.
(2) Different structural attitudes between Cambrian–Silurian and Silurian–Devonian series may indicate that the Silurian shales define a level of channelling of deformation, generating a disharmonic structure between series located below and above this level.
(3) Only the westernmost thrust cuts across Devonian rocks, and specifically the core of a syncline. This appears to show that this fold probably formed during the early emplacement of this thrust. Although difficult to interpret from outcrop observations, this kind of structure has been reproduced in analogue models with several detachment levels (Bonini 2001).
(4) The shallow westwards dip of thrusts in the southern sector of the studied area, generating small klippes, is more consistent with an upper detachment in the Silurian shales.

The origin of the forelandward dip of the northern thrusts may be due to two different processes: the internal dynamics of the stacking of thrust sheets (Butler 1987), or a later folding. In the study area the significant difference in dip between the thrust system (steep dips to the east) and the

folds involving the Devonian fold-and-thrust system (shallow dips to the west) is more consistent with the first hypothesis, because a later folding should have affected the entire system. However, a post-thrust folding has also been recognized, and is responsible for the folding that affects the Datos thrust and increases the eastward dips of the thrust system. The different attitude between the northern and southern sectors implies an important difference in shortening between the two sectors. In the northern sector several thrust sheets appear, accommodating a significant shortening, but towards the south these thrust sheets finish along oblique ramps, indicating less shortening in this direction. In the southern sector, deformation could have been accommodated by other structures towards the foreland, but the limited extent of outcrops does not allow for detailed observations to confirm this hypothesis.

The deformation phases proposed in this paper for the Iberian Chain contrast with the tectonic scenario proposed by other authors, probably because of the different scale of the interpreted structures. Tejero & Capote (1987), from structural analysis in the sector located to the NW of the area covered by us, interpreted three phases of deformation. The first phase is characterized by NW–SE-trending, east-verging hectometric folds, with axial surface cleavage. During the second phase thrust and faults developed, affecting mainly the overturned limbs of folds; the Datos thrust developed during this stage. Finally, the third phase is characterized by oblique folds (N 145°–N 100°) unequally distributed over the unit with associated crenulation cleavage; in the study area, these late structures have not been recognized.

Vílchez (1986) also proposed three deformation phases for the Herrera Unit. During the first stage, local, oblique folds (NE–SW) with SE vergence, and associated axial surface cleavage, developed. However, we did not recognize these structures in our study area. The second phase, according to this author, is the most important, and is characterized by east-verging, NNW–SSE-trending folds with associated cleavage. These structures correspond to the folds generated during the D_1 and D_2 phases proposed in this paper. Finally, during the third stage recognized by Vílchez (1986), thrusts (e.g. the Datos thrust) and back thrusts were generated; these back thrusts correspond to the folded thrust proposed by us. A similar deformation pattern was defined by Cardellach et al. (1988).

Relationship with other areas of the Variscan Orogen

The relationship between Palaeozoic rocks of the Iberian Range and other areas of the Iberian

Variscan Orogen has been a controversial matter. Different interpretations arise from the fact that the structural features of the Herrera unit that are intermediate between the CZ and the WALZ (i.e. the important thickness of the Cambro-Ordovician series and the lack of metamorphic facies) do not allow the area to be correlated directly with one or other domain. Julivert & Martínez (1983) considered all the Variscan rocks of the Iberian Range as belonging to the WALZ. Conversely, Gozalo & Liñán (1988) and Gutiérrez-Alonso (2004) propose the continuity of the Narcea antiform with the Precambrian rocks of the Iberian Chain. Accordingly, the Herrera unit would be a part of the Cantabrian Zone (CZ) and the Badules unit would then belong to the WALZ.

In spite of the existence of a pervasive cleavage set, the structural style of the Herrera Unit is similar to the CZ: it is characterized by thin-skinned tectonics and weak internal deformation; thrust- and fault-related folds are the main structures (Aller *et al.* 2004), with several duplex systems in the western sector (Alonso 1987, 1989*a*, *b*; Alonso *et al.* 1989; Gutiérrez-Alonso *et al.* 1990; Aller *et al.* 2004).

In their comparison between the Palaeozoic rocks of the Pyrenean area and the Iberian Massif, García-Sansegundo *et al.* (2011) concluded that the most deformed domains, affected by two cleavage families and metamorphism, correspond with the hinterland of the Variscan Orogen (WALZ or Ollo de Sapo Domain) whereas the less deformed domain, characterized by thrust and thrust-related folds with axial surface cleavage, corresponds with the foreland to the Variscan Orogen. However, the change of vergence and structural style makes it difficult to correlate the structural features of the Pyrenean area with those found in the Herrera Unit.

We interpret that a thrust system, with a foreland-dipping duplex geometry, can explain the structure of the Herrera unit. Associated with the different deformation events two cleavages can be distinguished. Therefore, the Herrera unit shows intermediate characteristics between the WALZ and the CZ. Nevertheless, significant variation in the degree of deformation to either side of the Precambrian outcrops has not been observed (Tejero & Capote 1987) in the way that occurs at the boundary between the WALZ and the CZ. We propose that the Precambrian outcrops of the Iberian Chain, and more specifically the Datos thrust, may be the boundary between the WALZ and the CZ (this would then include the Herrera unit) in the Iberian Range and that these Precambrian outcrops are equivalent to the core of the Narcea antiform (Gozalo & Liñán 1988; Gutiérrez-Alonso 2004). However, the shortening accumulated by the Datos thrust (at least in the area cropping out at present) would be much less than in the Narcea antiform, thus separating tectonic domains showing smaller differences than in the Iberian Massif.

Conclusions

The structure of the Herrera unit is characterized by an imbricate thrust system with a foreland-dipping geometry. Some of these structures can be interpreted as a duplex system, whose formation was favoured by the presence of several detachment levels. The lower one (floor detachment) is probably made up of Precambrian shales, the oldest rocks cropping out in the Iberian Range, or at the base of the Cambrian sequence. The upper level (roof detachment) is made up of Silurian shales which are about 1000 m thick, and constitute a preferential level of the deformation, generating a disharmonic structure between series located below and above.

The degree of deformation in the lowermost structural unit (Cambrian–Silurian) is higher than in the upper structural unit (Silurian–Devonian) and the superimposed generations of structures are better represented in the former. The upper unit is characterized by an east-verging fold-and-thrust system with axial surface cleavage.

Based on the deformation style and S_0-S_x relationships, structures related with two main stages of deformation can be distinguished: D_1 is associated with thrust emplacement; D_2 with the emplacement of the Datos thrust and wholesale folding. The tilting to the foreland of the duplex system was generated due to the internal dynamics of the thrust system (D_1), developed in a piggyback sequence and later by the emplacement of the Datos thrust (D_2). During D_1, horizontal, simple-shear, with top-to-the-east sense of movement was dominant, generating (originally in a westward direction), shallow-dipping structures, whereas during D_2, parallel-bed folding was dominant, with the subsequent formation of upright-to-overturned folds and related axial-surface cleavage.

The fold and thrust system in the Herrera unit may be correlated with the CZ of the Iberian Massif according to its structural features. However, the Herrera Unit has traditionally been included in the WALZ due to the important thickness of its Cambro-Ordovician series. We interpret that the NW–SE antiform associated with the Datos and Jarque faults can be correlated with the Narcea antiform of the Iberian Massif, and therefore we consider this antiform marks the boundary between the WALZ (towards the SW) and the CZ (towards the NE).

The authors thank the Geotransfer Research Group (Aragon Government and FEDER funds) for financial support for this research and are grateful for the use of Servicio General de Apoyo a la Investigación-SAI, Universidad de Zaragoza. The authors also acknowledge the careful and constructive revisions, comments and suggestion from Juan Luis Alonso and an anonymous reviewer, as well as the work done by the editors of this Special Publication.

References

ALLER, J., ÁLVAREZ-MARRÓN, J. *ET AL.* 2004. Estructura, deformación y metamorfismo. *In*: VERA, J. A. (ed.) *Geología de España*. SGE-IGME, Madrid, 42–49.

ALONSO, J. L. 1987. Sequences of thrust and displacement transfer in the superposed duplexes of the Esla Nappe Region (Cantabrian Zone, NW Spain). *Journal of Structural Geology*, **9**, 969–983, http://dx.doi.org/10.1016/0191-8141(87)90005-8

ALONSO, J. L. 1989a. Fold reactivation involving angular unconformable sequences: theoretical analysis and natural examples from the Cantabrian Zone (Northwest Spain). *Tectonophysics*, **170**, 57–77, http://dx.doi.org/10.1016/0040-1951(89)90103-0

ALONSO, J. L. 1989b. Síntesis cartográfica de la región del manto del Esla. *Trabajos de Geología*, **18**, 155–163.

ALONSO, J. L., ÁLVAREZ-MARRÓN, J. & PULGAR, J. A. 1989. Síntesis cartográfica de la parte sudoccidental del la Zona Cantábrica. *Trabajos de Geología*, **18**, 145–153.

ÁLVARO, M. 1991. *Memoria explicativa y mapa geológico de la Hoja 40 (Daroca) del Mapa Geológico de España 1:200.000 (1ª serie)*. ITGE, Madrid.

ÁLVARO, J. J., BAULUZ, B., GIL-IMÁZ, A. & SIMÓN, J. L. 2008. Multidisciplinary constraints on the Cadomian compression and early Cambrian extension in the Iberian Chains, NE Spain. *Tectonophysics*, **461**, 215–277, http://dx.doi.org/10.1016/j.tecto.2008.04.006

BANKS, C. J. & WARBURTON, J. 1986. Passive-roof duplex geometry in the frontal structures of the Kirthar and Sulaiman mountain belts, Pakistan. *Journal of Structural Geology*, **8**, 229–238, http://dx.doi.org/10.1016/0191-8141(86)90045-3

BONINI, M. 2001. Passive roof thrusting and forelandward fold propagation in scaled brittle-ductile physical models of thrust wedges. *Journal of Geophysical Research*, **106**, 2291–2311.

BOYER, S. E. & ELLIOT, D. 1982. Thrust systems. *American Association of Petroleum Geologists Bulletin*, **66**, 1196–1230.

BUTLER, R. 1987. Thrust sequences. *Journal of the Geological Society, London*, **144**, 619–634, http://dx.doi.org/10.1144/gsjgs.144.4.0619

CARDELLACH, E., BESTEIRO-RAFALES, J., OSÁCAR-SORIANO, C. & POCOVÍ, A. 1988. Aspectos geológicos y metalogenéticos del área de Herrera de los Navarros (Cadena Ibérica oriental. Zaragoza-Teruel). *Boletín Sociedad Española de Mineralogía*, **11-2**, 84–87.

CASAS-SAINZ, A. M. & CORTÉS-GRACIA, A. L. 1996. Cabalgamientos plegados en el macizo hercínico de la Sierra de Herrera (Cordillera Ibérica). *Geogaceta*, **19**, 3–6.

CASAS-SAINZ, A. M., CORTÉS-GRACIA, A. L. & MAESTRO-GONZÁLEZ, A. 2000. Intraplate deformation and basin formation during the Tertiary within the northern Iberian plate: Origin and evolution of the Almazán Basin. *Tectonics*, **19**, 258–289, http://dx.doi.org/10.1029/1999TC900059

CORTÉS-GRACIA, A. L. & CASAS-SAINZ, A. M. 1996. Deformación alpina de zócalo y cobertera en el borde norte de la Cordillera Ibérica (cubeta de Azuara–Sierra de Herrera). *Revista de la Sociedad Geológica de España*, **9**, 51–66.

COUZENS-SCHULTZ, B. A., VENDEVILLE, B. C. & WILTSCHKO, D. V. 2003. Duplex style and triangle zone formation: insights from physical modeling. *Journal of Structural Geology*, **25**, 1623–1644, http://dx.doi.org/10.1016/S0191-8141(03)00004-X

FERREIRO, E., RUIZ, V. *ET AL.* 1991. Mapa y memoria explicativa de la Hoja 40 (Daroca) del Mapa Geológico Nacional a escala 1:200.000. *ITGE*, 239.

GARCÍA-SANSEGUNDO, J., POBLET, J., ALONSO, J. L. & CLARIANA, P. 2011. Hinterland–foreland zonation of the Variscan orogen in the Central Pyrenees: comparison with the northern part of the Iberian Variscan Massif. *In*: POBLET, J. & LISLE, R. J. (eds) *Kinematic Evolution and Structural Styles of Fold-and-Thrust Belts*. Geological Society, London, Special Publications, **349**, 169–184, http://dx.doi.org/10.1144/SP349.9

GOZALO, R. & LIÑÁN, E. 1988. Los materiales hercínicos de la Cordillera Ibérica en el contexto el Macizo Ibérico. *Estudios geológicos*, **44**, 399–404.

GUTIÉRREZ-ALONSO, G. 1996. Strain partitioning in the footwall of the Somiedo Nappe: structural evolution of the Narcea Tectonic Window, NW Spain. *Journal of Structural Geology*, **18**, 1217–1229, http://dx.doi.org/10.1016/S0191-8141(96)00034-X

GUTIÉRREZ-ALONSO, G. 2004. La transición de la Zona Asturoccidental-Leonesa con la Zona Cantábrica: el Antiforme del Narcea. *In*: VERA, J. A. (ed.) *Geología de España*. SGE-IGME, Madrid, 52–54.

GUTIÉRREZ-ALONSO, G., VILLAR-ALONSO, P. & MARTÍN-PARRA, L. M. 1990. La estructura del Antiforme de Narcea. *Cuadernos Laboratorio Xeolóxico de Laxe*, **15**, 271–279.

JULIVERT, M. & MARTÍNEZ, F. J. 1983. Estructura de conjunto y visión global de la Cordillera Herciniana. *Geología de España. Libro Jubilar J. M. Ríos. IGME*, **1**, 612–630.

LAGO, M., ARRANZ, E., POCOVÍ, A., GALÉ, C. & GIL-IMAZ, A. 2004. Lower Permian magmatism of the Iberian Chain, Central Spain, and its relationship to extensional tectonics. *In*: WILSON, M., NEUMANN, E.-R., DAVIES, G. R., TIMMERMAN, M. J., HEEREMANS, M. & LARSEN, B. T. (eds) *Permo-Carboniferous Magmatism and Rifting in Europe*. Geological Society, London, Special Publications, **223**, 465–491, http://dx.doi.org/10.1144/GSL.SP.2004.223.01.19

MUÑOZ, J. A. 1992. Evolution of a continental collision belt: ECORS Pyrenees crustal balanced cross-section. *In*: MCCLAY, K. R. (ed.) *Thrust Tectonics*. Chapman & Hall, London, 235–246.

TEJERO, R. 1987. *Tectónica de los macizos paleozoicos al NE de Calatayud. Rama Aragonesa de la Cordillera Ibérica (Provincia de Zaragoza)*. PhD thesis, Complutense University of Madrid.

TEJERO, R. & CAPOTE, R. 1987. La deformación hercínica en los materiales paleozoicos nororientales de la Cordillera Ibérica. *Estudios geológicos*, **43**, 425–434.

VÍLCHEZ, J. F. 1986. *Rasgos geológicos y estructurales de la Unidad de Herrera (Cadena Ibérica). Resúmenes tesinas: curso 1983–1984*. Zaragoza University, Spain.

WILSON, T. H. & SCHUMAKER, R. C. 1992. Three-dimensional structural interrelationships within Cambro-Ordovician lithotectonic unit of the central Appalachians. *American Association of Petroleum Geologists Bulletin*, **74**, 1310–1324.

Crustal-scale folding: Palaeozoic deformation of the Mt Painter Inlier, South Australia

A. WEISHEIT[1], P. D. BONS[1]*, M. DANIŠÍK[2] & M. A. ELBURG[3]

[1]*Department of Geosciences, Eberhard-Karls University Tübingen, Wilhelmstraße 56, 72074 Tübingen, Germany*

[2]*Department of Earth and Ocean Sciences, The University of Waikato, Private Bag 3105, Hamilton 3240, New Zealand*

[3]*Geology Division, SAEES, University of KwaZulu-Natal, Private Bag X54001, Durban 4000, South Africa*

**Corresponding author (e-mail: paul.bons@uni-tuebingen.de)*

Abstract: The Mt Painter Inlier (Northern Flinders Ranges, South Australia) was exhumed in the Yankaninna Anticline, in the hanging wall of the major NE–SW-running Paralana Fault System. Regional north–south compression resulted in SE-directed oblique and lateral ramping on to the rigid Curnamona Province. Formation of the crustal-scale anticline caused exhumation of up to 15 km of basement rocks in the core of the anticline. Currently exposed rocks reached the surface in Permian times. Folding and thrusting commenced between *c.* 500 and *c.* 450 Ma and lasted until the Permian Period. Contrary to previous studies, only a minor part of the deformation can be attributed to the *c.* 500 Ma Delamerian Orogeny, with most of the tectonic activity being contemporaneous with the Alice Springs and Lachlan Orogenies. New Permo-Triassic zircon and apatite (U–Th)/He ages, as well as a titanite fission-track age point to a long-lived near-surface hydrothermal event that overprinted the basement and cover rocks. Hydrothermal reheating and burial below the Mesozoic Eromanga Basin sediments, to a depth of no more than 2 km, resulted in partial or full rejuvenation of the apatite (U–Th)/He system.

Crustal-scale folding and faulting is known from many orogenic belts, such as the Alpine Orogens (Fügenschuh *et al.* 1997; Schellart 2002; Ziv *et al.* 2010), the Himalayan Orogens (Burg *et al.* 1997; Maheo *et al.* 2004; Mukhopadhyay & Sharma 2010), the Variscan or the Pan-African belts (Konopasek *et al.* 2001; Passchier *et al.* 2002; Llana-Funez & Marcos 2007). Crustal deformation and thickening in these areas usually occurred during a restricted period of time, usually contemporaneous with, or followed by, exhumation. Phanerozoic continent–continent-collision deformation and exhumation usually lasted for several tens of millions of years (e.g. Suppe *et al.* 1992; Fügenschuh *et al.* 1997; Chen *et al.* 2002), while Precambrian collisions usually affect the crust for >100 myr (e.g. Gower *et al.* 2008; Hildebrand *et al.* 2010). Deformation in Archaean to Proterozoic intercratonic settings, however, may be long-lived, although probably intermittent, and can remain similar for >gyr (e.g. Holdsworth & Pinheiro 2000; Spaggiari 2007).

In the case of the Proterozoic Mt Painter region in the Northern Adelaide Fold Belt, South Australia (Fig. 1), crustal-scale folding and faulting have been interpreted as a result of intracrustal Delamerian (Pan-African) deformation, spanning *c.* 20 myr at *c.* 500 Ma (Preiss 1987, 2000; Teale 1993; Paul *et al.* 1999; Elburg *et al.* 2001; McLaren *et al.* 2006; Armit *et al.* 2012). It is generally assumed that the gross structure of the Mt Painter Inlier (MPI) had formed >400 Ma and was exhumed during Palaeozoic times (McLaren *et al.* 2002). This was recently challenged by Elburg *et al.* (2012) and Weisheit *et al.* (2013), who suggested that folding was not restricted to the Delamerian Orogeny, but lasted throughout almost the entire Palaeozoic Era.

Here we present new structural and thermochronological data from the Mt Painter region that confirms long-lived crustal-scale deformation in the Northern Flinders Ranges by far exceeded the duration of the Delamerian Orogeny as recorded in the southern Adelaide Fold Belt. In the following we first provide an overview of the geological setting and present our new mapping results and structural analyses. We then describe our new model of the inlier's tectonithermal evolution, which will be presented from higher to lower temperatures. We start with the reinterpretation of

From: LLANA-FÚNEZ, S., MARCOS, A. & BASTIDA, F. (eds) 2014. *Deformation Structures and Processes within the Continental Crust.* Geological Society, London, Special Publications, **394**, 53–77.
First published online November 22, 2013, http://dx.doi.org/10.1144/SP394.9

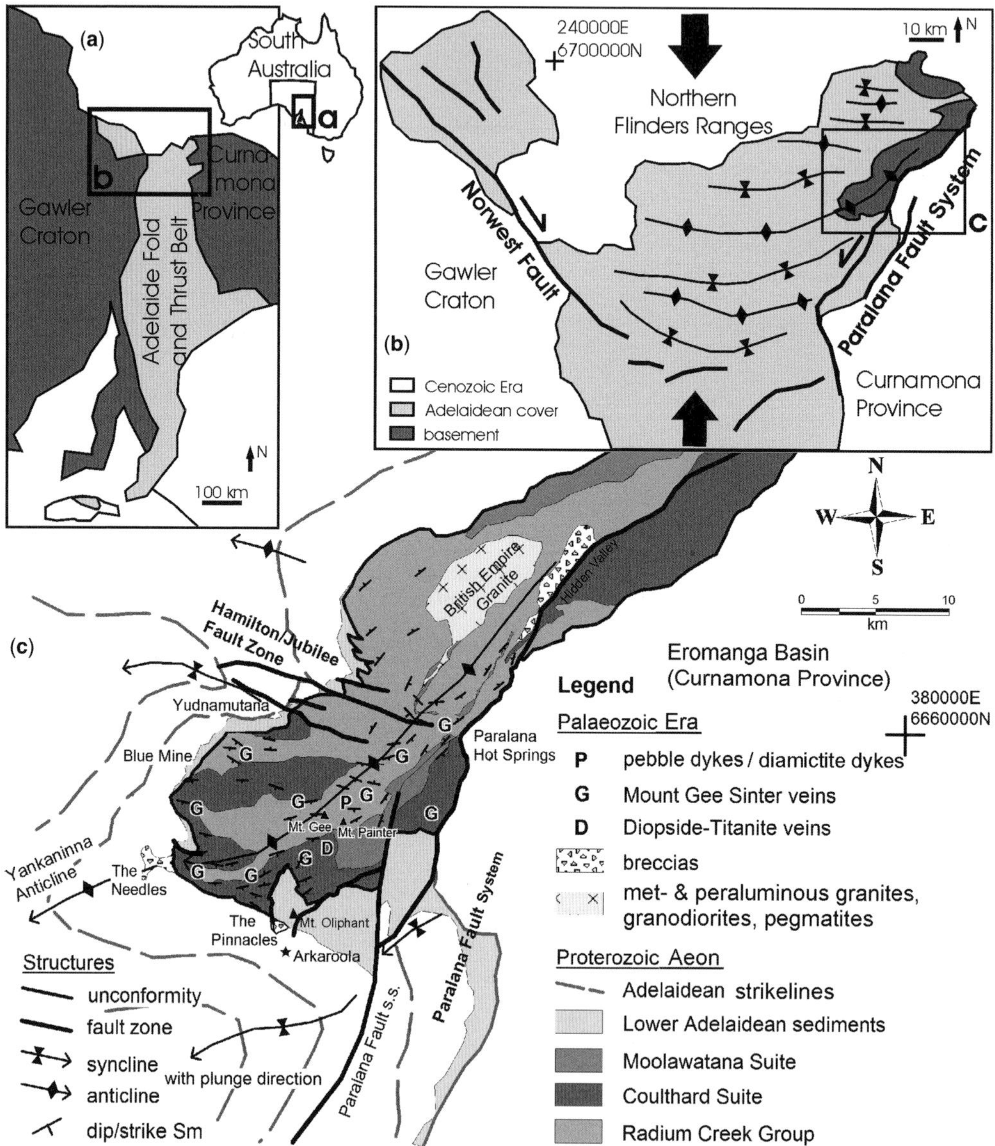

Fig. 1. Location and geological map of the Northern Flinders Ranges and the Mt Painter Inlier (MPI; after Coats & Blissett 1971 and Paul *et al.* 1999). (**a**) The Neoproterozoic Adelaide Rift Complex that is later deformed into the Adelaide Fold and Thrust Belt formed between the Archean Gawler Craton and the Palaeoproterozoic Curnamona Province. (**b**) The Northern Flinders Ranges are characterized by an arcuate pattern of 10 km-scale folds that formed between the dextral Norwest Fault and the sinistral Paralana Fault System. The MPI is exposed in the core of the NE–SW-trending Yankaninna Anticline. (**c**) Most of the MPI is formed by multiply deformed Mesoproterozoic basement, intruded by Palaeozoic magmatic intrusions (British Empire Granite) and overprinted by breccia zones. Locations of hydrothermal diopside-titanite veins and younger, Permian Mt Gee quartz sinter occurrences are shown.

published (higher-temperature) Ar/Ar data, and then work down to lower-temperature history based on published apatite fission-track data and new data from titanite fission-track, zircon, apatite and goethite (U–Th)/He analyses. Finally, we demonstrate that igneous and hydrothermal activity, 10 km-scale crustal folding and exhumation of the MPI occurred contemporaneously during *c.* 200 myr, spanning the Delamerian and Alice Springs Orogenies in central and southern Australia.

Geological setting

The Northern Flinders Ranges in South Australia are characterized by 10 km-scale folds that affect crystalline basement and the overlying Neoproterozoic (Adelaidean) cover succession (Fig. 1b; Preiss 1987, 2000; Paul *et al.* 1999). This area of crustal-scale folds is separated from the rigid Gawler Craton by the Norwest Fault System to the west and from the Curnamona Province by the Paralana Fault System to the east (Fig. 1b). The strike-slip components of these oblique-slip faults are dextral and sinistral, respectively. It formed during moderate north–south shortening of *c.* 10–20% between both cratons (Paul *et al.* 1999). So far, most authors have attributed the crustal-scale deformation to the Delamerian Orogeny at *c.* 500 Ma (Coats & Blissett 1971; Paul *et al.* 1999; Foden *et al.* 2006), while exhumation and refolding of the existing folds occurred during Palaeozoic times, possibly as an effect of the Alice Springs Orogeny (e.g. Teale 1993; Paul *et al.* 1999; McLaren *et al.* 2002, 2006; Armit *et al.* 2012). The Delamerian Orogeny is characterized by low pressure–high temperature metamorphism and the intrusion of syn- to post-tectonic granitoids in the southern Adelaide Fold and Thrust Belt (Preiss 1987; Foden *et al.* 2006). Style and intensity of deformation in this area (Flöttmann *et al.* 1994; Flöttmann & James 1997) differ from the structures in the northern Adelaide Fold and Thrust Belt (i.e. Northern Flinders Ranges; Paul *et al.* 1999). The major east–west-striking folds in the centre of the Northern Flinders Ranges fold–thrust belt turn to a NE–SW strike in sinistral transpression along the eastern Paralana Fault System. Here, Mesoproterozoic basement, the MPI, is exposed in one of the major anticlines (Yankaninna Anticline) in the hanging wall of the Paralana Fault System (Fig. 1b; Coats & Blissett 1971; Paul *et al.* 1999; Fraser & Neumann 2010).

Proterozoic eon

The oldest lithology in the MPI is a deltaic or shoalwater environment succession (Radium Creek Group; Coats & Blissett 1971; Cowley *et al.* 2012) with a maximum depositional age of *c.* 1590–1580 Ma (Figs 1c & 2; U–Pb detrital zircon analyses; Fanning *et al.* 2003; Wülser 2009; Fraser & Neumann 2010). At *c.* 1585–1575 Ma, a first suite of granitoids (Coulthard Suite dated by Pb/Pb and U–Pb of zircon; Elburg *et al.* 2001, 2012; Neumann 2001; Stewart & Foden 2003) intruded the (meta-) sediments, followed by the U-rich, high-heat-producing Moolawatana Suite at *c.* 1565–1555 Ma (U–Pb zircon, Pb/Pb K-feldspar and whole-rock data; Sheard & Cockshell 1992; McLaren *et al.* 2006; Fraser & Neumann 2010;

Elburg *et al.* 2012). The Moolawatana Suite appears to be associated with albitization of the older lithologies (Elburg *et al.* 2001, 2012; Neumann 2001). Metasomatism, metamorphism and deformation of the basement occurred prior to the deposition of the Adelaidean sedimentary cover (e.g. Teale 1993; Paul *et al.* 1999; Dipple *et al.* 2005; Elburg *et al.* 2001, 2012).

Intracratonic rifting during NE–SW extension resulted in the formation of NW–SE striking horst and graben systems in the southern MPI during Neoproterozoic times (Preiss 1987; Paul *et al.* 1999). The early graben structures were filled with sediments and flood basalts that extruded at *c.* 830 Ma, based on correlation with dolerite dykes dated by U–Pb on baddeleyite (Wingate *et al.* 1998). Ongoing extension and subsidence resulted in the formation of the large Adelaidean sedimentary cover basin along the full extent of the Adelaide Rift Complex that now forms the Adelaide Fold and Thrust Belt (Fig. 1a; Preiss 1987). Sedimentation of marine and terrestrial sediments continued until the onset of the Delamerian Orogeny at *c.* 500 Ma (Preiss 1987; Foden *et al.* 2006). The Adelaidean succession reaches a thickness of 10–12 km at the MPI (Coats & Blissett 1971; Paul *et al.* 1999). The geothermal gradient at that time is estimated to have been high at 40–50 °C km^{-1}, which was caused by the high concentration of heat-producing elements in the basement granitoids (U, Th, K; Mildren & Sandiford 1995; Sandiford *et al.* 1998; McLaren *et al.* 2002, 2006). Therefore, the temperature at the base of the Adelaidean basin reached *c.* 500 °C at the cessation of sedimentation at *c.* 500 Ma.

Palaeozoic magmatism and hydrothermal alteration

At *c.* 460–440 Ma, granites (metaluminous and peraluminous British Empire Granite), granodiorites and leucocratic pegmatites intruded the basement and adjacent cover sediments of the MPI (U–Pb ages of monazite and zircon; Elburg *et al.* 2003, 2013; McLaren *et al.* 2006; Wülser 2009). The inferred intrusion depth of the British Empire Granite (Fig. 1c) is about 13 km (McLaren *et al.* 2006). Pegmatitic diopside-titanite veins intruded the southern MPI at about the same time (Figs 1c & 3a; U–Pb age of titanite; Elburg *et al.* 2003). Based on fluid inclusions, Bakker & Elburg (2006) obtained conditions of 510 ± 20 °C at 130 ± 10 MPa (*c.* 5 km depth) for the formation of these veins, assuming a 100 °C/km geothermal gradient. A greater depth of 12–13 km results if the 40–50 °C km^{-1} gradient is applied instead. Considering that the vein-forming fluids were probably hotter

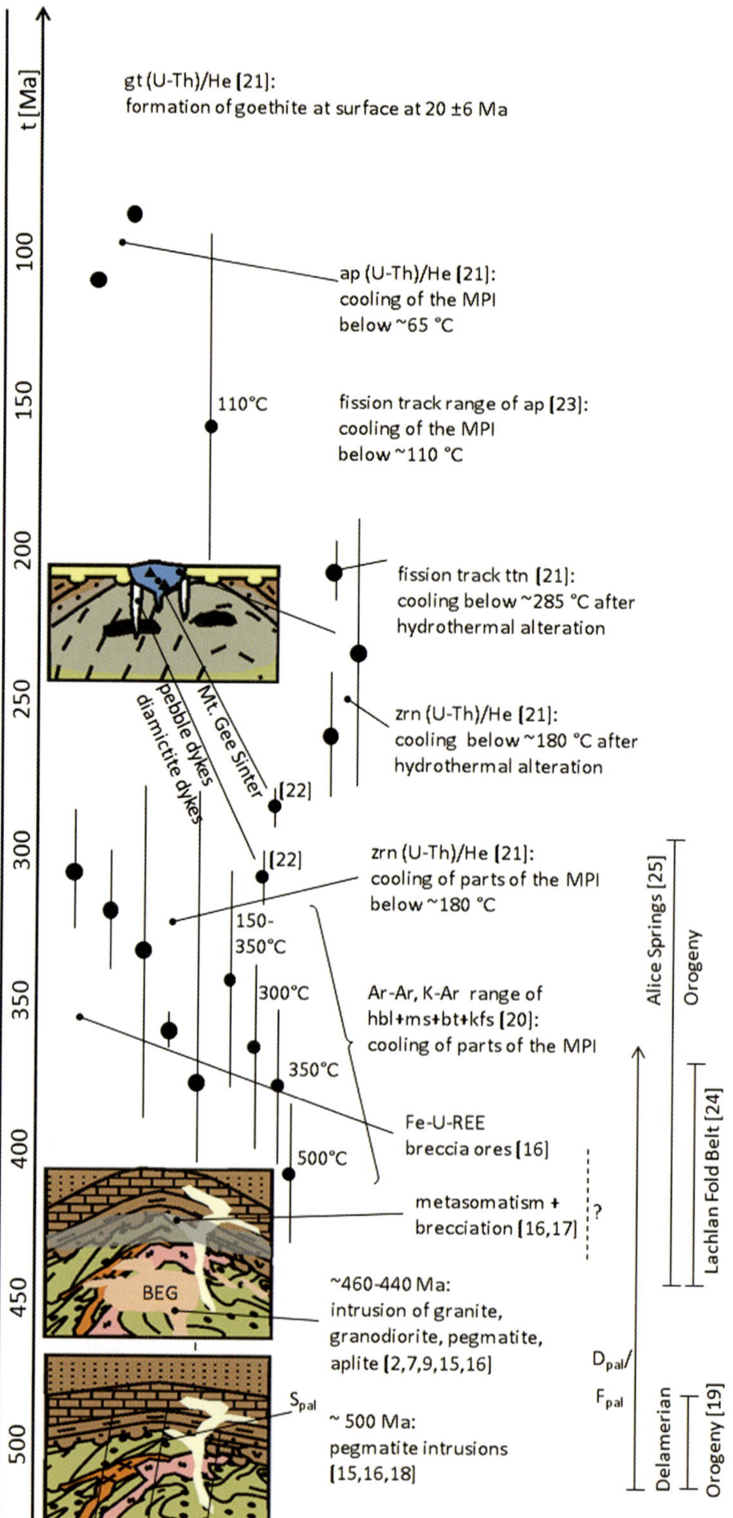
t [Ma]
gt (U-Th)/He [21]:
formation of goethite at surface at 20 ±6 Ma
ap (U-Th)/He [21]:
cooling of the MPI
below ~65 °C
110°C
fission track range of ap [23]:
cooling of the MPI
below ~110 °C
fission track ttn [21]:
cooling below ~285 °C after
hydrothermal alteration
zrn (U-Th)/He [21]:
cooling below ~180 °C after
hydrothermal alteration
pebble dykes
diamictite dykes
Mt. Gee Sinter
[22]
[22]
zrn (U-Th)/He [21]:
cooling of parts of the MPI
below ~180 °C
150-
350°C
300°C
Ar-Ar, K-Ar range of
hbl+ms+bt+kfs [20]:
cooling of parts of the MPI
350°C
Fe-U-REE
breccia ores [16]
500°C
metasomatism +
brecciation [16,17]
?
~460-440 Ma:
intrusion of granite,
granodiorite, pegmatite,
aplite [2,7,9,15,16]
BEG
S_pal
~ 500 Ma:
pegmatite intrusions
[15,16,18]
D_pal/
F_pal
Alice Springs [25] Orogeny
Lachlan Fold Belt [24]
Delamerian Orogeny [19]

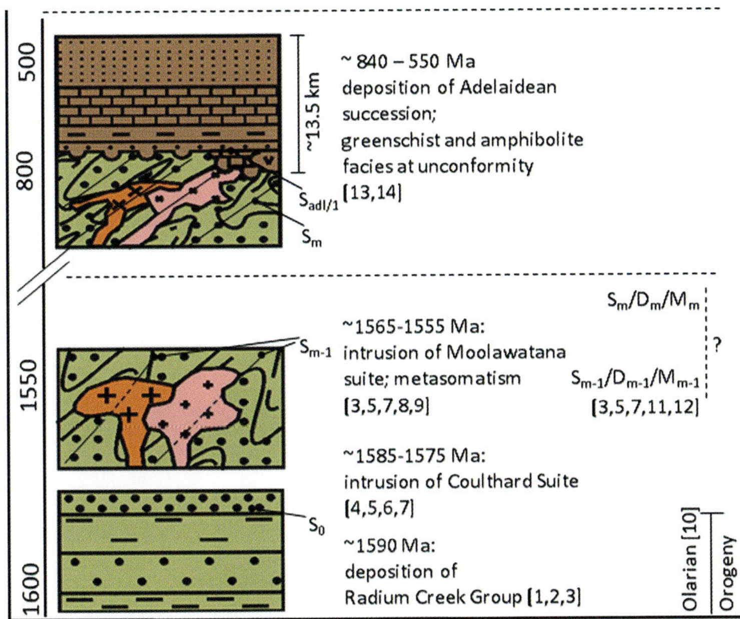

Fig. 2. Summary of the geological and structural evolution of the Mt Painter Province, based on published geochronological studies and low-temperature thermochronology of this study (bold circles). Differential exhumation of the basement between 500 and 300 Ma resulted in the formation of a crustal-scale anticline. The regional exhumation signal of the cooling ages is overprinted by local hydrothermal alteration. Australian regional events are shown for comparison. See text for description of the structural elements. Mineral abbreviations after Kretz (1983). MPI, Mount Painter Inlier; BEG, British Empire Granite. 1, Fanning *et al.* (2003); 2, Wülser (2009); 3, Fraser & Neumann (2010); 4, Neumann (2001); 5, Elburg *et al.* (2001); 6, Steward & Foden (2003); 7, Elburg *et al.* (2012); 8, Sheard & Cockshell (1992); 9, McLaren *et al.* (2006); 10, Rutherford *et al.* (2007); 11, Paul *et al.* (1999); 12, Armit *et al.* (2012); 13, Wingate *et al.* (1998); 14, Preiss (2000); 15, Elburg *et al.* (2003); 16, Elburg *et al.* (2013); 17, Weisheit *et al.* (2013); 18, Wülser *et al.* (2011); 19, Foden *et al.* (2006); 20, McLaren *et al.* (2002); 21, this study; 22, Brugger *et al.* (2011); 23, Mitchell *et al.* (2002); 24, Gray & Forster (2004); 25, Buick *et al.* (2008).

than their host rock, the veins formed at a depth of between *c.* 5 and 13 km.

The pegmatites and the diopside-titanite veins intrude and are overprinted by hydrothermal K-feldspar alteration and breccia zones (Coats & Blissett 1971; Drexel & Major 1987; Brugger *et al.* 2005). About 60 km^2 of K-feldspar-altered basement and cover rocks are now exposed (Weisheit *et al.* 2013). Palaeozoic fluid flow is also expressed by mineralizations of Fe-oxide–U–Cu–REE (rare earth element) ores (*c.* 355 Ma, U–Pb of monazite; Elburg *et al.* 2013).

Steep, decimetre to >metre-wide dykes, filled with rounded clasts of allochthonous and local (K-feldspar-altered) lithologies in a non-metamorphic fine-grained sandy matrix cut basement rocks near Mt Painter and Mt Gee. These diamictite dykes (Cowley *et al.* 2012), previously also referred to as 'pebble dykes' (Coats & Blissett 1971; Brugger *et al.* 2011), are interpreted as deep cracks filled with allochthonous and autochthonous (glacial) sediments. The youngest detrital zircons in the

matrix of the dykes are of Permian age (315 ± 8 Ma: U–Pb zircon; Brugger *et al.* 2011). The age of the dykes is also assumed to be Permian, based on the glacial nature of the dyke material and their likely Antarctic provenance, which indicates transport during the Late Carboniferous to Early Permian Glaciation (320–290 Ma; Eyles *et al.* 2002; Jones & Fielding 2004; Veevers 2006). The centre of the MPI was thus exhumed from its deepest burial at *c.* 500 Ma to near-surface levels at *c.* 300 Ma.

According to McLaren *et al.* (2002), the MPI cooled below *c.* 500 °C (Ar/Ar hornblende) after the 460–440 Ma intrusion of the British Empire Granite. Based on modelled K–Ar and Ar/Ar hornblende, mica and K-feldspar ages, McLaren *et al.* (2002) proposed three steps of fast cooling of the whole inlier at *c.* 430 Ma, *c.* 400 Ma and finally down to the closure temperature of K-feldspar between 300 and 150 °C at *c.* 325 Ma. These authors interpreted this cooling signal as a stepwise exhumation of the Delamerian Yankaninna Anticline.

(a)
Qtz+Cal
Di
Ttn
(b)
(c)
qtz-hem sinter vein
diamictite dyke
(d)
S_adl
quartzite
gneiss
S_m
(e)
Qtz
Bt
S_m-1
S_m
(f)
S_m
Palaeozoic crenulation
isoclinal fold
(g)
isoclinal S_m in migmatite
S_m
(h)
S_m
Palaeozoic crenulation
(i)
Qtz
ex-And
Kfs
Bt
S_pal
1mm

Outcrops of a near-surface quartz-hematite–calcite–fluorite sinter breccia and veins (Mt Gee Sinter; Fig. 3b; Coats & Blissett 1971; Brugger *et al.* 2011) can be found scattered throughout the centre of the MPI (Fig. 1c). These breccias and veins overprint the (K-feldspar-altered) basement rocks, as well as the diamictite dykes (Fig. 3c). They formed during near-surface boiling of an almost pure H_2O-fluid of 140–100 °C (Bakker & Elburg 2006) at *c.* 290 Ma (U–Pb age of a davidite; Brugger *et al.* 2011). This age is consistent with a formation of the diamictite dyke during the 320–290 Ma Carboniferous–Permian Glaciation (e.g. Jones & Fielding 2004). It should be noted that Bakker & Elburg (2006) and Brugger *et al.* (2005, 2011) regarded the various hydrothermal rocks (K-feldspar-altered breccias, Fe-oxide breccias, Mt Gee Sinter) as products of a single, but long-lasting event that may even be linked to the present-day hot springs (Paralana Hot Springs, Brugger *et al.* 2005). However, the cross-cutting relationships with the diamictite dykes allow a separation of middle- to upper-crustal pre-Permian, and near-surface post-Permian hydrothermal activity (Elburg *et al.* 2013; Weisheit *et al.* 2013). That the presently exposed rocks were near the surface at *c.* 290 Ma seems contradictory to apatite fission-track ages that indicate cooling below *c.* 110 °C between 260 and 175 Ma (Mitchell *et al.* 2002). As discussed below, these cooling ages can be interpreted as the waning stages of hydrothermal activity, rather than being related to exhumation as proposed by Mitchell *et al.* (2002).

Extension in Jurassic to mid-Cretaceous times (*c.* 200–90 Ma) led to the renewed formation of horst and graben structures to the east of the MPI. Up to *c.* 2.5 km of sediments were deposited in the centre of the newly formed Eromanga Basin, which thins to a few 100 m at its margins where the MPI lies (Fig. 1c; Krieg *et al.* 1995). Some Mesozoic sediments on top of the northern MPI (Coats & Blissett 1971) indicate burial of the inlier during that time as well. The final and recent exhumation of the inlier started in the Neogene Period along the Paralana Fault System (Célérier *et al.* 2005).

Methods

Detailed geological mapping and structural analyses were carried out in and around the southern and central MPI. Samples from basement and Adelaidean cover rocks were taken for microstructural analyses. The software move™2010 was used for 3D modelling of the Yankaninna Anticline.

A set of apatite, zircon, titanite and goethite samples was prepared for (U–Th)/He measurements and fission-track (FT) dating to reconstruct the cooling and exhumation history of the MPI. The closure temperatures of the used mineral systems are *c.* 310–265 °C (titanite FT; Coyle & Wagner 1998), *c.* 180 ± 20 °C (zircon (U–Th)/He; Reiners *et al.* 2004), *c.* 110 ± 10 °C (apatite FT; Green *et al.* 1986) and *c.* 75–55 °C (apatite (U–Th)/He; Schuster *et al.* 2006). The (U–Th)/He method applied to supergene goethite records the age of goethite formation at the surface (Shuster *et al.* 2005). Details on the (U–Th)/He method are given in Zeitler *et al.* (1987), Farley *et al.* (1996), Farley (2002), Meesters & Dunai (2002*a*, *b*), Reiners *et al.* (2004), Reiners (2005) and Schuster *et al.* (2006), and for the FT method in Gleadow & Lovering (1974), Wagner & Van den haute (1992), Galbraith & Laslett (1993), Reiners & Farley (1999) and Siebel *et al.* (2009).

Apatite and zircon (U–Th)/He dating was performed on clear, inclusion-free, euhedral grains, which were hand-picked under both binocular and petrographic microscopes. Single grains with a minimum diameter of 80 μm, were packed into Pt (apatite) and Nb (zircon) tubes and measured for He by isotope dilutions on the Patterson Ltd-designed helium-extraction lines with a quadrupole mass-spectrometer at the Department for Geosciences of the University of Tübingen (Germany), and at the John de Laeter Centre of Isotope Research in Perth

---

Fig. 3. (**a**) Pegmatitic diopside (Di) and titanite (Ttn) veins (south of Mt Painter) are overprinted by hydrothermal quartz–hematite Mt Gee Sinter and platy calcite that grew in open voids. (**b**) Typical hematite–quartz crusts of Mt Gee Sinter that can be found in veins and breccia zones throughout the southern MPI. (**c**) A diamictite dyke is cut by a Mt Gee Sinter vein, indicating that the Mt Gee event post-dated the diamictite dyke. (**d**) The contact between the sheared basement gneisses and the Adelaidean cover sediments has a low angle at the unconformity. (**e**) Microphotograph of a biotite schist NW of the Pinnacles. Biotite (Bt) and quartz (Qtz) form the main foliation S_m which is a crenulation cleavage that post-dates an older foliation S_{m-1}. (**f**) The main Mesoproterozoic deformation formed tight to isoclinal folds with the main foliation (S_m) parallel to the axial plane. The kink folding of S_m formed during the Palaeozoic deformation. (**g**) Leucosomes and melanosomes of a migmatite at Paralana Hot Springs are sheared and folded in the solid stage during the formation of S_m. (**h**) Typical Palaeozoic kink folds that developed mainly in mica-rich basement units. (**i**) Microphotograph of a phyllite in the Mt Oliphant Graben System east of the Pinnacles. The main Palaeozoic biotite–quartz (Bt, Qtz) foliation in the phyllite developed parallel to bedding. Crenulation of the main foliation occurred after the growth of intertectonic ?andalusite (ex-And) porphyroblasts that are now completely altered, while feldspar (Kfs) porphyroblasts grew post-tectonically.

(Australia). Apatites and zircons were degassed at *c.* 960 °C and *c.* 1250 °C, respectively, under ultra-high vacuum using a 960 nm diode (Tübingen) and Nd-YAG (Perth) lasers. To verify complete degassing in the first step, each grain was heated and measured again. The re-extractions generally revealed <1% of He of the first heating. Samples analysed for He in Tübingen were sent for U–Th measurements to the University of Arizona at Tucson, where they were spiked, dissolved and analysed by isotope dilution using inductively coupled plasma-mass spectrometry (ID-ICP-MS). Samples degassed in Perth were analysed in-house for U–Th using the same procedure (Evans *et al.* 2005). Uncertainty of the mean (U–Th)/He ages was calculated as the standard deviation of at least three replicates per sample.

A centimetre-sized, fresh, homogeneous, high-quality black goethite was used for (U–Th)/He dating. The sample was crushed in a steel mortar; shards were then hand-picked under a binocular microscope, ultrasonically cleaned in ethanol and left to dry. Single shards were loaded into Nb tubes. Helium and U–Th were measured using isotope-dilution mass spectrometry (quadrupole gas-MS and ICP-MS, respectively) at the John de Laeter Centre of Isotope Research (Perth). Helium was extracted at *c.* 500 °C under ultra-high vacuum. Following the He measurements, Nb tubes containing the samples were retrieved from the laser cell, spiked, dissolved in 200 μl of 50% HCl in Teflon vials loaded in Parr bombs and heated to 210 °C for 60 h. Besides sample solutions, each Parr bomb also contained one blank and two spiked standard solutions for blank correction and calibration purposes. All solutions were analysed for U–Th by ID-ICP-MS. The total analytical uncertainty (TAU) was calculated as the square root of the sum of squares of uncertainty on He and weighted uncertainties on U and Th measurements. Given the large size and polycrystalline character of the dated sample, (U–Th)/He ages were not corrected for alpha ejection (Farley *et al.* 1996). The ages were also not corrected for diffusive loss (Shuster *et al.* 2005).

For FT analysis, titanite crystals were embedded in epoxy and polished to 4π geometry. Spontaneous tracks were revealed by etching with a solution of $1HF:2HNO_3:3HCl:6H_2O$ at 23 °C for six min (Gleadow & Lovering 1974). FT dating was carried out using the external detector method (Gleadow 1981) with low-U muscovite sheets (Goodfellow mica[TM]). The IUGS-recommended zeta calibration approach was taken for age determination (Hurford & Green 1983). Zeta values (Hurford 1998) of 283.5 ± 6.2 (titanite) for dosimeter glass CN-5 have been derived from six titanite age standards from the Fish Canyon Tuff.

FT ages were calculated with TrackKey 4.1 (Dunkl 2002).

In the following we discuss the geological history of the MPI in chronological order. We first present and interpret the results of our detailed mapping and structural analysis; this is then followed by discussion of the timing of Palaeozoic folding and exhumation. We used robust Ar/Ar plateau ages of samples from basement and cover units published by McLaren *et al.* (2002). These plateau ages are defined by McLaren *et al.* (2002) as 'that part of an age spectrum where the gases of consecutive steps are within 2σ error of one another and together comprise at least 50% of the gas release'; the authors also reported that a plateau-like signal was observed 'if one or two steps in such a sequence are discordant'). Finally, we present the results of low-temperature thermochronology.

Results

Deformation history and structural reconstruction

The Mesoproterozoic Era. A pervasive to localized NE–SW–to east–west-striking main foliation S_m, mainly defined by biotite and sillimanite, is found throughout almost the whole MPI (S_{gneiss} in Elburg *et al.* 2003 and S_{2+3} in Armit *et al.* 2012; Fig. 4). The Adelaidean unconformity truncates this foliation at a low angle of about ten degrees (Fig. 3d). Biotite, sillimanite and cordierite parageneses in meta-pelites indicate low pressure–high temperature metamorphism during the S_m-forming deformation event (Paul 1998; Armit *et al.* 2012). The S_m formed axial-planar to closed and isoclinal folds and is a crenulation cleavage of an earlier foliation S_{m-1} (Fig. 3e, f). Hence, at least one deformation event predates the main foliation. The metamorphic grade in the basement increases to migmatitic conditions in the Paralana Hot Springs area. Isoclinal buckle folds of leucosomes with S_m as the axial-planar foliation indicate that the leucosomes solidified before folding. Partial melting therefore occurred pre- to syn-S_m (Fig. 3g).

Adelaidean cover sediments and the main basement foliation S_m were folded during the formation of the Yankaninna Anticline. The original orientation of S_m can be found by rotating the cover layers back to horizontal and passively rotating S_m along with these layers (Fig. 4; Table 1). Figure 4 shows that the main foliation S_m was shallowly dipping with a mean orientation of *c.* 10° towards 283° prior to the deposition of the Neoproterozoic Adelaidean cover. The large spread in orientations may be due to (1) variations in S_m prior to deposition of the Adelaidean succession, (2) tilting of

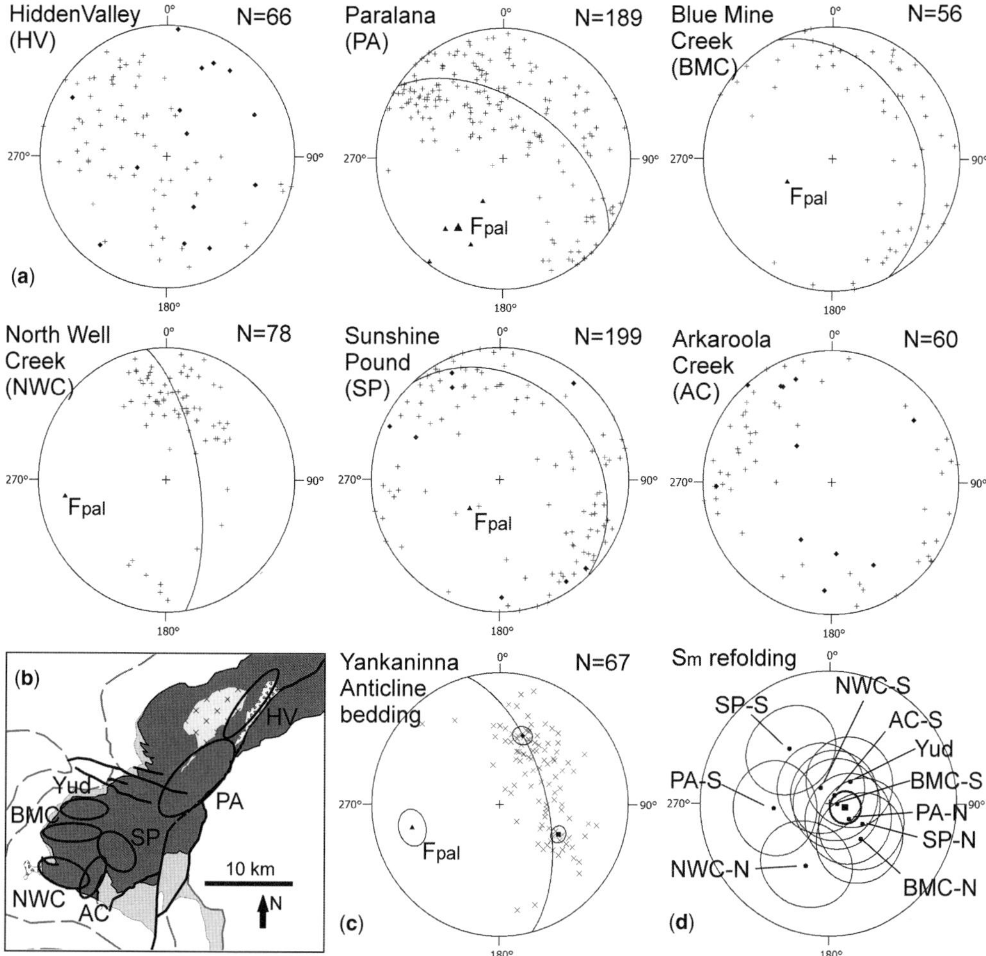

Fig. 4. (**a**) Stereoplots of poles (plus) to the main foliation S_m in selected areas (see (**b**)) in the southern and central MPI and (**c**) poles to bedding (cross) in the overlying Adelaidean succession. The fold axis of the Palaeozoic Yankaninna Anticline (triangle in (c)) in the Adelaidean succession plunges gently to the WSW. S_m in the basement is folded with fold axes (triangles in (a)) that change in plunge direction from west at North Well Creek to SW in the northwestern Paralana area. Poles of crenulation cleavage planes are plotted with diamonds. (**d**) Unfolding of the basement's S_m (round dot) shows that it made a small angle with the Adelaidean beds with a mean dip direction and dip (square) of 283/10 prior to the Palaeozoic folding.

basement during Neoproterozoic horst and graben formation and (3) ductile deformation within the basement during formation of the Yankaninna Anticline (see below).

The Neoproterozoic Era. Extensional tectonics in early Neoproterozoic times led to the formation of horst and graben structures in the exhumed and deformed Mesoproterozoic basement (Preiss 1987, 2000). The lowermost Adelaidean strata were deposited in NW–SE-running graben (Fig. 5). These graben deepened from the NW ($\leq$1 km) to

the SE where >2 km-thick graben sediments are found. They were passively folded during formation of the Yankaninna Anticline and now crop out on the western margin of the basement (graben at the Needles and at Yudnamutana) and in the SE (graben north of Arkaroola; Figs 1c & 5a). Whereas graben faults now strike east–west on the NW limb of the anticline (e.g. the Hamilton and Jubilee Faults, Fig. 1c), they run about north–south on the SE limb. This can be explained by the passive folding of an originally *c.* 310°-striking graben system by the Yankaninna Anticline with a

Table 1. *Structural elements*

	Fold axis Yankaninna Anticline	SE limb Yankaninna Anticline	NW limb Yankaninna Anticline	S_{main}	Paralana Fault s.s.	Paralana Fault System	Mt Oliphant Graben faults	Hamilton-Jubilee Fault System
Today	$30 \to 255 \pm 9°$	$200/46 \pm 6°$	$296/43 \pm 5°$	variably folded (see Fig. 4)	SE limb: 100/90* NW limb: 127/52 ± 8	270/90* to 315/90*	PiF: 087/16 NeF: 015/54	190/80
Unfolded	$00 \to 248$	000/00	000/00	283/ 10 ± 10°	280/65 ± 4°	270/90* to 315/90*	041/53 ± 8°	221/53 ± 8°

Orientations of structural elements today and after unfolding.
*Probable dip; PiF: Pinnacles Fault; NeF: Needles Fault.

SW-plunging fold axis (Fig. 5; Table 1). The graben structures on either side of the basement are thus in fact outcrops of one and the same graben structure, which we informally name the Mt Oliphant Graben System, after Mt Oliphant which is situated in one of the deepest parts of the graben north of Arkaroola Village (Fig. 5b). Graben faults, such as the Hamilton and Jubilee Faults cannot be traced all the way across the basement, probably because of heavy overprinting by hydrothermal activity.

Localized deposition in the lower Adelaidean graben was followed by widespread sedimentation in the first of a number of rift and sag cycles (Preiss 1987, 2000). While some of the early graben faults (e.g. at the Pinnacles) were deactivated before deposition of mid-Adelaidean succession, others remained active for longer. The Hamilton and Jubilee Faults still cause an offset of about 2 km in the mid-Adelaidean strata.

There is a distinct difference in metamorphic grade east and west of the Paralana Fault *sensu stricto* (Coats & Blissett 1971). The Paralana Fault *s.s.*, located east of Arkaroola Village, is the branch of the Paralana Fault System that gave it its name (Fig. 1c). Whereas the unconformity was buried up to *c.* 13.5 km just west of the Paralana Fault *s.s.* (Paul *et al.* 1999), it was only buried to *c.* 9.5 km east of it (Fig. 5c). This is confirmed by fluid inclusions in veins in shales east of the Paralana Fault *s.s.* that indicate a burial of 3–6 km of a horizon that is *c.* 4 km above the unconformity (Bons *et al.* 2007). The total thickness of the Adelaidean succession decreases to the east of the Paralana Fault System, where seismic surveys reveal a maximum thickness of *c.* 6 km (Korsch *et al.* 2010). The north–south-striking Paralana Fault *s.s.* thus became active during deposition of the Adelaidean succession and was one of several step faults with a west-down movement (and unknown horizontal offset) bounding a basin that deepened westwards.

The significant differences in the thickness of the sedimentary pile on either side of the Paralana Fault *s.s.* directly affected the metamorphic grade of the rocks at the unconformity. At an elevated geothermal gradient of 40–50 °C km^{-1} (Mildren & Sandiford 1995; Sandiford *et al.* 1998; McLaren *et al.* 2002, 2006) the lower Adelaidean sediments west of the fault experienced amphibolite-facies conditions close to 500 °C, while the same stratigraphic level east of the fault only reached depths of greenschist-facies conditions. The various metamorphic minerals, including cleavage-forming biotite, are aligned sub-parallel to the sedimentary bedding and the mineral-in isogrades are parallel to the unconformity (Coats & Blissett 1971; Mildren & Sandiford 1995). This suggests that the metamorphic grade of the Adelaidean units is a result of static burial of the unconformity and independent from the Delamerian Orogeny.

The Palaeozoic Era. The Adelaidean strata and crystalline basement of the MPI were folded into the 10 km-scale Yankaninna Anticline (Fig. 5a, b; Coats & Blissett 1971; Preiss 1987; Paul *et al.* 1999). The steep axial plane strikes NE–SW in the northeastern part of the inlier and turns towards east–west in the SW (Fig. 1). The crest of the anticline is located east of the British Empire Granite and north of Paralana Hot Springs. Here the fold strikes NE–SW with a gently NE-plunging fold axis in the northeastern part of the inlier. Towards the SW, the fold hinge plunges between 20 and 50° towards the SW to west (Fig. 4).

Apart from the large-scale folding and smaller, 100 m-scale parasitic folds of the main foliation S_{m} in the basement, few deformation structures developed during Palaeozoic folding. Within the basement, only centimetre-scale crenulation cleavages (S_{pal}) and kink folds in almost pure biotite schists can be attributed to this folding (Fig. 3h). Palaeozoic folding caused the formation of a steep, axial-planar

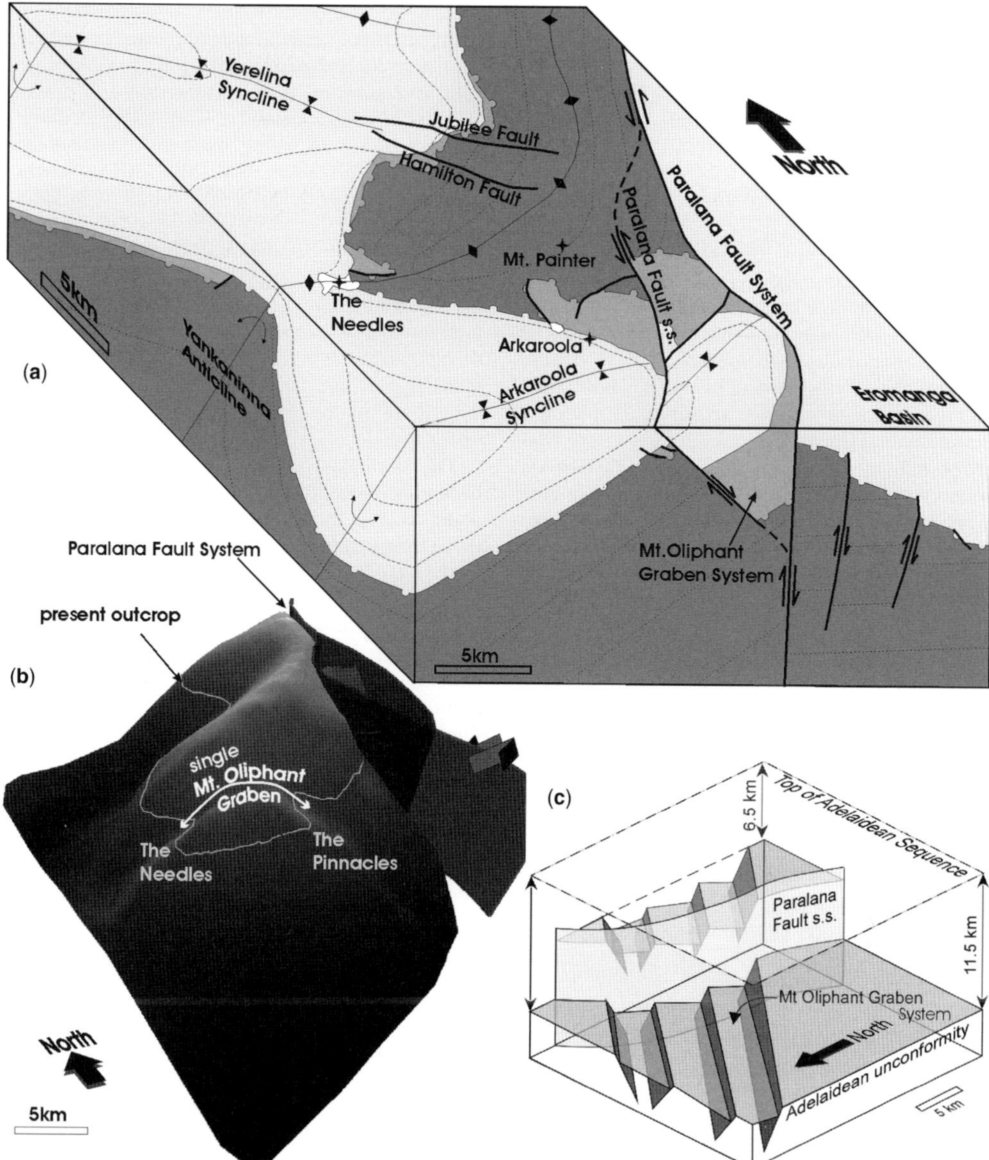

Fig. 5. Structural block-diagram of the present-day MPI and 3D reconstruction of the unconformity surface with contour line of the unconformity's current outcrop. (**a**) The MPI forms the core of the thick-skinned Yankanina Anticline at the hanging wall of the steep Paralana Fault System. It was exhumed from depths >15 km, while the cores of the adjacent synclines were exhumed by only about 2 km. (**b**) Three-dimensional model of the current Adelaidean unconformity surface, constructed with move[TM]2010. The surface in the centre of the anticlinal core reaches a maximum height of *c.* 3 km above the present land surface. Clearly visible is the folded Mt Oliphant Graben System between the Needles and the Pinnacles. (**c**) Schematic block-diagram of the Mt Oliphant Graben System prior to Palaeozoic folding.

crenulation cleavage (S_{pal}) in suitable lower Adelaidean units, in particular in phyllites north of Arkaroola Village (Elburg *et al.* 2003, fig. 1b). Minor crenulations locally overprint this dominant Palaeozoic foliation. Crenulations in the phyllites post-date the growth of porphyroblasts (Fig. 3i).

The Yankaninna Anticline developed during progressive north–south compression (Fig. 1b;

Paul *et al.* 1999). Structural profiles through the fault system show that the late Neoproterozoic Paralana Fault System became reactivated with a sinistral, NW-up movement. Maximum vertical displacement (>10 km) occurred in the NE–SW-striking section of the fault system NE of Paralana Hot Springs, where these faults formed oblique ramps. Vertical movement decreases to the south, where these faults turn to north–south-striking and become lateral ramps. Sinistral offset is of the order of a few kilometres. The deepest sections of these faults are now exposed in the Hidden Valley area, where mylonites are found with sinistral strike-slip and oblique-slip movement.

Basement and cover rocks within the Yankaninna Anticline were thrust up during formation of the anticline. Meanwhile, rocks in the cores of adjacent synclines maintained their approximate burial depth. Assuming that no high-elevation mountain range formed during this intracontinental deformation, the formation of the anticline resulted in exhumation of rocks in its core. Here we use the term 'exhumation' for a reduction in the absolute distance between a volume of rock and the surface (England & Molnar 1990). This opens up the possibility of determining the rate of formation of the Yankaninna Anticline by measuring the rate of exhumation. In the following section we evaluate published and new thermochronology data and show that the Yankaninna Anticline grew over a period of approximately 200 myr.

Timing of folding and exhumation

To reconstruct the timing of the crustal-scale folding, exhumation and fluid flow in the MPI, we use published Ar/Ar plateau cooling ages (McLaren *et al.* 2002) and several low-temperature thermochronology methods (Fig. 6; Mitchell *et al.* 2002; this study). The geothermal gradient down to the Adelaidean unconformity at the time of its maximum burial was estimated at $40–50\ °C\ km^{-1}$ (Mildren & Sandiford 1995; Sandiford *et al.* 1998; McLaren *et al.* 2002, 2006). The present-day heat flow in the region is about double the crustal average (McLaren *et al.* 2006), which indicates that the geothermal gradient is still high. For the interpretation of cooling ages, we therefore assumed a constant gradient of $40\ °C\ km^{-1}$ since 500 Ma. Deviations from this value may have occurred, but would not change the results significantly. Errors in maximum burial depth at the end of Adelaidean deposition are estimated to be ± 0.5 km. The surface temperature is assumed to be 20 °C.

Exhumation to near-surface levels

McLaren *et al.* (2002) interpret their modelled K–Ar and Ar/Ar cooling ages of hornblende, muscovite, biotite and K-feldspar in basement and cover samples as reflecting cooling and exhumation over a prolonged period during Palaeozoic times, with several stages of mid-Palaeozoic fast cooling (*c.* 430 Ma, *c.* 400 Ma and *c.* 325 Ma). Fault activity and a possible reheating during magmatism and metamorphism (Elburg *et al.* 2003, 2013) were not taken into account. This interpretation is based on the assumption that the Yankaninna Anticline was exhumed after its formation. With this assumption, samples retain their relative depth difference during exhumation. However, it is not feasible to first have a >10 km-amplitude fold which was then passively exhumed. Instead, as the fold grew, rocks in the core of the anticline exhumed faster than those in the limbs and their relative depth difference decreased. The data of McLaren *et al.* (2002) show that rocks in the MPI were exhumed over a prolonged time, and samples from greatest depth reached their closure temperatures later than those from shallower depth in the hinge of the anticline (Fig. 7).

The maximum burial depth of the Adelaidean unconformity at the onset of the Delamerian Orogeny (Fig. 6) and the current level and shape of this unconformity in the Yankaninna Anticline as shown in Figure 5a and b are based on our own mapping and that of Coats & Blissett (1971). This structure achieved its present shape by *c.* 290 Ma at the latest, based on the near-surface, <315 Ma diamictite dyke and the *c.* 290 Ma Mt Gee hydrothermal system. Using the Ar/Ar hornblende, muscovite and biotite plateau ages of McLaren *et al.* (2002) we constrain the shape, and hence the average growth rate of the anticline over time (Figs 7, 8 & 9). K-feldspar Ar/Ar cooling ages are not taken into account because of the strong dependence of the closure temperature on the (poorly known) size of diffusion domains (300–150 °C, Lovera *et al.* 1989). With the limited data available, we need to make the assumption that the anticline maintained its shape, but increased its amplitude over time.

McLaren *et al.* (2002) report two hornblende plateau ages (closure temperature *c.* 500 ± 50 °C; Harrison 1981): 446 ± 1 Ma for sample MP31 of a flood basalt at the base of the Mt Oliphant Graben System with a maximum burial depth of *c.* 12.4 km, and 432 ± 1 Ma for sample MP28 of an amphibolite dyke intruding basement quartzites near the British Empire Granite with a maximum burial depth of *c.* 15 km. An exhumation rate of 0.19 ± 0.11 km per myr (1.4–3.6 km per 12–16 myr) is obtained when identical exhumation rates for the two samples are assumed (Fig. 7a). Slightly different rates result if the different exhumation rates as a function of burial depth are taken into account. However, as the samples were

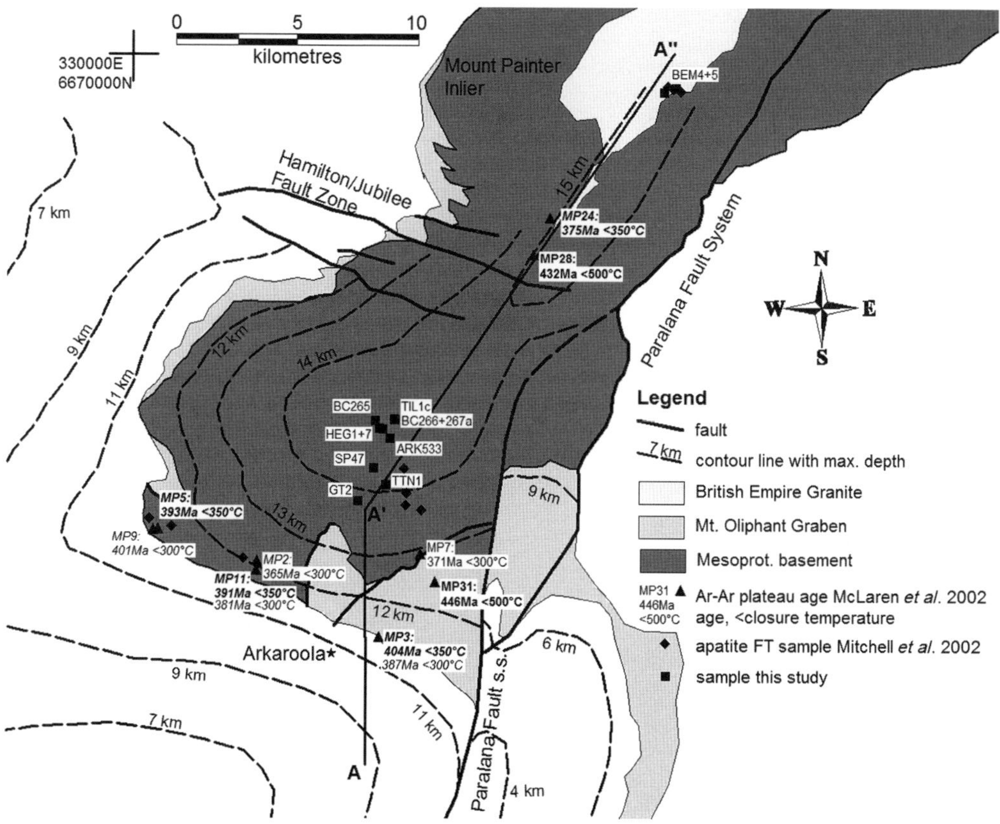

Fig. 6. Sample location of published analyses (McLaren *et al.* 2002; Mitchell *et al.* 2002) and of this study. The simplified map of the MPI shows the contour lines of maximum burial depth after deposition of the Adelaidean succession, as well as the location of the profile of Figure 8.

still close to their maximum burial depth, the difference is not significant. If constant rates are assumed for the initial exhumation, both samples would still have been at their original burial depth at *c.* 450 Ma (Fig. 7b). Earlier, Delamerian, initiation of exhumation would have resulted in a larger difference in cooling ages between the two samples.

By the time of the intrusion of the British Empire Granite (460–440 Ma), the core of the anticline had risen by *c.* 1.7 km ± 1.0 km, according to the structural reconstructions (Fig. 9). The British Empire Granite thus intruded at *c.* 13.3 ± 1.5 km, which is in accordance with earlier, petrologic and thermal estimates (McLaren *et al.* 2006).

Muscovites from a quartzite sample in the vicinity of the British Empire Granite (*c.* 3 km below the unconformity) closed their Ar/Ar system at *c.* 375 Ma (*c.* 350 ± 50 °C; Hames & Bowring 1994), while basement gneisses, pegmatite, biotite schist and Neoproterozoic phyllite samples at the unconformity in the southern MPI went through the closure temperature of biotites (*c.* 300 ± 50 °C; Harrison *et al.* 1985) at about this time. At 375 Ma the unconformity above the central MPI reached a depth of *c.* 6 km, while it rose more slowly in the southern limb of the anticline (Fig. 9). The mean exhumation rate in the centre of the inlier between 440 and 375 Ma was *c.* 0.08 ± 0.04 km per myr (at a depth of 13.3 ± 1.5 km at 440 Ma).

The inlier reached the surface at the latest at *c.* 290 Ma with the formation of the Mt Gee Sinter (Fig. 8; Brugger *et al.* 2011). This implies an average exhumation at the hinge of the anticline of 0.08–0.11 km per myr since 375 Ma. About half of the total exhumation, and hence folding, thus occurred after 375 Ma. Only a very minor part of the exhumation can potentially be attributed to the Delamerian Orogeny and even this is doubtful as the hornblende ages indicate commencement of exhumation at *c.* 450 Ma.

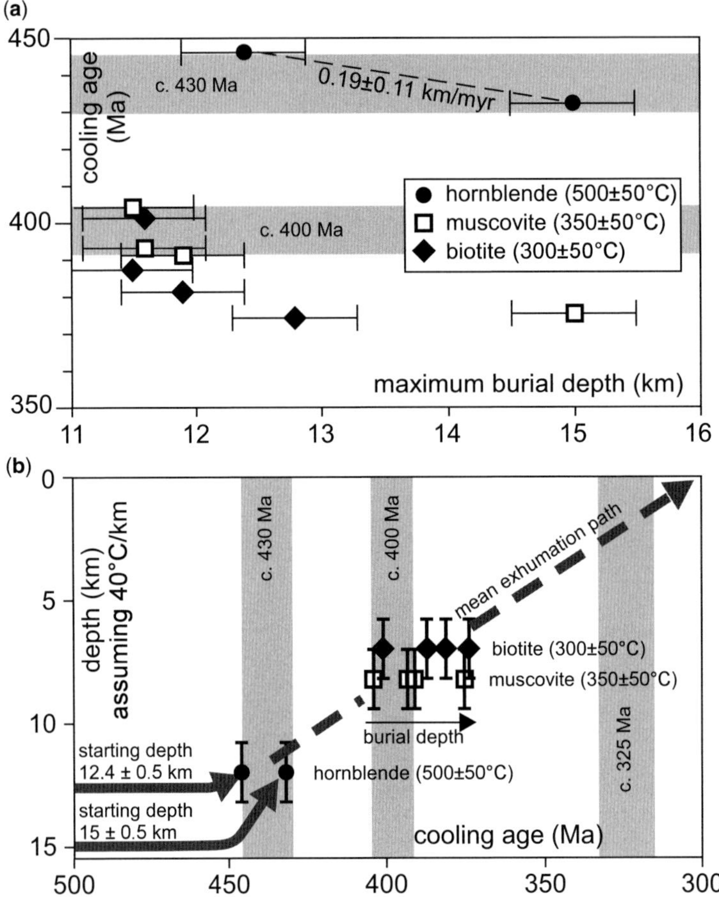

Fig. 7. (**a**) Ar/Ar plateau ages from McLaren *et al.* (2002) as a function of maximum burial depth, showing that more deeply buried samples exhumed later than shallower ones. Exhumation rate during passage through the hornblende closing temperature is *c.* 0.2 km per myr. (**b**) Same ages plotted at the depth at their closure temperature (at 40 °C km^{-1} and a surface temperature of 20 °C). Initial exhumation at 0.2 km per myr appears to have commenced later than the 500 Ma Delamerian Orogeny. Dashed arrow shows the mean exhumation rate with full exhumation at *c.* 300 Ma. Periods of fast exhumation according to McLaren *et al.* (2002) are shown as grey bands in both graphs.

Final exhumation and hydrothermal overprint

Results of low-temperature thermochronology. A total of 12 samples, collected from the southern and central MPI, were used for thermochronology analyses. BEM4 is a sample of an undeformed metaluminous variety of the British Empire Granite (BEG), taken at the eastern side near the base of the Palaeozoic granite intrusion (Fig. 6; Table 2). BEM5 was collected close to BEM4 and represents the peraluminous variety of the granite. Both samples are from an intrusion in metasediments that were at a maximum depth of *c.* 15 km after the deposition of the Adelaidean sediments, as inferred from the structural reconstructions. The

monazite U–Pb ages of these samples form a cluster at *c.* 440 Ma (Elburg *et al.* 2003). Single zircon (U–Th)/He ages of BEM4 ($n = 3$) and BEM5 ($n = 4$) are scattered between 245 ± 16 and 11 ± 0.7 Ma (Table 3). The scatter is likely to be related to the radiation damage and zoning of parent nuclides, since all zircons have high-U cores of Mesoproterozoic age (Elburg *et al.* 2012, 2013). Therefore, their average age was not calculated and they are not included in the interpretation. Apatite (U–Th)/He ages of BEM4 ($n = 3$), in contrast, reproduce very well (all replicates overlap within one-sigma error) and give an average age of 110 ± 3 Ma.

Sample GT2 is from a shear zone in the basement south of Mt Painter, which had a maximum burial

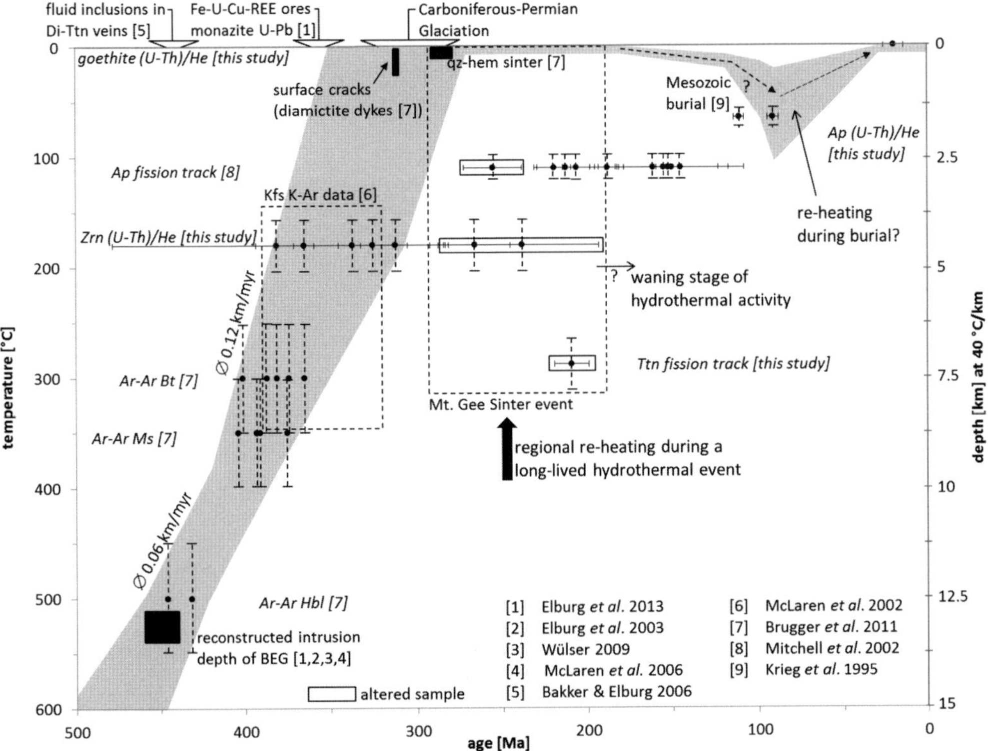

Fig. 8. Temperature (depth) v. age diagram of published Ar/Ar data (McLaren *et al.* 2002), apatite FT samples (Mitchell *et al.* 2002) and new low-temperature thermochronology analyses. The depth is calculated assuming a geothermal gradient of 40 °C km^{-1} and 20 °C at the surface. The mean exhumation rate of the central and southern MPI is ≤0.1 km per myr between <500 Ma and *c.* 300 Ma when the currently outcropping rocks reached the surface. Here they remained until the end of the Triassic Period (*c.* 200 Ma). Zircon, apatite and titanite (U–Th)/He ages are overprinted by the hydrothermal Mt Gee Sinter event. A Mesozoic burial of ≤2.5 km may have affected the low-temperature thermochronology samples. A goethite sample formed at the surface in early Miocene times.

depth of *c.* 14 km. Monazite U–Pb analyses revealed mixed ages between 530 and 440 Ma (Elburg *et al.* 2003; Buick, pers. comm.). Zircon (U–Th)/He ages (*n* = 2) of GT2 reproduce within one-sigma error and give an average age of 313 ± 3 Ma.

A goethite sample collected on the surface of a gneiss (SP47) revealed an average (U–Th)/He age of 20.0 ± 6 Ma (*n* = 8). This age gives the time of ferruginization, oxidation, or gossan formation when the basement must have been at the surface.

TIL1c and BC267a are volcanic and granitic pebbles in a sedimentary diamictite dyke; BC265 and BC266 were collected from the wall rock of these dykes to test whether the pebbles and the wall rock experienced different exhumation. Zircon (U–Th)/He ages of TIL1c (*n* = 4) gave an average age of 323 ± 19 Ma. Two similar ages of *c.* 344 and *c.* 307 Ma, and one much older age of *c.* 491 Ma were obtained from three zircons in sample BC267a (Table 3). Zircons of BC265 gave an average age of 239 ± 45 Ma (*n* = 3). The scatter in ages is even greater for the single apatite measurements of BC265 and BC266, which show a wide spread between *c.* 179 and *c.* 23 Ma. The (U–Th)/He zircon age of the host rock is clearly younger than the (U–Th)/He zircon age of the pebble boulders. The ages of the boulders rather resemble the results of the shear zone GT2 with ages around 320 Ma. The younger zircon age of the wall rock of 238 ± 45 Ma could be a result of an overprint by the quartz-hematite Mt Gee Sinter.

ARK533 (*n* = 7) is a meta-granite clast within the Permian quartz-hematite Mt Gee Sinter. The mean zircon (U–Th)/He age of 264 ± 21 Ma overlaps within the error with that of BC265 (239 ± 45 Ma). Apatite (U–Th)/He ages of ARK533 give an average age of 91 ± 3 Ma (*n* = 4), which is close to the apatite age of BEM4 (110 ± 3 Ma).

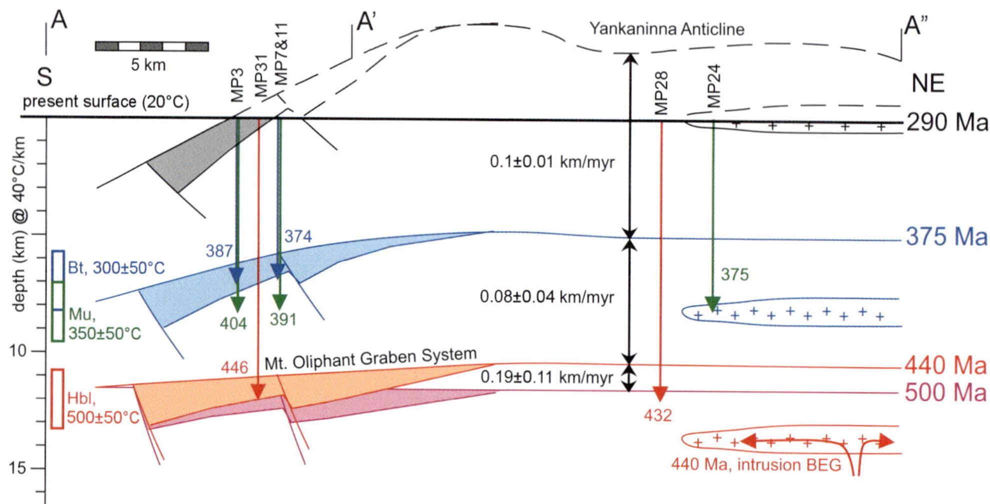

Fig. 9. Profile along the hinge of the Yankaninna Anticline (line A–A′–A″ in Fig. 6), showing the exhumation of the Adelaidean unconformity and Mt Oliphant Graben System at different points in time: deepest burial at *c.* 500 Ma (purple), *c.* 440 Ma when the BEG intruded (red), *c.* 375 Ma and at *c.* 300 Ma, when the current outcropping surface had reached the surface. Profiles are based on plateau Ar–Ar ages of McLaren *et al.* (2002) on hornblende (red arrows), muscovite (green arrows) and biotite (blue arrows), assuming a geothermal gradient of 40 °C km^{-1} and a surface temperature of 20 °C. Sample numbers from McLaren *et al.* (2002).

At the base of the Mt Gee Sinter, wall-rock samples of meta-granites were sampled from different distances to the sinter. HEG1, closest to the sinter, gave a scattered zircon (U–Th)/He signal with an average age of 337 ± 57 Ma (*n* = 3), which is similar to 345 ± 6 Ma (*n* = 2) for sample HEG7, 50 m away from the sinter. The (U–Th)/He ages of the two samples do not appear to have been reset by the Mt Gee Sinter event. Instead, they, and sample GT2, probably record cooling during exhumation.

Finally, TTN1 is a titanite sample from a diopside-titanite vein (*c.* 440 Ma, U–Pb titanite; Elburg *et al.* 2003), whose FT age of 209 ± 10 Ma is appreciably younger than the rest of the samples (Table 4).

Interpretation of low-temperature thermochronology. The presented zircon (U–Th)/He ages that show good reproducibility (closure temperature *c.* 180 ± 20 °C at *c.* 3–6 km depth) cluster around 350–300 Ma and fit well into the general integrated exhumation trend of *c.* 0.1 km per myr derived from the Ar/Ar data of McLaren *et al.* 2002 (Fig. 8). An exhumation rate for these samples cannot, however, be calculated from their age-elevation relationship, as the difference in the recent elevation is low (*c.* 250 m; Table 2) and the isotherms need not have been horizontal.

The MPI reached the surface in the early Permian Period (Brugger *et al.* 2011), which is consistent with the integrated exhumation rate of *c.* 0.1 km per myr. However, two zircon (U–Th)/ He samples record a highly scattered signal and plot well outside of the Palaeozoic exhumation trend (ARK533 and BC265). Also the apatite FT data (closure temperature *c.* 110 °C, equating to 2–3 km depth) reported by Mitchell *et al.* (2002) show a wide range of Mesozoic ages (Fig. 8). The titanite FT age (closure temperature *c.* 310–265 °C, equating to 6–7 km) depth and the apatite (U–Th)/He ages (closure temperature *c.* 75–55 °C or *c.* 1 km) are also younger than the regional exhumation trend. An unpublished apatite FT age of the diopside-titanite vein is, at 127 ± 7 Ma, well outside of the regional trend (M. Elburg & S. Glorie, unpublished data).

The low-temperature thermochronology data were potentially affected by the burial of the MPI during the formation of the Mesozoic Eromanga Basin. This burial was certainly less than the maximum of 2.5 km that is recorded in the centre of the basin (Krieg *et al.* 1995). However, even with a stable geothermal gradient of 40 °C km^{-1} and a regionally constant burial of 2.5 km, the MPI would be heated only to ≤120 °C. This burial alone does not explain the zircon (U–Th)/He apatite and titanite FT ages that indicate higher temperatures in late Palaeozoic and Mesozoic times. Only the apatite (U–Th)/He ages may have been affected, even if the MPI had been buried by at least 1 km.

Table 2. *Low-temperature thermochronology samples*

Sample	Description	E (GDA94)	N (GDA94)	Elevation (m)	Max burial depth (km)	Method	Age (Ma)
GT2	Shear zone in meta-granite (Elburg *et al.* 2003)	339381	6651990	410	14	Zircon (U–Th)/He	313 ± 2.5
TIL1c	Volcanic pebble in sedimentary pebble dyke	340856	6655291	490	14.5	Zircon (U–Th)/He	323 ± 19
BEM4	Metaluminous BEG (Elburg *et al.* 2003)	351886	6668428	650	15	Zircon & apatite (U–Th)/He	n.c.* & 110 ± 3.4
BEM5	Peraluminous BEG (Elburg *et al.* 2003)	352372	6668618	650	15	Zircon (U–Th)/He	n.c.
ARK533	Boulder of meta-granite in Qtz-hem sinter	340682	6654508	440	14.3	Zircon & apatite (U–Th)/He	264 ± 20.7 & 90.8 ± 3.4
SP47	Goethite in gneiss	340017	6653328	430	14.1	Goethite (U–Th)/He	20 ± 5.7
TTN1	Ttn in a Di-Ttn vein, overprinted by Qzt-hem sinter	340523	6652638	370	14.1	Titanite fission-track	209.2 ± 9.9
BC265	K-fsp rich breccia: wall rock of pebble dyke	340083	6655225	520	14.3	Zircon & apatite (U–Th)/He	238.7 ± 45.2 & n.c.
BC266	K-fsp rich meta-granite: wall rock of pebble dyke	340853	6655295	490	14.5	Apatite (U–Th)/He	n.c.
BC267a	Granite pebble in pebble dyke	340895	6655268	490	14.5	Zircon (U–Th)/He	380.7 ± 97.5
HEG1	Bt-rich gneiss at contact to Qtz-hem sinter	340397	6654887	600	14.3	Zircon (U–Th)/He	337 ± 56.5
HEG7	K-fsp rich gneiss, 60 m distance to Qtz-hem sinter	340265	6654900	530	14.3	Zircon & apatite (U–Th)/He	345 ± 5.6 & n.c.

Sample locations, maximum Palaeozoic burial depth of the sample and low-temperature thermochronology age.
*Not calculated

Table 3. *Apatite and zircon (U–Th)/He measurements*

Sample	N_c	Th (ng)	± (%)	U (ng)	± (%)	Sm	± (%)	He (ncc)	± (%)	TAU (%)	Th/U	unc. age (Ma)	±1σ (Ma)	F_t	cor. age (Ma)	±1σ (Ma)
Zircon																
ARK533-1	1	5.784	4.0	10.736	4.1	n/a	n/a	254.518	0.8	4.2	0.53	170.4	7.1	0.70	243.8	15.9
ARK533-2	1	1.119	4.0	5.046	4.1	n/a	n/a	144.044	0.8	4.1	0.22	218.6	9.1	0.78	280.5	18.2
ARK533-3	1	2.908	4.0	7.101	4.1	n/a	n/a	196.388	0.8	4.1	0.41	203.6	8.4	0.80	253.1	16.4
ARK533-4	1	0.424	4.1	1.088	4.1	n/a	n/a	37.524	0.5	4.2	0.39	253.8	10.6	0.84	301.1	19.6
ARK533-5	1	0.651	4.0	3.029	4.1	n/a	n/a	70.213	0.5	4.1	0.21	178.5	7.3	0.73	245.9	15.9
ARK533-6	1	0.787	4.0	2.334	4.1	n/a	n/a	68.585	0.5	4.1	0.33	219.4	9.0	0.80	274.1	17.7
ARK533-7	1	2.095	4.0	3.045	4.1	n/a	n/a	84.739	0.5	4.1	0.68	193.6	7.9	0.72	268.5	17.3
Average age (Ma) ± Std dev. (Ma) (all replicates)																**264 ± 21**
BEM4-1	1	0.954	4.2	5.069	4.1	n/a	n/a	102.348	0.9	4.2	0.19	156.8	6.6	0.84	187.3	12.2
BEM4-2	1	3.564	4.2	10.243	4.2	n/a	n/a	93.374	0.9	4.3	0.35	68.9	2.9	0.78	87.8	5.8
BEM4-3	1	2.875	4.2	4.852	4.1	n/a	n/a	142.718	0.9	4.2	0.59	208.5	8.8	0.85	244.7	16.0
Average age (Ma) ± Std dev. (Ma) (all replicates)																**Not calculated**
BEM5-1	1	163.842	4.2	5.679	4.2	n/a	n/a	749.127	0.9	4.3	28.64	138.4	6.0	0.90	154.5	10.2
BEM5-2	1	594.440	4.4	36.781	4.3	n/a	n/a	196.237	0.6	4.4	16.05	9.1	0.4	0.82	11.0	0.7
BEM5-3	1	239.028	4.2	21.141	4.2	n/a	n/a	1842.909	0.9	4.3	11.22	194.0	8.4	0.86	225.6	14.9
BEM5-4	1	791.540	4.4	70.699	4.4	n/a	n/a	1103.965	0.6	4.4	11.12	35.2	1.6	0.84	41.8	2.8
Average age (Ma) ± Std dev. (Ma) (all replicates)																**Not calculated**
GT2-1	1	0.645	4.2	2.100	4.1	n/a	n/a	77.991	0.9	4.2	0.30	277.8	11.7	0.88	314.4	20.5
GT2-2	1	1.121	4.2	3.525	4.1	n/a	n/a	123.958	0.9	4.2	0.32	262.8	11.0	0.85	310.9	20.3
Average age (Ma) ± Std dev. (Ma) (all replicates)																**313 ± 3**
TIL1c-1	1	2.257	4.2	4.031	4.1	n/a	n/a	158.673	1.2	4.3	0.56	279.2	12.1	0.92	304.4	20.1
TIL1c-2	1	26.100	4.2	12.279	4.1	n/a	n/a	622.566	0.9	4.2	2.11	272.4	11.5	0.87	312.8	20.5
TIL1c-3	1	12.925	4.2	7.231	4.1	n/a	n/a	388.005	0.9	4.2	1.77	303.6	12.8	0.89	340.3	22.3
TIL1c-4	1	9.725	4.2	20.939	4.1	n/a	n/a	875.171	0.9	4.2	0.46	301.7	12.8	0.88	341.6	22.4
Average age (Ma) ± Std dev. (Ma) (all replicates)																**323 ± 19**
BC267a_1*	1	1.463	1.9	1.194	2.1	n/a	n/a	72.252	0.9	2.8	1.22	377.7	10.7	0.76	491.2	13.9
BC267a_2*	1	1.152	1.9	1.315	2.0	n/a	n/a	46.751	0.9	2.8	0.88	239.4	6.6	0.78	306.8	8.5
BC267a_3*	1	1.851	1.8	1.416	2.1	n/a	n/a	60.082	0.9	2.8	1.31	263.3	7.3	0.76	344.2	9.5
Average age (Ma) ± Std dev. (Ma) (all replicates)																**381 ± 98***
BC265a_1*	1	0.483	1.8	2.575	2.0	n/a	n/a	70.615	0.9	2.7	0.19	213.3	5.7	0.78	270.7	7.3
BC265a_2*	1	0.878	1.8	2.197	2.0	n/a	n/a	55.744	0.9	2.6	0.40	188.9	5.0	0.73	257.8	6.8
BC265a_3*	1	0.764	1.9	3.477	2.1	n/a	n/a	64.151	0.9	2.8	0.22	143.4	4.0	0.77	186.7	5.2
Average age (Ma) ± Std dev. (Ma) (all replicates)																**239 ± 45***
HEG1-1*	1	12.406	1.3	8.764	2.0	n/a	n/a	415.125	0.9	2.4	1.42	287.9	6.8	0.83	347.4	8.3
HEG1-2*	1	1.980	1.2	5.154	2.8	n/a	n/a	212.014	0.9	3.1	0.38	304.1	9.4	0.78	387.5	11.9
HEG1-3*	1	1.265	1.1	3.606	1.9	n/a	n/a	96.137	0.9	2.2	0.35	214.5	4.8	0.77	276.0	6.2

Average age (Ma) ± Std dev. (Ma) (all replicates)																**337 ± 57***
HEG7-1*	1	1.486	1.1	5.598	1.8	n/a	n/a	205.638	0.9	2.1	0.27	279.2	6.0	0.81	341.1	7.3
HEG7-2*	1	1.098	1.1	3.637	1.5	n/a	n/a	136.677	0.9	1.9	0.30	283.3	5.4	0.81	349.0	6.6
Average age (Ma) ± Std dev. (Ma) (all replicates)																**345 ± 6***
Apatite																
ARK533-1	1	0.272	3.8	0.188	3.8	n/a	n/a	2.142	0.9	3.9	1.43	69.5	2.7	0.80	86.9	5.5
ARK533-2	1	0.041	3.8	0.029	3.9	n/a	n/a	0.345	0.9	4.0	1.39	72.7	2.9	0.79	92.4	5.9
ARK533-3	1	0.049	3.8	0.030	3.9	n/a	n/a	0.307	0.9	3.9	1.65	60.9	2.4	0.68	89.9	5.7
ARK533-4 †	1	0.060	3.8	0.031	3.9	n/a	n/a	0.420	0.9	4.0	1.94	113.2	4.5	0.81	139.6	8.9
ARK533-5	1	0.008	4.1	0.011	3.9	n/a	n/a	0.138	1.1	4.1	0.77	76.9	3.1	0.81	94.9	6.1
Average age (Ma) ± Std dev. (Ma) (all replicates)																**91 ± 3**
BEM4-1	1	0.020	2.7	0.038	3.1	0.7	10.1	0.443	0.9	4.6	0.52	85.4	3.9	0.79	108.1	7.3
BEM4-2	1	0.028	2.6	0.046	2.8	0.5	10.3	0.589	0.9	4.0	0.62	91.8	3.7	0.83	110.6	7.1
BEM4-3	1	0.015	2.7	0.016	12.9	0.1	10.5	0.229	0.9	11.8	0.90	94.1	11.1	0.82	114.8	14.7
Average age (Ma) ± Std dev. (Ma) (all replicates)																**110 ± 3**
BC266_1*	1	0.059	3.2	0.036	2.4	1.4	1.0	0.665	0.9	4.0	1.64	106.3	4.2	0.78	136.4	5.4
BC266_2*	1	0.077	1.8	0.100	1.9	0.8	1.0	0.907	0.9	2.6	0.77	62.8	1.6	0.80	78.3	2.0
BC266_3*	1	0.117	1.8	0.088	2.0	2.3	1.0	1.201	0.9	2.7	1.33	83.4	2.2	0.81	102.5	2.7
BC266_4*	1	0.090	1.8	0.130	2.0	0.2	1.0	0.327	0.9	2.7	0.69	17.9	0.5	0.78	23.0	0.6
Average age (Ma) ± Std dev. (Ma) (all replicates)																**Not calculated***
BC265_1*	1	0.900	1.7	0.562	1.9	10.8	1.0	8.763	0.9	2.5	1.60	91.6	2.3	0.87	104.8	2.6
BC265_2*	1	0.402	1.7	0.424	1.9	9.6	1.0	5.398	0.9	2.6	0.95	83.8	2.2	0.87	95.9	2.5
BC265_3*	1	0.833	1.7	0.458	1.9	13.5	1.0	12.944	0.9	2.5	1.82	158.0	4.0	0.88	178.5	4.5
Average age (Ma) ± Std dev. (Ma) (all replicates)																**Not calculated***
HEG7-1*	1	0.015	2.2	0.045	1.9	1.4	1.0	0.382	0.9	2.9	0.33	63.0	1.8	0.77	82.1	2.4
HEG7-2*	1	0.395	1.5	0.951	1.8	3.6	1.0	16.052	0.9	2.3	0.42	125.4	2.9	0.82	153.3	3.6
HEG7-3*	1	0.096	1.6	0.039	1.9	3.9	1.0	0.861	0.9	2.5	2.45	106.1	2.6	0.80	131.8	3.3
Average age (Ma) ± Std dev. (Ma) (all replicates)																**Not calculated***
Goethite																
SP47-1	1	0.005	17.8	7.358	1.7	n/a	n/a	18.954	1.0	1.9	0.0006	21.1	1.1	1.00	21.1	1.1
SP47-2	1	0.002	44.7	18.253	2.4	n/a	n/a	46.790	1.0	2.6	0.0001	21.0	1.2	1.00	21.0	1.2
SP47-3	1	0.003	23.9	3.163	1.6	n/a	n/a	12.897	1.0	1.9	0.0010	33.4	1.8	1.00	33.4	1.8
SP47-4	1	0.002	34.7	5.797	1.8	n/a	n/a	17.988	1.0	2.1	0.0004	25.5	1.4	1.00	25.5	1.4
SP47-5	1	0.007	10.0	28.970	1.9	n/a	n/a	61.748	1.2	2.2	0.0002	17.5	1.0	1.00	17.5	1.0
SP47-6	1	0.025	3.2	22.209	1.9	n/a	n/a	51.608	1.3	2.3	0.0011	19.1	1.0	1.00	19.1	1.0
SP47-7	1	0.035	3.2	26.989	2.3	n/a	n/a	49.931	1.3	2.6	0.0013	15.2	1.1	1.00	15.2	1.1
SP47-8	1	0.006	7.3	37.622	1.8	n/a	n/a	85.582	1.3	2.2	0.0002	18.7	1.0	1.00	18.7	1.0
Average age (Ma) ± Std dev. (Ma) (all replicates)																**20 ± 6**

*Samples measured at the Department for Geosciences, University of Tübingen (Germany). Samples not marked with an asterisk were measured at the John de Laeter Centre of Isotope Research in Perth (Australia).

†Sample considered as 'flier' was excluded from calculation of the average.

N_c, number of dated crystals/shards; Th, ^{232}Th; U, ^{238}U; Sm, ^{147}Sm; He, ^{4}He at STP; TAU, total analytical uncertainty; unc. age, uncorrected He age; F_t, alpha recoil correction factor after Farley *et al.* (1996); cor. age, corrected He age. Average age was calculated as error weighted mean. n/a, not analysed.

Table 4. *Titanite fission-track measurements*

Sample	N	ρ_s	N_s	ρ_i	N_i	ρ_d	N_d	P($\chi2$) (%)	age (Ma)	$\pm 1\sigma$ (Ma)
TTN1	20	137.669	2474	56.759	1020	6.184	2801	>95	209.0	10.0

N, number of dated titanite crystals; ρs (ρi), spontaneous (induced) track densities ($\times 10^5$ tracks/cm^2); Ns (Ni), number of counted spontaneous (induced) tracks; ρd, dosimeter track density ($\times 10^5$ tracks/cm^2); Nd, number of tracks counted on dosimeter; P($\chi2$), probability obtaining Chi-square value ($\chi2$) for n degrees of freedom (where n = No. of crystals − 1); age $\pm$ 1σ, central FT age $\pm$ 1 standard error (Galbraith & Laslett 1993). Age was calculated using zeta calibration method (Hurford & Green 1983), glass dosimeters CN-5, and zeta values of 283.5 $\pm$ 6.2 a cm^{-2}. Samples were measured at the John de Laeter Centre of Isotope Research in Perth (Australia).

There are another two processes that can influence FT and (U–Th)/He ages of the zircons and apatites in the MPI: (1) lowering of the closure temperatures of the minerals due to radiation damage (Flowers *et al.* 2009); and (2) change of the geothermal gradient due to equilibration and reheating during burial and regional–local hydrothermal fluid flow (Timar-Geng *et al.* 2004; Filip *et al.* 2007). The second of these processes in particular could have resulted in the scattered ages that are recorded in our data (Fig. 8). We believe that the partial reheating of the MPI has been caused by the Mesozoic burial, as well as by regional–local hydrothermal activity during the quartz-hematite Mt Gee Sinter event.

Sinter formation in the southern MPI started in the early Permian Period (Brugger *et al.* 2011). According to our titanite FT age of the overprinted diopside-titanite veins (Fig. 3a), this sinter event was active at least until the late Triassic Period. With Mesozoic burial of the MPI the sinter activity may have ceased, but this long-lasting hydrothermal event probably resulted in a locally elevated geothermal gradient that equilibrated with time (Fig. 8). This could explain the scattered and relatively young apatite FT and apatite (U–Th)/He ages.

After the Mesozoic burial the MPI must have again reached the surface at the latest at *c.* 20 Ma, when the goethite sample formed (Fig. 8). This interpretation is supported by the style of sedimentation in the Eromanga Basin that changed from low-energy, paralic conditions to alluvial fans in the Miocene during the initial uplift of the Northern Flinders Ranges (Célérier *et al.* 2005, and references therein). The exhumation path of the MPI since 200 Ma cannot be constrained from geological observations. A slow mean exhumation rate of *c.* 0.01 km per myr in Tertiary times could be possible, as it is indicated from apatite FT ages all over Australia (Kohn *et al.* 2002).

Discussion

The Yankaninna Anticline is a large fold in which all structural elements present at the onset of the Palaeozoic Era were passively folded (Fig. 5b). The fold pattern of S_m in the basement closely mimics, but does not exactly match, that of the bedding in the overlying Adelaidean succession, because of the small angle between S_m and bedding before folding (Fig. 3d). From the current map pattern, it is not immediately obvious that the east–west-striking faults, such as the Hamilton and Jubilee Faults, active in the Neoproterozoic Era, belong to the same set as those in the southeastern limb of the anticline (Fig. 5a). Considering that these faults were formed well before the folding, it becomes clear that they originally formed a single *c.* 310°-striking Mt Oliphant Graben System (Fig. 5c).

The north–south- to NE–SW-striking Paralana Fault System developed after the Mt Oliphant Graben System, but still during deposition of the Late Proterozoic Adelaidean succession. During Palaeozoic times, this fault system was reactivated as a set of lateral and oblique ramps against which the MPI was pushed up and towards the south (Fig. 5a). Older faults on the NW limb of the developing anticline were passively rotated to an east–west strike, but not reactivated. Graben faults on the SW limb, however, rotated towards a north–south strike, which is closer to that of the Paralana Fault System. This allowed these faults to be reactivated, as can be observed NE of Mt Oliphant. Here the old graben faults have thick quartz veins with slickensides and striations that indicate sinistral, transpressive movement. The formation of the Yankaninna Anticline, and, by inference, folds throughout the Northern Flinders Ranges (Fig. 1b), is here explained by a single, but prolonged event. However, the above shows that structural elements progressively changed their position and orientation within the growing anticline and could have become deactivated or reactivated at various points in time. The overall single folding event may therefore locally appear as a series of distinct events. This observation applies especially to the phyllites north of Arkaroola Village that were affected by a complex interference of local small-scale faulting and folding.

The data of McLaren *et al.* (2002) show that exhumation took place over a *c.* 200 myr period in

Palaeozoic times. We have argued that this exhumation is not the result of passively exhuming a Delamerian fold, but represents the actual growth of the Yankaninna Anticline. It was suggested by Elburg *et al.* (2013) that the data of McLaren *et al.* (2002) may have been affected by igneous and hydrothermal heat pulses. Although potentially of local importance, we regard heat pulses of minor overall importance, based on the following argument. A conservative estimate for the heat production of rocks in the MPI is $5 \pm 1 \times 10^{-6}$ W m^{-3} (Mildren & Sandiford 1995). With an area of 500 km^2 and a crustal section of 10 km the basement units would have produced a total of 1.5×10^{23} J within 200 myr. One intrusion such as the British Empire Granite at an estimated volume of 20 km^3 would have contributed *c.* 0.01% to the total heat budget if it cooled by 500 °C after intrusion. Hydrothermal fluids could also have caused thermal pulses. However, cooling 30 km^3 of hot fluids (Weisheit *et al.* 2013) by 300 °C would have contributed <0.05% to the total heat budget. Furthermore, the data from various locations in the central and southern MPI show a strong consistency between the various mineral systems and their structural position in the Yankaninna Anticline (Fig. 7). We therefore infer that the data mostly record cooling by exhumation with only minor local effects of thermal pulses.

McLaren *et al.* (2002) proposed stepwise cooling and exhumation of the MPI at *c.* 430, *c.* 400 and *c.* 325 Ma. In their model (their fig. 7), folding was assumed to have taken place during the Delamerian Orogeny, after which all rocks experienced the same Palaeozoic cooling history as the fold was passively exhumed. This model is in contrast to our model, where we interpret the same data as reflecting different exhumation rates as a function of the structural position of the samples within the developing anticline. This allows us to constrain both the development of the anticline at different points in time and average exhumation rates during the intervening periods (Fig. 9). Rates most probably varied, as proposed by McLaren *et al.* (2002). This is not contradictory to our model. The only difference is that we interpret their exhumation rates as reflecting growth of the Yankaninna fold itself, rather than its passive exhumation.

Low-temperature thermochronology data (Mitchell *et al.* 2002; this study) cannot be explained by cooling resulting from exhumation as these postdate the time that the presently outcropping rocks reached the near surface. Evidence for this are the diamictite dykes and the near-surface Mt Gee Sinter deposit that was active from at least *c.* 290 Ma (Brugger *et al.* 2011). Burial under Mesozoic Eromanga Basin sediments cannot explain the various Permian to Jurassic apatite, zircon and titanite FT ages. Only the *c.* 100 Ma apatite (U–Th)/He ages could be the result of burial under at least 1 km of Eromanga Basin sediments and their subsequent exhumation in Cretaceous–Tertiary times (Fig. 8).

Other low-temperature thermochronology data can only be explained with resetting by hydrothermal fluids of the Mt Gee System. The low closure temperatures are easily reset by hydrothermal and ore-forming fluids (Bojar *et al.* 1998; Arnaud & Eide 2000; Balogh & Dunkl 2005; Dempster & Persano 2006; Kelley *et al.* 2006; Marton *et al.* 2010). A *c.* 210 Ma titanite FT age indicates that these hydrothermal fluids must have reached temperatures above *c.* 265 °C. The sample comes from a 440 Ma diopside-titanite vein in the centre of the MPI (Bakker & Elburg 2006), which was overprinted by Mt Gee Sinter deposits (Fig. 3a). Our low-temperature thermochronology data of hydrothermally overprinted samples are all older than *c.* 190 Ma (except apatite (U–Th)/He), indicating that hydrothermal activity decreased after this time (Fig. 8). However, the active Paralana Hot Springs indicate that hydrothermal activity may have lasted from the Permian Period until today (Brugger *et al.* 2005).

Most of the exhumation of the MPI occurred during mid-Palaeozoic times at a time of major metamorphic, magmatic and deformation activities in the Harts Ranges in central Australia (Alice Springs Orogeny at 450–300 Ma; Buick *et al.* 2008) and subduction and terrane amalgamation in the Lachlan Fold Belt in SE Australia (450–370 Ma; Chappell *et al.* 1988; Turner *et al.* 1996; Soesoo *et al.* 1997; Gray & Foster 2004; Fig. 2). The *c.* 500 Ma Delamerian Orogeny that is recorded in the south of the Adelaide Fold Belt in the Mt Lofty and Southern Flinders Ranges (Foden *et al.* 2006) only played a minor role in the crustal-scale deformation in the MPI, contrary to previous interpretations that the refolding of the mainly Delamerian structure was only a far-field effect of the Alice Springs Orogeny (e.g. Paul *et al.* 1999; McLaren *et al.* 2002; Armit *et al.* 2012). Instead, the Yankaninna Anticline grew over a long period of time by south-directed ramping on to the Curnamona Province. The Yankaninna Anticline is, however, not an isolated structure, but is only one of many similar folds in the Northern Flinders Ranges. This implies that the whole region experienced a long period of north–south shortening of the previous Adelaidean basin between the rigid Gawler and Curnamona Provinces. It remains to be investigated what the implications of this newly recognized regional intracontinental Palaeozoic deformation are for the tectonic evolution of central and southern Australia and how the folding and concomitant erosion relate to Palaeozoic sedimentary basins (e.g. Gravestock 1995).

Conclusions

The Mt Painter Inlier forms the core of the Yankaninna Anticline, which is part of the Northern Adelaide Fold and Thrust Belt. Our mapping and structural analysis allowed construction of the present-day fold shape and reconstruction of the system before folding. The Mesoproterozoic basement is truncated by the *c.* 800 Ma-Adelaidean unconformity. The angle between the unconformity and the main foliation in the basement was found to be about ten degrees. Initial Adelaidean succession deposition occurred in *c.* 310°-striking graben, followed by widespread deposition. The NE–SW- to north–south-trending Paralana Fault System became active during Adelaidean succession deposition. At the end of deposition, by *c.* 500 Ma, the Adelaidean unconformity was buried *c.* 11.5 km west of the Parallana Fault *sensu stricto*, and 1–2 km deeper in the early Adelaidean graben.

Folding commenced with the *c.* 500 Ma Delamerian Orogeny at the earliest, but possibly as late as, *c.* 450 Ma. In the study area, south-directed ramping up the Curnamona Province along the Paralana Fault System resulted in the formation of the Yankaninna Anticline in which the core rose by *c.* 15 km. This folding passively rotated all previous structural elements, including the early Adelaidean graben, which are now found exposed at nearly right angles to each other on the NW and SE limb of the fold.

Folding in the Northern Flinders Ranges has traditionally been interpreted as Delamerian in age, with subsequent Palaeozoic exhumation of the folds. We show that very limited folding occurred during the Delamerian Orogeny. Instead, the Yankaninna Anticline grew over a prolonged period of *c.* 200 myr up to the Permian Period. Younger low-temperature thermochronological data indicate a major near-surface hydrothermal event (Mt Gee event) that lasted at least until *c.* 180 Ma.

We kindly thank the Sprigg family for access to their land and continuing support during the field trips. S. Hore and W. Preiss are thanked for their inspiring discussions and cooperation during our projects. J. Holzäpfel, C. Kieslinger, C. Kling, S. Kocher, J. Lang, M. Lindhuber, J. Müller, T. Rehder and J. Roessiger contributed to the fieldwork. K. Warber is thanked for the 3D modelling. We are grateful to E. Enkelmann, who carried out the low-temperature analyses and interpretations in Tübingen. We thank W. Preiss and G. Mahéo for their helpful comments. This project was funded by Marathon Ltd and the German Science Foundation (DFG, grant BO-1776/8-1).

References

ARMIT, R. J., BETTS, P. G., SCHAEFER, B. F. & AILLERES, L. 2012. Constraints on long-lived Mesoproterozoic and Palaeozoic deformational events and crustal architecture in the northern Mt. Painter Province, Australia. *Gondwana Research*, **22**, 207–226.

ARNAUD, N. O. & EIDE, E. 2000. Brecciation-related argon redistribution in alkali feldspars; an in naturo crushing study. *Geochimica et Cosmochimica Acta*, **64**, 3201–3215.

BAKKER, R. J. & ELBURG, M. A. 2006. A magmatic-hydrothermal transition in Arkaroola (Northern Flinders Ranges, South Australia): from diopside-titanite pegmatites to hematite-quartz growth. *Contributions to Mineralogy and Petrology*, **152**, 541–569.

BALOGH, K. & DUNKL, I. 2005. Argon and fission track dating of Alpine metamorphism and basement exhumation in the Sopron Mts. (Eastern Alps, Hungary); thermochronology or mineral growth? *Mineralogy and Petrology*, **83**, 191–218.

BOJAR, A.-V., NEUBAUER, F. & FRITZ, H. 1998. Cretaceous to Cenozoic thermal evolution of the southwestern South Carpathians; evidence from fission-track thermochronology. *Tectonophysics*, **297**, 229–249.

BONS, P. D., MONTENARI, M., BAKKER, R. J. & ELBURG, M. 2007. Potential evidence of fossilised Neoproterozoic deep life: SEM observations on calcite veins from Oppaminda Creek, Arkaroola, South Australia. *International Journal of Earth Sciences*, **98**, 327–343.

BRUGGER, J., LONG, N., MCPHAIL, D. C. & PLIMER, I. 2005. An active amagmatic hydrothermal system: the Paralana hot springs, Northern Flinders Ranges, South Australia. *Chemical Geology*, **222**, 35–64.

BRUGGER, J., WÜLSER, P. A. & FODEN, J. 2011. Genesis and preservation of a uranium-rich Paleozoic epithermal system with a surface expression (Northern Flinders Ranges, South Australia): radiogenic heat driving regional hydrothermal circulation over geological timescales. *Astrobiology*, **11**, 499–508.

BUICK, I. S., STORKEY, A. & WILLIAMS, I. S. 2008. Timing relationships between pegmatite emplacement, metamorphism and deformation during the intra-plate Alice Springs Orogeny, central Australia. *Journal of Metamorphic Geology*, **26**, 915–936.

BURG, J.-P., DAVY, P., NIEVERGELT, P., OBERLI, F., SEWARD, D., DIAO, Z. & MEIER, M. 1997. Exhumation during crustal folding in the Namche-Barwa syntaxis. *Terra Nova*, **9**, 53–56.

CÉLÉRIER, J., SANDIFORD, M., HANSEN, D. L. & QUIGLEY, M. 2005. Modes of active intraplate deformation, Flinders Ranges, Australia. *Tectonics*, **24**, TC6006, http://dx.doi.org/10.1029/2004TC001679.

CHAPPELL, B. W., WHITE, A. J. R. & HINE, R. 1988. Granite provinces and basement terranes in the Lachlan Fold Belt, southeastern Australia. *Australian Journal of Earth Sciences*, **35**, 505–521.

CHEN, J., BURBANK, D. W., SCHARER, K. M., SOBEL, E., YIN, J., RUBIN, C. & ZHAO, R. 2002. Magnetochronology of the Upper Cenozoic strata in the Southwestern Chinese Tian Shan: rates of Pleistocene folding and thrusting. *Earth and Planetary Science Letters*, **195**, 113–130.

COATS, R. P. & BLISSETT, A. H. 1971. Regional and economic geology of the Mount Painter Province. Geological Survey of South Australia, Bulletin, **43**, 425.

COWLEY, W. M., HORE, S. B., PREISS, W. V., SHEARD, M. J. & WADE, C. 2012. A revised stratigraphic scheme for the Mount Painter and Mount Babbage Inliers. *SAREIC 2012 Technical forum poster.* http://minerals.dmitre.sa.gov.au/press_and_events/events/sareic_2012_technical_forum

COYLE, D. A. & WAGNER, G. A. 1998. Positioning the titanite fission-track partial annealing zone. *Chemical Geology*, **149**, 117–125.

DEMPSTER, T. J. & PERSANO, C. 2006. Low-temperature thermochronology; resolving geotherm shapes or denudation histories? *Geology*, **34**, 73–76.

DIPPLE, G., BONS, P. D. & OLIVER, N. H. S. 2005. A vector of high-temperature paleo-fluid flow deduced from mass transfer across permeability barriers (quartz veins). *Geofluids*, **5**, 67–82.

DREXEL, J. F. & MAJOR, R. B. 1987. Geology of the uraniferous breccia near Mt Painter, South Australia and revision of rock nomenclature. *Geological Survey of South Australia, Quarterly Notes*, **104**, 14–24.

DUNKL, I. 2002. TRACKKEY: a Windows program for calculation and graphical presentation of fission track data. *Computer Geosciences*, **28**, 3–12.

ELBURG, M. A., BONS, P. D., DOUGHERTY-PAGE, J., JANKA, C. E., NEUMANN, N. & SCHAEFER, B. 2001. Age and metasomatic alteration of the Mt Neill Granite at Nooldoonooldoona Waterhole, Mt Painter Inlier, South Australia. *Australian Journal of Earth Sciences*, **48**, 721–730.

ELBURG, M. A., BONS, P. D., FODEN, J. & BRUGGER, J. 2003. A newly defined Late Ordovician magmatic-thermal event in the Mt Painter Province, Northern Flinders Ranges, South Australia. *Australian Journal of Earth Sciences*, **50**, 611–631.

ELBURG, M. A., ANDERSEN, T., BONS, P. D., WEISHEIT, A., SIMONSEN, S. L. & SMET, I. 2012. Metasomatism and metallogeny of A-type granites of the Mt Painter – Mt Babbage Inliers, South Australia. *Lithos*, **151**, 83–104.

ELBURG, M. A., ANDERSEN, T., BONS, P. D., SIMONSEN, S. L & WEISHEIT, A. 2013. New constraints on Phanerozoic magmatic and hydrothermal events in the Mt Painter Province, South Australia. *Gondwana Research*, **24**, 700–712, http://dx.doi.org/10.1016/j.gr.2012.12.017

ENGLAND, P. & MOLNAR, P. 1990. Surface uplift, uplift of rocks, and exhumation of rocks. *Geology*, **18**, 1173–1177.

EVANS, N. J., BYRNE, J. P., KEEGAN, J. T. & DOTTER, L. E. 2005. Determination of uranium and thorium in zircon, apatite, and fluorite: application to laser (U–Th)/He thermochronology. *Journal of Analytical Chemistry*, **60**, 1159–1165.

EYLES, N., MORY, A. J. & BACKHOUSE, J. 2002. Carboniferous–Permian palynostratigraphy of West Australian marine rift basins; resolving tectonic and eustatic controls during Gondwanan glaciations. *Palaeogeography, Palaeoclimatology, Palaeoecology*, **184**, 305–319.

FANNING, C. M., TEALE, G. S. & ROBERTSON, R. S. 2003. Is there a Willyama supergroup sequence in the Mount Painter Inlier? *In*: PELJO, M. *Broken Hill Exploration Initiative: Abstracts from the July 2003 Conference. Geoscience Australia Record*, **2003/13**, 38–41.

FARLEY, K. A. 2002. (U–Th)/He dating: techniques, calibrations, and applications. *Reviews in Mineralogy and Geochemistry*, **47**, 819–843.

FARLEY, K. A., WOLF, R. A. & SILVER, L. T. 1996. The effects of long alpha-stopping distances on (U–Th)/He dates. *Geochimica et Cosmochimica Acta*, **60**, 4223–4230.

FILIP, J., ULRYCH, J., ASAMOVIC, J. & BALOGH, K. 2007. Apatite fission track implications for timing of hydrothermal fluid flow in Tertiary volcanics of the Bohemian Massif. *Journal of Geosciences*, **52**, 211–220.

FLÖTTMANN, T. & JAMES, P. 1997. Influence of basin architecture on the style of inversion and fold-thrust belt tectonics – the southern Adelaide Fold-Thrust Belt, South Australia. *Journal of Structural Geology*, **19**, 1093–1110.

FLÖTTMANN, T., JAMES, P., ROGERS, J. & JOHNSON, T. 1994. Early Paleozoic foreland thrusting and basin reactivation at the palaeo-Pacific margin of the Southeastern Australian Precambrian Craton – a reappraisal of the structural evolution of the Southern Adelaide Fold-Thrust Belt. *Tectonophysics*, **234**, 95–116.

FLOWERS, R. M., KETCHAM, R. A., SHUSTER, D. L. & FARLEY, K. A. 2009. Apatite (U–Th)/He thermochronometry using a radiation damage accumulation and annealing model. *Geochimica et Cosmochimica Acta*, **73**, 2347–2365.

FODEN, J., ELBURG, M. A., DOUGHERTY-PAGE, J. & BURTT, A. 2006. The timing and duration of the Delamerian Orogeny: correlation with the Ross Orogen and implications for Gondwana assembly. *Journal of Geology*, **114**, 189–210.

FRASER, G. L & NEUMANN, N. L. 2010. New SHRIMP U–Pb zircon ages from the Gawler Craton and Curnamona Province, South Australia, 2008–2010. *Geoscience Australia Record*, 2010/16.

FÜGENSCHUH, B., SEWARD, D. & MANCKTELOW, N. 1997. Exhumation in a convergent orogen; the western Tauern Window. *Terra Nova*, **9**, 213–217.

GALBRAITH, R. F. & LASLETT, G. M. 1993. Statistical-models for mixed fission-track ages. *Nuclear Tracks and Radiation Measurements*, **21**, 459–470.

GLEADOW, A. J. W. 1981. Fission track dating methods: what are the real alternatives? *Nuclear Tracks and Radiation Measurements*, **5**, 3–14.

GLEADOW, A. J. W. & LOVERING, J. F. 1974. The effect of weathering on FT dating. *Earth and Planetary Sciences Letters*, **22**, 163–168.

GOWER, C. F., KAMO, S. L., KWOK, K. & KROGH, T. E. 2008. Proterozoic southward accretion and Grenvillian orogenesis in the interior Grenville Province in eastern Labrador; evidence from U/Pb geochronological investigations. *Precambrian Research*, **165**, 61–95.

GRAVESTOCK, D. I. 1995. Western Warburton basin. *In*: DREXEL, J. F. & PREISS, W. V. (eds) *The Geology of South Australia, Vol. 2, The Phanerozoic.* South Australia Geological Survey, Bulletin, **54**, 41–43.

GRAY, D. R. & FOSTER, D. A. 2004. Tectonic evolution of the Lachlan Orogen, southeast Australia: historical review, data synthesis and modern perspectives. *Australian Journal of Earth Sciences*, **51**, 773–817.

GREEN, P. F., DUDDY, I. R., GLEADOW, A. J. W., TINGATE, P. T. & LASLETT, G. M. 1986. Thermal annealing of fission tracks in apatite: 1. A qualitative description. *Chemical Geology: Isotopic Geoscience Section*, **59**, 237–253, http://dx.doi.org/10.1016/0168-9622(86)90074-6

HAMES, W. E. & BOWRING, S. A. 1994. An empirical evaluation of the argon diffusion geometry in muscovite. *Earth and Planetary Science Letters*, **124**, 161–169.

HARRISON, T. M. 1981. Diffusion of 40Ar in hornblende. *Contributions to Mineralogy and Petrology*, **78**, 324–331.

HARRISON, T. M., DUNCAN, I. & MCDOUGALL, I. 1985. Diffusion of ^{40}Ar in biotite: temperature, pressure and compositional effects. *Geochimica et Cosmochimica Acta*, **49**, 2461–2468.

HILDEBRAND, R. S., HOFFMAN, P. F. & BOWRING, S. A. 2010. The Calderian Orogeny in Wopmay Orogen (1.9 Ga), northwestern Canadian Shield. *Geological Society of America Bulletin*, **122**, 794–814.

HOLDSWORTH, R. E. & PINHEIRO, R. V. L. 2000. The anatomy of shallow-crustal transpressional structures; insights from the Archaean Carajas fault zone, Amazon, Brazil. *Journal of Structural Geology*, **22**, 1105–1123.

HURFORD, A. J. 1998. ZETA: the ultimate solution to fission-track analysis calibration or just an interim measure. *In*: VAN DEN HAUTE, P. & DE CORTE, F. (eds) *Advances in Fission-Track Geochronology*. Kluwer Academic, Dordrecht, 19–32.

HURFORD, A. J. & GREEN, P. F. 1983. The zeta age calibration of fission-track dating. *Chemical Geology*, **41**, 285–312.

JONES, A. T. & FIELDING, C. R. 2004. Sedimentological record of the late Paleozoic glaciation in Queensland, Australia. *Geology*, **32**, 153–156.

KELLEY, D. L., KELLEY, K. D., COKER, W. B., CAUGHLIN, B. & DOHERTY, M. E. 2006. Beyond the obvious limits of ore deposits; the use of mineralogical, geochemical, and biological features for the remote detection of mineralization. *Economic Geology and the Bulletin of the Society of Economic Geologists*, **101**, 729–752.

KOHN, B. P., GLEADOW, A. J. W., BROWN, R. W., GALLAGHER, K., O'SULLIVAN, P. B. & FOSTER, D. A. 2002. Shaping the Australian crust over the last 300 million years: insights from fission track thermo-tectonic imaging and denudation studies of key terranes. *Australian Journal of Earth Sciences*, **49**, 697–717.

KONOPASEK, J., SCHULMANN, K. & LEXA, O. 2001. Structural evolution of the central part of the Krusne hory (Erzgebirge) Mountains in the Czech Republic; evidence for changing stress regime during Variscan compression. *Journal of Structural Geology*, **23**, 1373–1392.

KORSCH, R. J., PREISS, W. V. *ET AL.* 2010. Geological interpretation of deep seismic reflection and magneto-telluric line 08GA-C1: Curnamona Province, South Australia. *In*: KORSCH, R. J. & KOSITCIN, N. (eds) *South Australian Seismic and MT Workshop*, Extended Abstracts. Geoscience Australia Record, 2010/10, 42–53.

KRETZ, R. 1983. Symbols for rock-forming minerals. *American Mineralogist*, **68**, 277–279.

KRIEG, G., ALEXANDER, E. M. & ROGERS, P. A. 1995. Eromanga Basin. *In*: DREXEL, J. F. & PREISS, W. V. (eds) *The Geology of South Australia, vol. 2, The Phanerozoic*. South Australia Geological Survey, Bulletin, **54**, 101–105.

LLANA-FUNEZ, S. & MARCOS, A. 2007. Convergence in a thermally softened thick crust; Variscan intracontinental tectonics in Iberian Plate rocks. *Terra Nova*, **19**, 393–400.

LOVERA, O. M., RICHTER, F. M. & HARRISON, T. M. 1989. The ^{40}Ar/^{39}Ar thermochronometry for slowly cooled samples having a distribution of diffusion domain sizes. *Journal of Geophysical Research*, **94**, 17 917–17 935.

MAHEO, G., PECHER, A., GUILLOT, S., ROLLAND, Y. & DELACOURT, C. 2004. Exhumation of Neogene gneiss domes between oblique crustal boundaries in south Karakorum, northwest Himalaya, Pakistan. *In*: WHITNEY, D. L., TEYSSIER, C. & SIDDOWAY, C. S. (eds) *Gneiss domes in orogeny*. Geological Society of America, Special Papers, **380**, 141–154.

MARTON, I., MORITZ, R. & SPIKINGS, R. 2010. Application of low temperature thermochronology to hydrothermal ore deposits; formation, preservation and exhumation of epithermal gold systems from the eastern Rhodopes, Bulgaria. *Tectonophysics*, **483**, 240–254.

MCLAREN, S. N., DUNLAP, W. J., SANDIFORD, M. & MCDOUGALLN, I. 2002. Thermochronology of high heat-producing crust at Mount Painter, South Australia: implications for tectonic reactivation of continental interiors. *Tectonics*, **21**, http://dx.doi.org/10.1029/2000TC001275

MCLAREN, S., SANDIFORD, M., POWELL, R., NEUMANN, N. & WOODHEAD, J. 2006. Palaeozoic intraplate crustal anatexis in the Mount Painter Province, South Australia; timing, thermal budgets and the role of crustal heat production. *Journal of Petrology*, **47**, 2281–2302.

MEESTERS, A. G. C. A. & DUNAI, T. J. 2002*a*. Solving the production-diffusion equation for finite diffusion domains of various shapes – Part II. Application to cases with alpha-ejection and nonhomogeneous distribution of the source. *Chemical Geology*, **186**, 57–73.

MEESTERS, A. G. C. A. & DUNAI, T. J. 2002*b*. Solving the production-diffusion equation for finite diffusion domains of various shapes – Part 1. Implications for low-temperature (U–Th)/He thermochronology. *Chemical Geology*, **186**, 333–344.

MILDREN, S. D. & SANDIFORD, M. 1995. Heat refraction and low-pressure metamorphism in the Northern Flinders Ranges, South Australia. *Australian Journal of Earth Sciences*, **42**, 241–247.

MITCHELL, M. M., KOHN, B. P., O'SULLIVAN, P. B., HARTLEY, M. J. & FOSTER, D. A. 2002. Low-temperature thermochronology of the Mt Painter Province, South Australia. *Australian Journal of Earth Sciences*, **49**, 551–563.

MUKHOPADHYAY, S. & SHARMA, J. 2010. Crustal scale detachment in the Himalayas; a reappraisal. *Geophysical Journal International*, **183**, 850–860.

NEUMANN, N. 2001. *Geochemical and Isotopic Characteristics of South Australian Proterozoic Granites:*

Implications for the Origin and Evolution of High Heat-Producing Terrains. PhD thesis. Department of Geology and Geophysics, University of Adelaide, Adelaide.

PASSCHIER, C. W., TROUW, R. A. J., RIBEIRO, A. & PACIULLO, F. V. P. 2002. Tectonic evolution of the southern Kaoko Belt, Namibia. *Journal of African Earth Sciences*, **35**, 61–75.

PAUL, E. 1998. *Geometry and Controls on Basement-Involved Deformation in the Adelaide Fold Belt, South Australia*. PhD thesis. Department of Geology and Geophysics. University of Adelaide, Adelaide.

PAUL, E., FLÖTTMANN, T. & SANDIFORD, M. 1999. Structural geometry and controls on basement-involved deformation in the northern Flinders Ranges, Adelaide Fold Belt, South Australia. *Australian Journal of Earth Sciences*, **46**, 343–354.

PREISS, W. V. 1987. The Adelaide Geosyncline – Late Proterozoic stratigraphy, sedimentation, paleontology, and tectonics. *South Australian Geological Survey Bulletin*, **53**, 438.

PREISS, W. V. 2000. The Adelaide Geosyncline of South Australia and its significance in Neoproterozoic continental reconstruction. *Precambrian Research*, **100**, 21–63.

REINERS, P. W. 2005. Zircon (U–Th)/He thermochronometry. *Reviews in Mineralogy and Geochemistry*, **58**, 151–176.

REINERS, P. W. & FARLEY, K. A. 1999. Helium diffusion and (U–Th)/He thermochronometry of titanite. *Geochimica et Cosmochimica Acta*, **63**, 3845–3859.

REINERS, P. W., SPELL, T. L., NICOLESCU, S. & ZANETTI, K. A. 2004. Zircon (U–Th)/He thermochronometry: He diffusion and comparisons with 40Ar/39Ar dating. *Geochimica et Cosmochimica Acta*, **68**, 1857–1887.

RUTHERFORD, L., HAND, M. & BAROVICH, K. 2007. Timing of Proterozoic metamorphism in the southern Curnamona Province: implications for tectonic models and continental reconstructions. *Australian Journal of Earth Sciences*, **54**, 65–81.

SANDIFORD, M., HAND, M. & MCLAREN, S. 1998. High geothermal gradient metamorphism during thermal subsidence. *Earth and Planetary Science Letters*, **163**, 149–165.

SCHELLART, W. P. 2002. Alpine deformation at the western termination of the Axial Zone, southern Pyrenees. *Journal of the Virtual Explorer*, **8**, 35–55.

SCHUSTER, D. L., FLOWERS, R. M. & FARLEY, K. A. 2006. The influence of natural radiation damage on helium diffusion kinetics in apatite. *Earth and Planetary Science Letters*, **249**, 148–161, http://dx.doi.org/10.1016/j.epsl.2006.07.028

SHEARD, M. & COCKSHELL, C. D. 1992. *Seismic interpretation of Mt. Hopeless Line 1*. Department of Energy and Science, SA Report book, **92/17**.

SHUSTER, D. L., VASCONCELOS, P. M., HEIM, J. A. & FARLEY, K. A. 2005. Weathering geochronology by (U–Th)/He dating of goethite. *Geochimica et Cosmochimica Acta*, **69**, 659–673.

SIEBEL, W., DANIŠÍK, M. & CHEN, F. 2009. From emplacement to unroofing: thermal history of the Jiazishan gabbro, Sulu UHP terrane, China. *Mineralogy and Petrology*, **96**, 163–175.

SOESOO, A., BONS, P. D., GRAY, D. R. & FOSTER, D. A. 1997. Divergent double subduction: tectonic and petrologic consequences. *Geology*, **25**, 55–758.

SPAGGIARI, C. V. 2007. The Jack Hills greenstone belt, Western Australia; Part 1, Structural and tectonic evolution over >1.5 Ga. *Precambrian Research*, **155**, 204–228.

STEWART, K. & FODEN, J. 2003. *Mesoproterozoic granites of South Australia*. Department of Primary Industries and Resources, Report Book **2003/15**.

SUPPE, J., CHOU, G. T. & HOOK, S. C. 1992. *Rates of Folding and Faulting Determined from Growth Strata. Thrust Tectonics*. Chapman & Hall London, UK.

TEALE, G. S. 1993. Mount Painter and Mount Babbage Inliers. *In*: DREXEL, J. F., PREISS, W. V. & PARKER, A. J. (eds) *The Geology of South Australia, Vol. 1 The Precambrian*. Geological Survey of South Australia, Bulletin, **54**, 93–100.

TIMAR-GENG, Z., FÜGENSCHUH, B., SCHALTEGGER, U. & WETZEL, A. 2004. The impact of the Jurassic hydrothermal activity on zircon fission track data from the southern Upper Rhine Graben area. *Schweizerische Mineralogische und Petrographische Mitteilungen*, **84**, 257–269.

TURNER, S. P., KELLEY, S. P., VANDENBERG, A. H. M., FODEN, J. D., SANDIFORD, M. & FLÖTTMANN, T. 1996. Source of the Lachlan Fold Belt flysch linked to convective removal of the lithospheric mantle and rapid exhumation of the Delamerian-Ross Fold Belt. *Geology*, **24**, 941–944.

VEEVERS, J. J. 2006. Updated Gondwana (Permian–Cretaceous) Earth history of Australia. *Gondwana Research*, **9**, 231–260.

WAGNER, G. A. & VAN DEN HAUTE, P. 1992. *Fission-Track Dating*. Enke Verlag, Stuttgart.

WEISHEIT, A., BONS, P. D. & ELBURG, M. A. 2013. Long-lived crustal fluid-flow: the hydrothermal mega-breccia of Hidden Valley, Mt. Painter Inlier, South Australia. *International Journal of Earth Sciences*, **102**, 1219–1236, http://dx.doi.org/10.1007/s00531-013-0875-7

WINGATE, M. T. D., CAMPBELL, I. H., COMPSTON, W. & GIBSON, G. M. 1998. Ion microprobe U–Pb ages for Neoproterozoic basaltic magmatism in south-central Australia and implications for the breakup of Rodinia. *Precambrian Research*, **87**, 135–159.

WÜLSER, P. A. 2009. *Uranium metallogeny in North Flinders Ranges region of South Australia*. PhD thesis, Adelaide University, Adelaide.

WÜLSER, P.-A., BRUGGER, J., FODEN, J. & PFEIFER, H.-R. 2011. The sandstone-hosted Beverley uranium deposit, Lake Frome basin, South Australia; mineralogy, geochemistry, and a time-constrained model for its genesis. *Economic Geology and the Bulletin of the Society of Economic Geologists*, **106**, 835–867.

ZEITLER, P. K., HERCZEG, A. L., MCDOUGALL, I. & HONDA, M. 1987. U–Th–He dating of apatite – a potential thermochronometer. *Geochimica et Cosmochimica Acta*, **51**, 2865–2868.

ZIV, A., KATZIR, Y., AVIGAD, D. & GARFUNKEL, Z. 2010. Strain development and kinematic significance of the Alpine folding on Andros (western Cyclades, Greece). *Tectonophysics*, **488**, 248–255.

Constraints on the movement history of the Carboneras Fault Zone (SE Spain) from stratigraphy and ^{40}Ar–^{39}Ar dating of Neogene volcanic rocks

E. H. RUTTER[1]*, R. BURGESS[1] & D. R. FAULKNER[2]

[1]*University of Manchester, School of Earth, Atmospheric and Environmental Sciences, Manchester M13 9PL, UK*

[2]*University of Liverpool, Earth and Ocean Sciences, School of Environmental Sciences, Liverpool L69 3GP, UK*

**Corresponding author (e-mail: e.rutter@manchester.ac.uk)*

Abstract: The Carboneras Fault Zone (CFZ) of SE Spain separates the volcanic Cabo de Gata terrain to the SE (accumulated over 18 to 6 Ma BP) from the tract of uplifted Alpine metamorphic basement blocks and post-orogenic basins that comprise the Betic Cordilleras to the NW. The CFZ cuts metamorphic basement and folded post-orogenic sediments and volcanic rocks, and acted as a conduit for upper Miocene calc-alkaline volcanic rocks to the surface. NW of the CFZ, unconformities and deformation episodes affect successive sedimentary formations of upper Miocene age (upper Serravallian through Messinian). CFZ movements are constrained stratigraphically by an unconformity within the volcanic sequence. Older, pre-faulting volcanic rocks are uptilted against the CFZ whilst the youngest volcanic rocks step across it. ^{40}Ar–^{39}Ar age determinations on amphibole crysts from the volcanic rocks show the main CFZ movements began 11–12 Ma BP and ended about 6 Ma. Lesser movement rates continue today. The CFZ is interpreted as part of a transform fault system separating NE–SW stretched and NW–SE shortened crust deformed above a southwestward-retreating subducted slab, from a less deformed terrain lying to the SE. Locally, total offset on the CFZ may be up to 40 km but is at least 15 km.

The Carboneras Fault Zone (CFZ) of SE Spain separates the Miocene Gabo de Gata volcanic terrain to the SE from metamorphic basement rocks and post-orogenic sedimentary rocks of the internal zone of the Betic Cordilleras, an Alpine orogenic belt in which tectonometamorphic activity peaked during early Miocene time (25–18 Ma, Alonso-Chaves *et al.* 2004; Vissers 2012). The CFZ is only exposed cutting metamorphic basement rocks in a 15 km long tract near the town of Carboneras. Elsewhere on the Spanish mainland it is largely covered by upper Miocene through Recent sedimentary rocks. The CFZ forms part of a segmented system of left-lateral strike-slip faults that extends 150 km NE towards Alicante, as the Palomares and Alhama de Murcia fault systems (Fig. 1). To the SW the Carboneras fault extends continuously 100 km into the Alborán Sea (Moreno Mota 2010), beyond which it may continue as a soft-linked network to include the Nekor and Jebha faults in Morocco. The concept of the Trans-Alborán shear zone (Leblanc & Olivier 1984; de Larouzière *et al.* 1988; Gràcia *et al.* 2006) can be extended usefully to include the whole of this system of left-lateral strike-slip faults.

To the NW of the CFZ it is well established that NE–SW-directed extensional faulting of late Miocene age has affected crystalline basement rocks (García-Dueñas *et al.* 1992; Johnson *et al.* 1997; Augier *et al.* 2005; Martínez-Martínez *et al.* 2006; Meijninger & Vissers 2006; Giaconia *et al.* 2011; Vissers 2012; Sanz de Galdeano *et al.* 2012) and allowed them to become exposed as metamorphic core complexes (Sierra Nevada, Sierra de Los Filabres, Sierra Alhamilla and Sierra Cabrera). This produced accommodation space for the accumulation of post-orogenic sediments, and suggests that the faults of the Trans-Alborán shear zone act as lateral walls or ramps to allow the extensional faulting to occur (Fig. 2). At the same time, there is evidence for widespread co-eval NW–SE-directed shortening causing deformation within the internal Betic zone and folding and thrusting in the external zones (Galindo-Zaldívar *et al.* 2003; Platt *et al.* 2003b; Vissers 2012), making the CFZ a transpressional fault. Basement blocks uplifted during late Miocene time typically take the form of large-scale antiformal structures with a box-fold geometry, often bounded by NE–SW-trending high-angle reverse faults, and form the mountain

From: LLANA-FÚNEZ, S., MARCOS, A. & BASTIDA, F. (eds) 2014. *Deformation Structures and Processes within the Continental Crust.* Geological Society, London, Special Publications, **394**, 79–99.
First published online November 22, 2013, http://dx.doi.org/10.1144/SP394.5

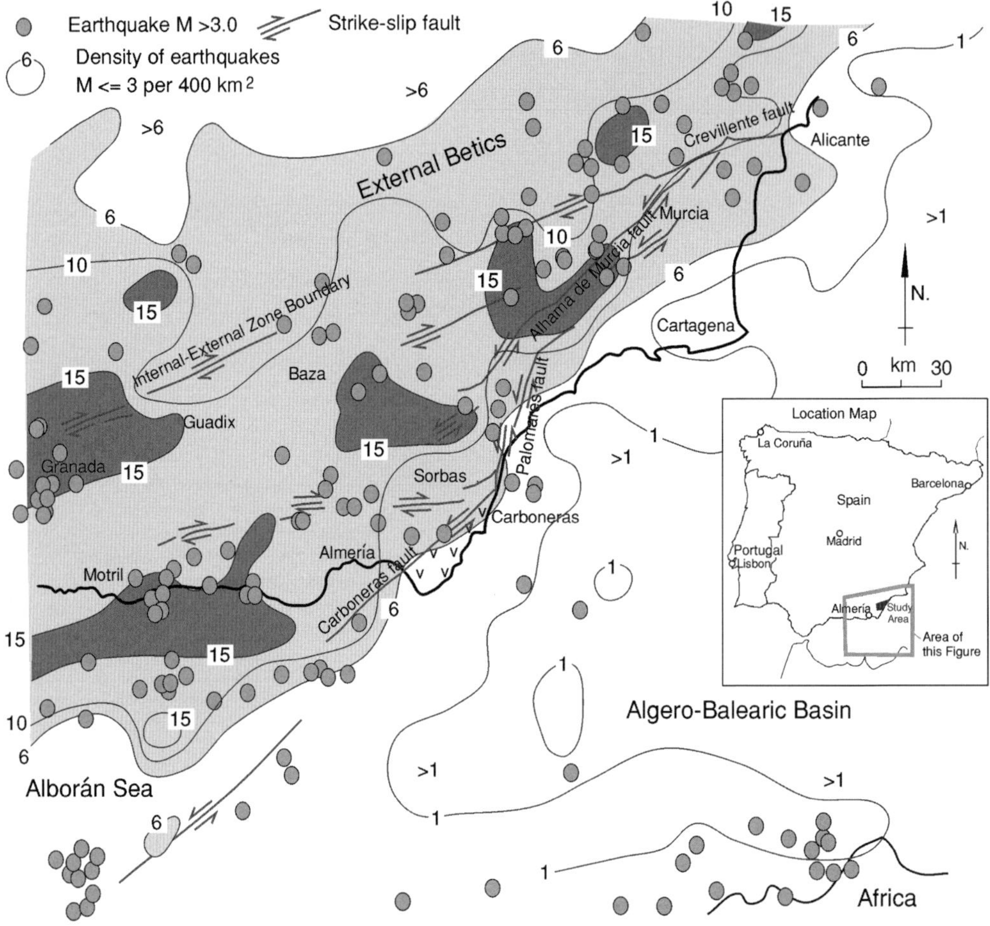

Fig. 1. Location of the Carboneras Fault Zone (CFZ) in SE Spain, and the linked Palomares and Alhama de Murcia faults, comprising the NE end of the Trans-Alborán shear zone. Seismic activity shown between years 2000 and 2010, is based on Martínez-Díaz *et al.* (2012). Discrete events of local magnitude M > 3 shown as filled circles, activity of M < 3 is density-contoured as number of such earthquakes per 400 km². The Betic internal zone is wedge shaped, between the left-lateral Trans-Alborán system and the right-lateral Internal-External Zone Boundary–Crevillente fault system. The wedge is most active, with a marked seismic activity decrease to the SE across the trend of the Trans-Alborán shear zone, which is interpreted to be an upper mantle flow velocity discontinuity.

ranges of the Betic Cordilleras. They testify to upper Miocene through Recent convergence of North Africa with Iberia.

Rutter *et al.* (2012) suggested that the Trans-Alborán fault system has acted as a stretching fault (Means 1989), accommodating a velocity discontinuity (Odé 1960) in the upper mantle flow regime. The fast polarization directions from observations of teleseismic shear-wave splitting throughout the region show a consistent NE–SW trend, that has been interpreted as indicating the contemporary and/or Neogene upper mantle principal extension direction (Díaz *et al.* 2010), and this is consistent with the orientation and

movement picture of extensional faults in upper Miocene rocks that partially cover the internal Betic zone (Fig. 2) on *both* sides of the Trans-Alborán shear zone (Dabrio & Polo 1991). Following Martínez-Díaz *et al.* (2012), Figure 1 also shows the distribution of earthquakes in the region during the ten years since 2000. The Trans-Alborán shear zone appears to delimit the largely onshore tract of more frequent earthquakes to the NW from the lower frequency of tremors to the SE. This probably implies a greater degree of active deformation in the region to the NW and supports the notion of the shear zone acting as a velocity discontinuity.

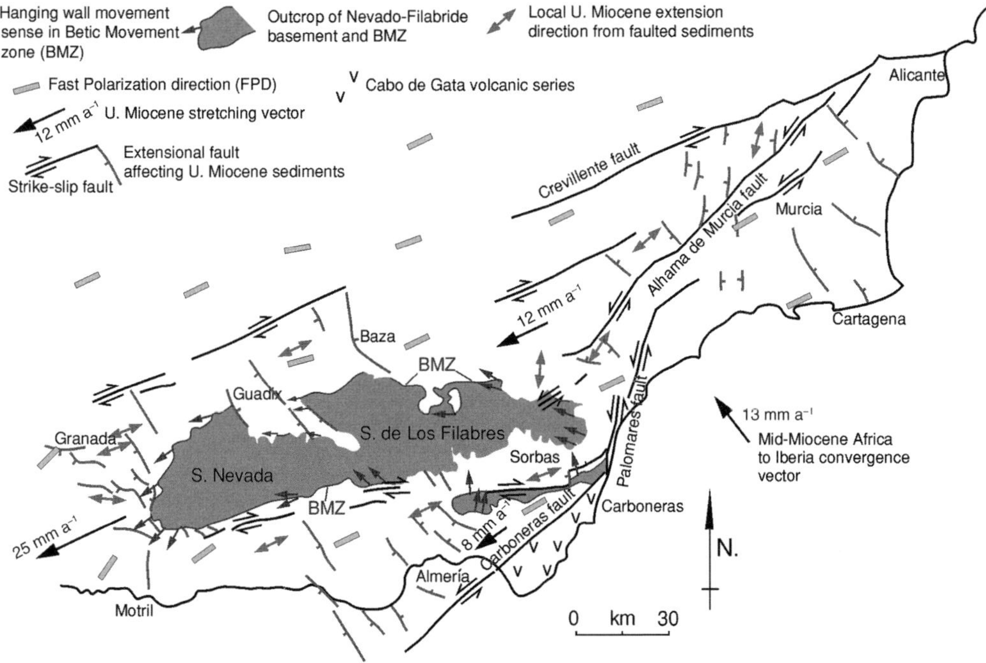

Fig. 2. Tortonian NE–SW-directed stretching in the back-arc wedge that now comprises the Betic internal zone. Pre-Tortonian stretching is revealed by the displacements on the Betic Movement Zone (BMZ), separating the Alpujárride nappe from the underlying Nevado-Filábride metamorphic rocks (after Vissers 2012). Traces of generally NW–SE-trending extensional faults that generate the Tortonian extensional basins and cut Tortonian sedimentary rocks are linked by the network of SW–NE-trending strike-slip faults (compiled from Martínez-Díaz & Hernández-Enrile 2004; Martínez-Martínez *et al.* 2006; Meijninger & Vissers 2006). Stretching velocities shown based on Gutscher (2012). Mid-Miocene Africa–Iberia convergence rate after Vissers (2012). Upper mantle S-wave fast polarization directions from Díaz *et al.* (2010).

Rocks of the Cabo de Gata calc-alkaline volcanic terrain cover much of the onshore area to the SE of the CFZ, between Cabo de Gata and a few kilometres north from Carboneras, and include a number of individual eruptive volcanic centres (e.g. Rodalquilar, Majada Redondo, Los Frailes). The ages of these rocks are dominantly Serravallian through Messinian (Serrano & Gonzáles Donoso 1989; Comas *et al.* 1999; Turner *et al.* 1999). In the Carboneras area, rather older volcanic rocks also occur (Scotney *et al.* 2000). The age of the oldest *in-situ* volcaniclastic rocks at the exposed stratigraphic base of the sequence is 18 Ma, but evidence of older volcanic activity extending to 21 Ma is found as amphibole crysts in water-lain sediments (Rutter *et al.* 2012). Hardly any volcanic rocks of the Cabo de Gata suite outcrop to the north and west of the CFZ, suggesting that displacements on the fault zone have been sufficient to remove volcanic rocks originally lying to the NW. The fault zone hosts a swarm of vertical intrusive andesitic dykes that have apparently used the fault zone to reach the surface (Rutter *et al.* 2012),

testifying to a link between faulting and volcanic activity, and also suggesting that the CFZ cuts right through the continental crust and into the upper mantle (Rutter *et al.* 2012). Stratigraphic relationships between the Cabo de Gata volcanic rocks and the CFZ are of particular importance to dating movements on the CFZ, and the presence of K-bearing amphiboles in the volcanic rocks means they are particularly well suited to ^{40}Ar–^{39}Ar dating. The main aim of this paper is to establish these relative and absolute age relationships, to constrain the movement history, and to consider the implications for the geodynamic evolution of the region.

Summary of the geology of the CFZ near Carboneras

Only near Carboneras town is the CFZ extensively exposed, cutting basement rocks over a distance of 9 km and juxtaposing them against volcanic rocks of the Cabo de Gata complex. Further SW the fault zone disappears beneath uppermost Miocene and Pliocene rocks of the NE corner of the Níjar basin, lying between the towns of Almería and Carboneras (Fig. 1). Figures 3–5 show that the geology of the area may be considered as group of lithotectonic units bounded by unconformities and/or major faults (Rutter *et al.* 2012).

Metamorphic basement rocks of the Nevado-Filábride and Alpujárride groups outcrop extensively to the NW of the CFZ and form fault-bounded blocks within it (Fig. 4). To the SE of the CFZ near Carboneras only Maláguide and Alpujárride (Vissers 2012) basement rocks are found at the current level of exposure, and these are overlain unconformably by Burdigalian through Serravallian post-orogenic volcanic rocks of the Cabo de Gata complex that are tilted to the vertical against the

fault zone and hence are probably older than the time of onset of the faulting. Tortonian volcaniclastic rocks lie unconformably above the older volcanic rocks and overstep their base northwards, only to become truncated against the CFZ (Figs 3 & 4). Messinian volcaniclastic rocks overlie these, and overstep the CFZ northwards but are slightly affected by the later fault movements (Fig. 3). Intrusive andesitic dykes outcrop in the CFZ, including two that are some 500 m wide.

To the NW of the CFZ the oldest post-orogenic rocks are Serravallian/Tortonian marine marls and sands of the Saltador Formation (Fig. 3). These are preserved from erosion in tight synformal inliers in the basement rocks. In the CFZ they are cut and offset by the faults and often are intruded by andesitic dykes. They are equivalent to the Chozas

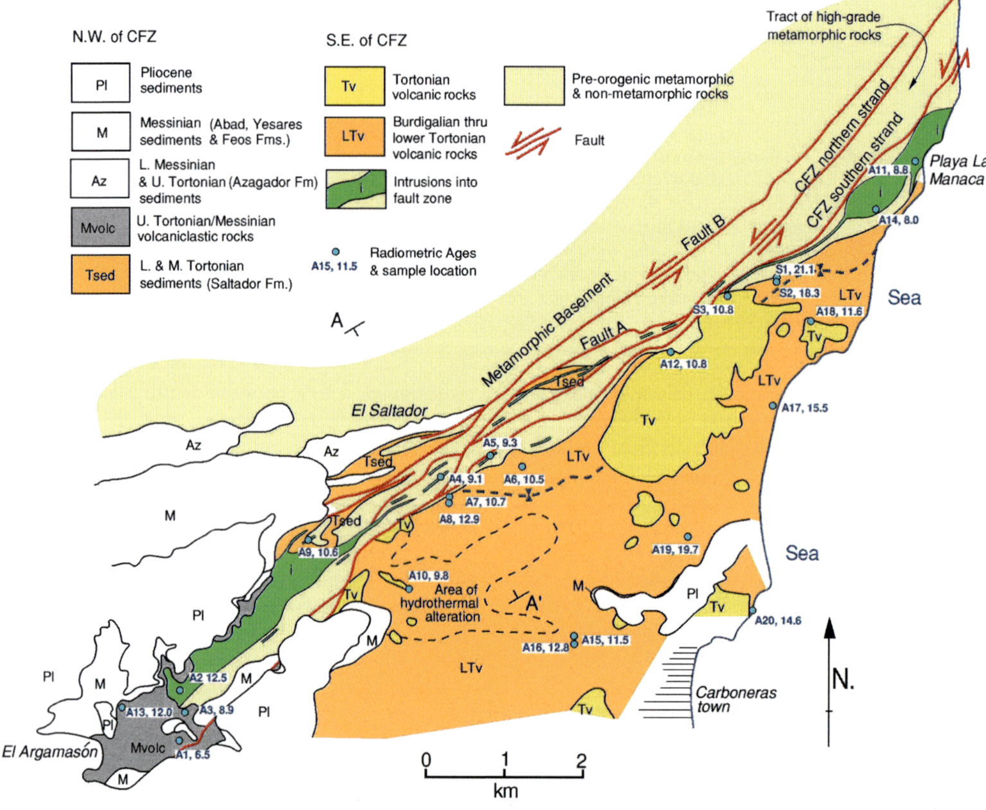

Fig. 3. Geological outline of the CFZ near Carboneras town, where the fault system can be seen cutting metamorphic basement rocks. Outlines of the succession of main unconformities affecting the post-orogenic sedimentary and volcanic rocks are shown, together with andesitic dykes and minor intrusions into the fault zone. The outcrop area of basement rocks is shown not divided into lithotectonic units. The intrusive rocks cut only basement rocks and lower Tortonian sediments. Locations of dated volcanic rocks are shown, together with mean ages in Ma. Uncertainties ($\pm 2\sigma$) on the mean ages are shown in Table 1. A–A′ marks the position of the line of the geological cross-section in Figure 4, which is close to the trend of the valley of the Saltador river.

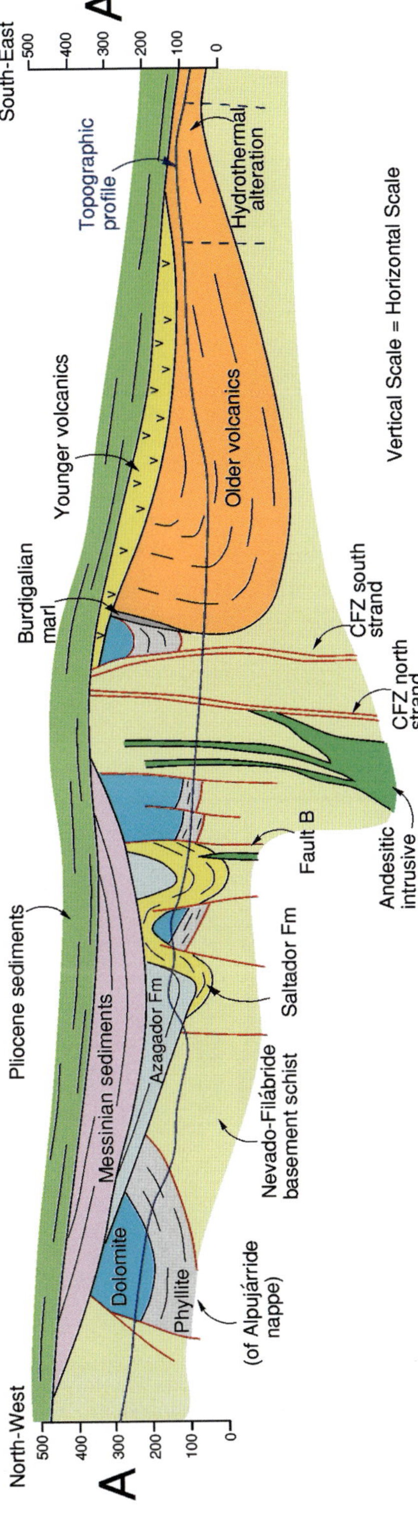

Fig. 4. Simplified vertical geological cross-section along line A–A' on Figure 3, emphasizing the succession of unconformable sedimentary and volcanic sequences deposited upon the basement rocks affected by the orogenic activity.

formation of the Sorbas basin, which is strongly affected by NE–SW-directed extensional faulting at its eastern margin with uplifted basement rocks of the Sierra Cabrera (Giaconia *et al.* 2011).

The rocks of the Saltador Fm are overlain unconformably (Figs 4 & 5) by the less strongly folded rocks (beach sands and marls) of the Azagador Fm (upper Tortonian – lower Messinian, Van der Poel 1992; Fortuin & Krijgsman 2003). These are not cut by igneous dykes but are affected by faulting. This unit is in turn overlain unconformably by the three Messinian formations (Abad Fm (marls), Yesares Fm (selenitic and alabastrine gypsum), Feos Fm (marls)) (Fortuin & Krijgsman 2003), and in turn unconformably by Pliocene formations. Pliocene formations are gently arched over the trace of the CFZ, are only minimally affected by fault movements and testify to continued transpression or to localized uplift of the rocks of the CFZ. They are also tilted gently against the southern side of the Sierra Cabrera and each of the progressively older units is tilted at a steeper angle against the basement block, testifying to continued progressive uplift of the Sierra Cabrera from Tortonian through Quaternary time. This succession of basement uplifts is responsible for the overall shape of the geological units in the CFZ area.

The CFZ consists of four main fault strands labelled on Figure 3 as follows:

(a) The northern and southern strands of the CFZ. These seem to accommodate the largest displacements of the fault system. Between them lies a belt of distinctly higher metamorphic grade Nevado-Filábride basement schists (Westra 1969), including sillimanite-bearing gneisses with evidence of partial melting (e.g. at GR 597550 4099510) dated at 20 Ma BP (Platt & Whitehouse 1999).

(b) Faults labelled A and B in Figure 3. These lie immediately to the north of the main faults of the CFZ and collectively have an estimated strike-slip offset of almost 2 km on the basis of the displacement of the axial zone of a synform in the Saltador Fm. Fragments of the Saltador Fm are strung out along the northern strand of the CFZ over at least 6 km. Fault B cuts the Azagador Fm but is unconformably overlain by the Messinian Abad Fm. Hence it was active until lowermost Messinian time.

The fault cores consist of typically several metres thickness of clay-bearing fault gouge derived from graphitic mica schist. The clay minerals are produced by breakdown of mechanically granulated

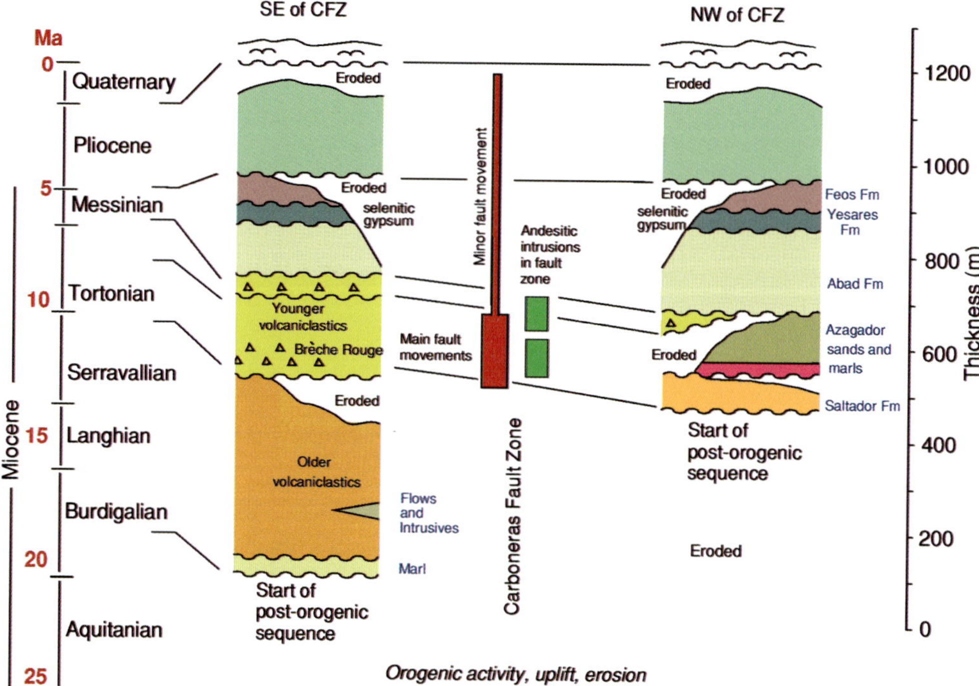

Fig. 5. Comparative stratigraphic successions for the Neogene rocks on either side of the CFZ, in relation to the main period of movement on the CFZ. Formations within the sediments of Pliocene age are undifferentiated.

feldspars and micas (Rutter & White 1979; Solum & van der Pluijm 2009). The fault rocks characteristically display a foliation that is transected by Riedel shear surfaces (Rutter *et al.* 1986, 2012; Faulkner *et al.* 2003). Displacement indicators within the fault rocks and stratigraphic constraints show that the fault displacement pattern is close to pure strike-slip (Rutter *et al.* 2012). The country rocks between the fault cores are typically variably damaged by crack networks.

The depth of burial during the formation of the fault gouges now seen at the surface is not well constrained, but does not seem likely to have been much more than 1 km, according to the likely maximum thickness of the Saltador Fm (bearing in mind it is always truncated at the top by an unconformity) or of the thickness of the volcanic deposits. Absence of a dip-slip component of faulting explains the lack of higher grade rock products of faulting. The formation of chlorite as a secondary mineral in the fault gouge may have been aided by elevation of the geothermal gradient under the influence of the igneous activity.

^{40}Ar–^{39}Ar dating of the volcanic rocks

Here we report nine new ^{40}Ar–^{39}Ar ages from amphibole crysts in the volcanic series, to augment a further 11 measurements that were reported in Rutter *et al.* (2012). In Table 1 and elsewhere in the text sample locations are given as grid references in the UTM (Universal Transverse Mercator, zone 30) coordinate system that is used on the IGN (Instituto Geográfico Nacional) topographic maps, European 1950 datum.

Amphibole separates were prepared by lightly crushing bulk rocks, followed by hand-picking grains under a binocular microscope. Samples were washed in deionized water and acetone and dried under an infra-red heating lamp. Between 0.032–0.058 g of samples were weighed, wrapped in aluminium foil and sealed in quartz ampoules together with the flux monitor Hb3gr ($t = 1073.6 \pm 5.3$ Ma, Jourdan *et al.* 2007). Nuclear irradiation was carried out in the RODEO facility of the High Flux Reactor, Petten, the Netherlands, with a fast fluence of *c.* 2×10^{18} n cm^{-2}. Argon was extracted from the samples using a Ta resistance furnace over the temperature interval 400–1400 °C using 30-minute heating steps. The argon was purified using a SEAES NP10 getter at 250 °C and transferred to the MS1 mass spectrometer for isotopic analysis. Blank levels were determined at three temperatures during the course of the experiments, in units of 10^{-15} moles ^{40}Ar: these were 22 ± 4 (600 °C); 40 ± 9 (1000 °C); and 97 ± 20 (1400 °C). Blank corrections were typically *c.* 10% of most

argon release steps. Raw data were corrected for blanks, mass discrimination (calibrated using atmospheric argon), radioactive decay and neutron interference. The neutron interference corrections were determined from pure CaF_2 and K_2SO_4 salts included in the ampoule with the following values: $(^{40}Ar/^{39}Ar)_K = 0.0.027$; $(^{38}Ar/^{39}Ar)_K = 0.0201$; $(^{39}Ar/^{37}Ar)_{Ca} = 0.000694$; and $(^{36}Ar/^{37}Ar)_{Ca} = 0.000267$. Full details of the ^{40}Ar–^{39}Ar stepped heating data and age spectrum diagrams are given in Table 2 and Figure 6. All ages are reported at the 2σ level of uncertainty. Age spectrum and isotope correlation diagrams were produced using the Isoplot/Ex3.23 program (Ludwig 2003).

Stratigraphy and dating results

Table 1 summarizes the age determinations from volcanic rocks and they are shown in map view on Figure 3. The age determinations coupled with stratigraphic relations show that the volcanic rocks north and west of Carboneras, but lying to the SE of the CFZ, can be divided into two groups: an older group formed between 18 and 11 Ma (Burdigalian–upper Serravallian) which were deposited largely before the CFZ was active, and a younger, unconformable group formed between 11 and 6 Ma (upper Serravallian–Messinian), during the main stage of activity of the CFZ. An angular unconformity separates these two groups of volcanic rocks.

The older volcanic rocks

The older group sits with unconformity on non-metamorphic marls, dated as upper Burdigalian by Serrano (1990) using microfossils. These in turn rest upon carbonate rocks and red siltstones of the Maláguide unit, the highest tectonic slice within the basement complex. These stratigraphic relationships are exposed in the northeastern extremity of the outcrop of the volcanic rocks (on either side of the Granatilla valley, GR 599950 4101050, at the location of sample number S1 in Fig. 3), and the sequence is tilted to the vertical against the southern margin of the CFZ. Thus the deposition of the volcanic rocks predates the origins of the CFZ but must have been taking place at the same time as tectonically driven cooling and uplift of metamorphic basement was occurring by deep-seated extensional faulting (Platt & Whitehouse 1999; Platt *et al.* 2003a; Vissers 2012) that would eventually lead to exposure and erosion of Nevado-Filábride metamorphic rocks in Tortonian times (Johnson *et al.* 1997; Augier *et al.* 2005). The onset of volcanism is marked at the stratigraphic top of the Burdigalian marl by the incorporation of small

     E. H. RUTTER *ET AL.*

Table 1. $^{40}Ar-^{39}Ar$ *radiometric ages on amphiboles from igneous rocks of the Cabo de Gata volcanic series near Carboneras*

Spec. no.	Location (GR) Easting/Northing	Age BP (Ma)	Source rock
A1	592488 4095381	6.5 ± 2.2	Volcaniclastic 500 m east of El Argamasón
A2	592463 4095938	12.5 ± 1.9	Centre of thick andesite dyke
A3	592502 4095718	8.9 ± 0.8	Volcaniclastic rock south of CFZ near el Argamasón
A4	**595750 4098590**	**9.1 ± 1.0**	**Andesite dyke intruding mica schist**
A5	**596410 4098870**	**9.3 ± 1.1**	**Fault-damaged andesite dyke**
A6	596850 4098730	10.5 ± 3.9	Volcanic breccia
A7	595860 4098350	10.7 ± 2.9	Volcanic breccia
A8	**595860 4098310**	**12.9 ± 0.8**	**Breccia 40 m south from A7**
A9	594090 4097840	10.6 ± 0.8	Andesite dyke cutting basal Saltador Fm
A10	595390 4097240	9.8 ± 1.8	Outlier of red (younger) volcanics near Minas El Palaín
A11	601800 4102550	8.8 ± 2.7	Massive intrusive andesite near La Manaca beach
A12	598600 4100150	10.8 ± 1.6	Andesitic breccia overstepping CFZ southern strand
A13	591640 4095820	12.0 ± 1.9	Andesitic breccia beneath gypsum, El Argamasón
A14	**601260 4101870**	**8.0 ± 0.8**	**Andesite intrusive near La Manaca beach**
A15	597420 4096650	11.5 ± 0.9	Cañada de Don Rodrigo, near Carboneras
A16	**597420 4096650**	**12.8 ± 1.3**	**Cañada de Don Rodrigo, near Carboneras**
A17	**599960 4099495**	**15.5 ± 1.0**	**Amphibole-phyric lava**
A18	**600450 4100550**	**11.6 ± 0.7**	**Amphibole-phyric fragments in breccia**
A19	**589851 4097885**	**19.7 ± 1.4**	**Flow-banded amphibole-phyric lava**
A20	**599700 4096975**	**14.6 ± 1.7**	**Flow-banded dacite beneath Brèche Rouge**
SP1	599950 4101040	21.1 ± 1.2	Amphibole crystals in tuff, R. Granatilla
SP2	600010 4100997	18.3 ± 1.0	Volcaniclastic breccia, R. Granatilla
SP3	599350 4100870	10.8 ± 1.9	Volcaniclastic breccia, Younger volcanic rocks on Cerro de Don Fernando

Uncertainty estimates are given as ±2 standard deviations. The nine new age determinations reported in this paper are in bold. Age determinations for samples SP1, SP2 and SP3 were described fully in Scotney *et al.* (2000) and the remaining eleven samples in Rutter *et al.* (2012).

(10 mm) clasts of pyroxenite and cumulate gabbro, thought to have been ripped from the walls of the fresh volcanic conduits where they passed through the lower continental crust. These are mixed with clasts of dolomite and red siltstone, testifying to the nearby exposure of basement rocks of the Maláguide complex at the time. There are also thin (10 mm) layers of water-lain sands bearing reworked amphibole crysts of volcanic origin. These yielded a $^{40}Ar-^{39}Ar$ age of 21.1 ± 1.2 Ma (sample SP1, Table 1), whereas amphiboles from lava fragments in breccia at the base of the immediately overlying main volcanic series (sample SP2, Table 1) yielded 18.3 ± 1.0 Ma. This is taken to be more representative of the age of the immediately underlying marls, and dates the onset of the older volcanic episode as represented by *in-situ* eruptive rocks. The detrital amphibole crysts are inferred to represent the remains of even older volcanic materials that have been completely eroded.

It is difficult to ascertain the general attitude of the older volcanic rocks owing to irregularity of contacts within the volcanic pile as a result of deposition on irregular surfaces and erosion of surfaces, but over most of the outcrop area such layering as can be discerned dips gently towards the NW. In the northeastern part of the outcrop, however, the layering is abruptly tilted up to the vertical or even overturned against the south side of the CFZ, but the axial trace of this asymmetrical syncline is truncated against the base of the deposit on the east side of the Saltador valley (Fig. 3), which lies close to the cross-section line of Figure 4.

The coastal tract about 1 km wide between Playa La Manaca and Carboneras town (Fig. 3) bears the oldest of the volcanic rocks that lie beneath the Tortonian unconformity, ranging in measured ages between 19.7 and 14.4 Ma. This tract includes massive andesitic and dacitic lavas and minor intrusive rocks, layered tuffs (including evidence of locally subaqueous deposition), subaerial mudflow deposits and coarse, clast-supported volcanic breccias. Whereas most of the dating results display a high degree of consistency in terms of their spatial distribution and accord with lithostratigraphic constraints, there are occasional apparent discrepancies. Samples A19 (19.7 Ma) and A20

Table 2. $^{40}Ar-^{39}Ar$ analytical data

Temp. ($^{\circ}$C)	^{36}Ar ($\times 10^{-15}$ moles)	Cl* ($\times 10^{-9}$ moles)	Ca† ($\times 10^{-9}$ moles)	K‡ ($\times 10^{-9}$ moles)	^{40}Ar ($\times 10^{-15}$ moles)	^{40}Ar*§ (%)	Cum. ^{39}Ar (%)	^{40}Ar*/^{39}Ar	Age$^{\parallel}$ (Ma)
A4 mass = 0.04435 g									
600	0.31 ± 0.03	2.9 ± 0.0	1094 ± 9	45.6 ± 0.5	102.3 ± 0.1	9.6	1.8	2.32 ± 2.23	31.6 ± 30.0
800	0.08 ± 0.02	18.1 ± 0.3	2508 ± 15	91.1 ± 1.0	28.1 ± 0.3	11.7	5.3	0.39 ± 0.72	5.3 ± 9.9
1000	0.12 ± 0.03	702.1 ± 9.7	10662 ± 57	525.7 ± 5.3	66.5 ± 0.1	47.1	25.5	0.64 ± 0.16	8.8 ± 2.2
1100	1.25 ± 0.05	4001.2 ± 55.3	57161 ± 344	1638.7 ± 16.8	469.2 ± 0.4	21.3	88.4	0.66 ± 0.09	9.0 ± 1.2
1200	0.35 ± 0.02	739.8 ± 10.2	6052 ± 74	194.6 ± 2.4	115.2 ± 0.1	11.1	95.9	0.70 ± 0.27	9.6 ± 3.7
1400	0.04 ± 0.02	183.6 ± 2.5	3935 ± 28	106.6 ± 1.3	14.3 ± 0.1	27.7	100.0	0.40 ± 0.51	5.5 ± 7.0
Total	2.15 ± 0.07	5647.7 ± 57.2	81413 ± 358	2602.4 ± 17.9	795.7 ± 0.6			0.67 ± 0.08	9.1 ± 1.1
	48.4 ± 1.6 $\times 10^{-15}$ mol/g	4514.3 ± 45.7 ppm	7.35 ± 0.03 wt%	2294.3 ± 15.8 ppm	17941.1 ± 13.5 $\times 10^{-15}$ mol/g				
A5 mass = 0.04367 g									
600	0.08 ± 0.02	5.7 ± 0.1	1228 ± 9	55.7 ± 0.6	33.4 ± 0.2	28.1	2.5	1.81 ± 0.94	24.7 ± 12.7
800	0.04 ± 0.02	0.2 ± 0.0	5940 ± 43	157.6 ± 1.6	22.6 ± 0.0	53.6	9.4	0.83 ± 0.37	11.3 ± 5.0
1000	0.12 ± 0.03	32.2 ± 0.4	7664 ± 57	204.7 ± 2.1	48.8 ± 0.0	28.9	18.4	0.74 ± 0.46	10.1 ± 6.2
1100	0.84 ± 0.04	1930.2 ± 26.7	40473 ± 263	1552.7 ± 15.7	343.3 ± 0.4	27.9	86.8	0.66 ± 0.08	9.1 ± 1.1
1200	0.31 ± 0.03	669.3 ± 9.3	4958 ± 31	201.1 ± 2.0	108.8 ± 0.1	16.0	95.7	0.93 ± 0.42	12.7 ± 5.7
1400	0.01 ± 0.02	171.2 ± 2.4	2271 ± 24	97.6 ± 1.2	13.3 ± 0.0	81.8	100.0	1.20 ± 0.53	16.3 ± 7.2
Total	1.39 ± 0.06	2808.9 ± 28.4	62534 ± 276	2269.4 ± 16.1	570.2 ± 0.5			0.76 ± 0.09	10.3 ± 1.2
	31.83 ± 1.46 $\times 10^{-15}$ mol/g	2280.2 ± 23.0 ppm	5.73 ± 0.02 wt%	2031.9 ± 14.5 ppm	13057.4 ± 10.8 $\times 10^{-15}$ mol/g				
A8 mass = 0.03360 g									
600	0.42 ± 0.04	3.8 ± 0.1	1736 ± 21	40.8 ± 0.5	120.1 ± 0.1	−3.8	2.2	−1.20 − ± 2.89	nd
800	0.33 ± 0.02	3.6 ± 0.1	7498 ± 85	235.4 ± 2.9	122.0 ± 0.5	19.6	14.8	1.10 ± 0.23	14.9 ± 3.1
1000	0.09 ± 0.01	90.4 ± 1.3	5996 ± 45	197.2 ± 2.3	45.1 ± 0.2	40.1	25.3	0.99 ± 0.22	13.5 ± 3.0
1100	1.22 ± 0.02	2602.8 ± 36.0	43274 ± 452	1115.5 ± 12.4	457.6 ± 0.6	21.1	84.9	0.93 ± 0.07	12.7 ± 0.9
1200	0.20 ± 0.03	389.4 ± 5.4	6330 ± 48	160.7 ± 1.9	70.3 ± 0.1	16.7	93.5	0.79 ± 0.49	10.8 ± 6.7
1400	0.00 ± 0.02	264.7 ± 3.7	3583 ± 37	121.5 ± 1.5	7.9 ± 0.1	96.8	100.0	0.68 ± 0.44	9.3 ± 6.0
Total	2.27 ± 0.06	3354.7 ± 36.6	68417 ± 466	1871.2 ± 13.2	823.0 ± 0.8			0.88 ± 0.10	12.1 ± 1.3
	67.42 ± 1.70 $\times 10^{-15}$ mol/g	3539.4 ± 38.6 ppm	8.15 ± 0.05 wt%	2177.5 ± 15.4 ppm	24494.4 ± 23.3 $\times 10^{-15}$ mol/g				

(*Continued*)

Table 2. *Continued*

Temp. (°C)	^{36}Ar ($\times 10^{-15}$ moles)	Cl* ($\times 10^{-9}$ moles)	$Ca^†$ ($\times 10^{-9}$ moles)	$K^‡$ ($\times 10^{-9}$ moles)	^{40}Ar ($\times 10^{-15}$ moles)	$^{40}Ar*^§$ (%)	Cum. ^{39}Ar (%)	$^{40}Ar*/^{39}Ar$	$Age^{\|}$ (Ma)
A14 mass = 0.03606 g									
600	0.04 ± 0.01	4.8 ± 0.1	287 ± 3	10.0 ± 0.1	16.9 ± 0.0	33.4	0.4	6.06 ± 4.12	81.3 ± 54.0
800	0.05 ± 0.02	11.8 ± 0.2	3586 ± 24	120.0 ± 1.3	18.1 ± 0.0	23.7	5.2	0.39 ± 0.50	5.3 ± 6.8
1000	0.03 ± 0.02	35.4 ± 0.5	12284 ± 70	377.8 ± 3.8	29.3 ± 0.0	74.1	20.5	0.62 ± 0.20	8.5 ± 2.7
1100	0.59 ± 0.03	1604.5 ± 22.2	33217 ± 204	1584.3 ± 16.1	259.8 ± 0.2	32.9	84.3	0.58 ± 0.07	7.9 ± 0.9
1200	0.18 ± 0.01	225.2 ± 3.1	8008 ± 83	262.1 ± 2.7	68.4 ± 0.1	20.8	94.9	0.58 ± 0.14	8.0 ± 1.9
1400	0.03 ± 0.02	100.3 ± 1.4	3759 ± 27	126.3 ± 1.3	15.4 ± 0.1	50.7	100.0	0.67 ± 0.41	9.1 ± 5.6
Total	*0.91 ± 0.05*	*1982.1 ± 22.5*	*61141 ± 234*	*2480.4 ± 16.8*	*407.9 ± 0.2*			*0.60 ± 0.07*	*8.2 ± 0.9*
	25.23 ± 1.43 *×10⁻¹⁵ mol/g*	*1948.6 ± 22.1* *ppm*	*6.79 ± 0.02* *wt%*	*2689.5 ± 18.2* *ppm*	*11311.9 ± 6.5* *×10⁻¹⁵ mol/g*				
A16 mass = 0.03529 g									
600	0.32 ± 0.02	1.9 ± 0.0	2054 ± 25	48.2 ± 0.6	82.6 ± 0.1	− 14.6	2.6	− 2.70 ± 1.63	nd
1000	0.07 ± 0.01	68.3 ± 9.4	7264 ± 59	157.8 ± 1.7	40.6 ± 0.1	52.2	11.3	1.45 ± 0.29	19.7 ± 4.0
1100	1.33 ± 0.04	3564.4 ± 493.0	35182 ± 239	1278.6 ± 13.0	499.9 ± 0.6	21.6	81.2	0.91 ± 0.10	12.4 ± 1.3
1200	0.29 ± 0.02	734.0 ± 101.7	10309 ± 105	253.3 ± 3.1	111.8 ± 0.2	24.5	95.0	1.16 ± 0.29	15.9 ± 3.9
1400	0.20 ± 0.03	279.1 ± 38.6	4301 ± 59	90.9 ± 1.3	65.7 ± 0.1	8.4	100.0	0.65 ± 0.89	8.9 ± 12.1
Total	*2.20 ± 0.06*	*4647.7 ± 504.9*	*59110 ± 275*	*1828.9 ± 13.6*	*800.7 ± 0.6*			*0.88 ± 0.10*	*12.1 ± 1.4*
	62.38 ± 1.69 *×10⁻¹⁵ mol/g*	*4668.8 ± 507.2* *ppm*	*6.70 ± 0.03* *wt%*	*2026.3 ± 15.1* *ppm*	*22689.8 ± 18.3* *×10⁻¹⁵ mol/g*				
A17 mass = 0.03314 g									
600	0.34 ± 0.02	16.5 ± 0.2	1378 ± 10	30.4 ± 0.4	103.2 ± 0.1	3.9	2.2	1.40 ± 1.98	19.2 ± 26.9
800	0.25 ± 0.02	20.8 ± 0.3	1980 ± 21	43.0 ± 0.6	76.9 ± 0.1	5.3	5.3	1.01 ± 1.14	13.9 ± 15.5
1000	0.15 ± 0.02	39.2 ± 0.5	933 ± 18	27.7 ± 0.6	46.8 ± 0.1	3.5	7.3	0.64 ± 2.44	8.8 ± 33.2
1100	1.03 ± 0.03	2800.3 ± 38.7	46820 ± 303	992.1 ± 10.4	408.7 ± 0.3	25.6	78.5	1.13 ± 0.08	15.5 ± 1.1
1200	0.22 ± 0.01	536.3 ± 7.4	4873 ± 44	185.8 ± 2.1	84.3 ± 0.1	24.0	91.8	1.17 ± 0.24	16.0 ± 3.2
1400	0.03 ± 0.02	305.3 ± 4.2	4441 ± 29	114.2 ± 1.3	22.0 ± 0.1	63.4	100.0	1.31 ± 0.46	17.9 ± 6.3
Total	*2.01 ± 0.05*	*3718.5 ± 39.7*	*60424 ± 309*	*1393.2 ± 10.7*	*638.7 ± 0.4*			*1.15 ± 0.11*	*15.6 ± 1.4*
	60.61 ± 1.40 *×10⁻¹⁵ mol/g*	*3977.7 ± 42.4* *ppm*	*7.30 ± 0.03* *wt%*	*1643.7 ± 12.6* *ppm*	*19271.7 ± 10.8* *×10⁻¹⁵ mol/g*				

A18 mass = 0.05760 g

600	0.73 ± 0.02	6.7 ± 0.1	3370 ± 31	40.9 ± 0.5	218.8 ± 0.2	1.8	2.5	1.04 ± 1.38	14.3 ± 18.7
800	0.27 ± 0.02	7.2 ± 0.1	21018 ± 166	290.2 ± 3.1	101.8 ± 0.1	22.9	20.4	0.87 ± 0.19	11.8 ± 2.5
1000	0.20 ± 0.02	48.0 ± 0.7	11213 ± 97	137.2 ± 1.5	68.3 ± 0.1	13.6	28.8	0.73 ± 0.41	10.0 ± 5.6
1200	1.53 ± 0.02	3497.1 ± 48.4	87563 ± 450	1145.1 ± 11.6	542.2 ± 0.4	16.7	99.4	0.85 ± 0.05	11.6 ± 0.7
1400	0.08 ± 0.03	0.9 ± 0.0	718 ± 15	10.0 ± 0.3	25.1 ± 0.0	8.6	100.0	2.32 ± 10.96	31.5 ± 147.6
Total	*2.80 ± 0.05*	*3559.8 ± 48.4*	*123883 ± 491*	*1623.4 ± 12.1*	*956.2 ± 0.5*			*0.86 ± 0.10*	*11.7 ± 1.3*
	48.59 ± 0.85	2190.9 ± 29.8	8.61 ± 0.03	1102.0 ± 8.2	16601.1 ± 8.1				
	$\times 10^{-15}$ mol/g	ppm	wt%	ppm	$\times 10^{-15}$ mol/g				

A19 mass = 0.03177 g

600	0.65 ± 0.03	5.5 ± 0.1	5745 ± 47	136.6 ± 1.5	202.4 ± 0.2	5.7	8.0	0.91 ± 0.65	12.5 ± 8.8
800	0.28 ± 0.02	5.5 ± 0.1	5802 ± 67	132.7 ± 1.6	89.5 ± 0.1	6.8	15.8	0.50 ± 0.39	6.8 ± 5.4
1000	0.13 ± 0.01	64.2 ± 0.9	6823 ± 68	112.0 ± 1.2	50.5 ± 0.1	23.9	22.4	1.16 ± 0.23	15.8 ± 3.1
1100	1.22 ± 0.03	2932.7 ± 40.6	42927 ± 329	962.3 ± 10.1	489.5 ± 0.5	26.4	79.0	1.44 ± 0.09	19.7 ± 1.2
1200	0.36 ± 0.03	887.5 ± 12.3	12043 ± 116	259.3 ± 2.7	139.8 ± 0.1	24.3	94.2	1.41 ± 0.40	19.2 ± 5.4
1400	0.23 ± 0.02	330.2 ± 4.6	3890 ± 32	98.7 ± 1.4	102.6 ± 0.1	34.4	100.0	3.84 ± 0.50	51.9 ± 6.7
Total	*2.86 ± 0.06*	*4225.5 ± 42.6*	*77230 ± 366*	*1701.7 ± 10.9*	*1074.3 ± 0.6*			*1.44 ± 0.10*	*19.7 ± 1.4*
	90.14 ± 1.76	4715.0 ± 47.6	9.73 ± 0.04	2094.3 ± 13.4	33816.5 ± 19.4				
	$\times 10^{-15}$ mol/g	ppm	wt%	ppm	$\times 10^{-1}$ mol/g				

A20 mass = 0.04158 g

600	0.61 ± 0.02	21.6 ± 0.3	872 ± 11	34.3 ± 0.4	189.4 ± 0.2	5.2	1.5	3.08 ± 1.68	41.7 ± 22.6
800	0.39 ± 0.02	27.8 ± 0.4	3526 ± 31	148.3 ± 1.6	128.5 ± 0.2	10.3	8.2	0.96 ± 0.39	13.1 ± 5.3
1000	0.32 ± 0.02	51.3 ± 0.7	6568 ± 34	294.2 ± 3.0	123.1 ± 0.1	22.9	21.5	1.03 ± 0.27	14.1 ± 3.7
1100	1.64 ± 0.02	3855.1 ± 53.3	35683 ± 277	1273.1 ± 13.2	613.1 ± 0.5	20.8	78.9	1.08 ± 0.05	14.7 ± 0.7
1200	0.28 ± 0.02	696.3 ± 9.6	10963 ± 74	346.7 ± 3.9	119.1 ± 0.1	29.4	94.5	1.08 ± 0.17	14.8 ± 2.3
1400	0.13 ± 0.03	454.4 ± 6.3	4517 ± 37	120.9 ± 1.6	43.3 ± 0.1	9.5	100.0	0.37 ± 0.81	5.0 ± 11.1
Total	*3.38 ± 0.06*	*5106.6 ± 54.6*	*62129 ± 293*	*2217.5 ± 14.3*	*1216.5 ± 0.6*			*1.06 ± 0.08*	*14.4 ± 1.1*
	81.30 ± 1.32	4353.7 ± 46.5	5.98 ± 0.02	2085.3 ± 13.4	29257.2 ± 13.9				
	$\times 10^{-15}$ mol/g	ppm	wt%	ppm	$\times 10^{-15}$ mol/g				

Errors quoted are ± 2 standard deviations.

*Determined from ^{37}Cl(n, γ, β)^{38}Ar:Cl $= 9.0159 \times 10^4 \cdot \dfrac{^{38}\text{Ar}}{\beta \cdot \text{J}} \cdot$ (mole/mole); $\beta = \left(\dfrac{\text{K}}{\text{Cl}} \cdot \dfrac{^{38}\text{Ar}}{^{39}\text{Ar}} \right)_{Hb3gr}$; $(\text{K/Cl})_{Hb3gr} = 5.242$; $\beta = 5.29 \pm 0.20$.

†Determined from ^{40}Ca(n, α)^{37}Ar:Ca $= 7.975 \times 10^4 \cdot \dfrac{^{37}\text{Ar}}{\alpha \cdot \text{J}} \cdot$ (mole/mole); $\alpha = \left(\dfrac{\text{K}}{\text{Ca}} \cdot \dfrac{^{37}\text{Ar}}{^{39}\text{Ar}} \right)_{Hb3gr}$; $(\text{K/Ca})_{Hb3gr} = 0.1674$; $\alpha = 0.472 \pm 0.02$.

‡Determined from ^{39}K(n, p)^{39}Ar:K $= \dfrac{^{39}\text{Ar}}{\text{J}} \cdot \dfrac{\text{K}}{^{40}\text{K}} \cdot \dfrac{\lambda}{\lambda_e}$ (mole/mole); $\lambda/\lambda_e = 9.54$; ^{40}K/K $= 1.167 \times 10^{-4}$; J $= 0.007599 \pm 0.000038$.

§^{40}Ar* $= {}^{40}$Ar$_{\text{total}} - 295.5 \times {}^{36}$Ar.

‖Includes uncertainties on the J value.

(14.6 Ma) could from a lithological point of view be derived from the same flow-banded lava or minor intrusive, in which case more closely corresponding ages would be expected. Although it should be noted that A19 did not yield a plateau age (only a total age is quoted in Table 1), if they are indeed different rock units, then A19 represents the oldest surviving igneous-textured rock in the area.

Younger rocks of the older volcanic group (lying beneath the Tortonian unconformity) also outcrop extensively further SW in Figure 3, but are always

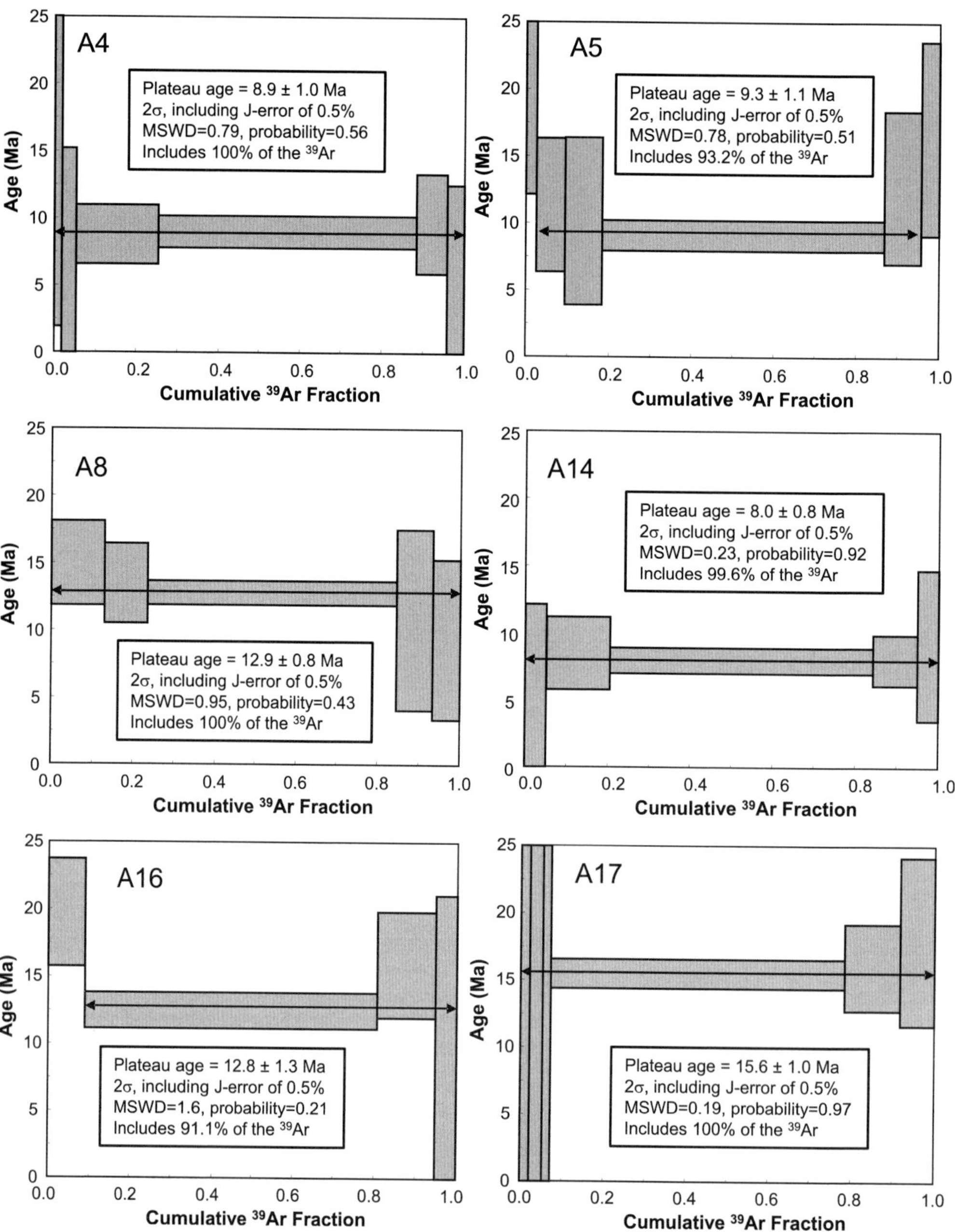

Fig. 6.

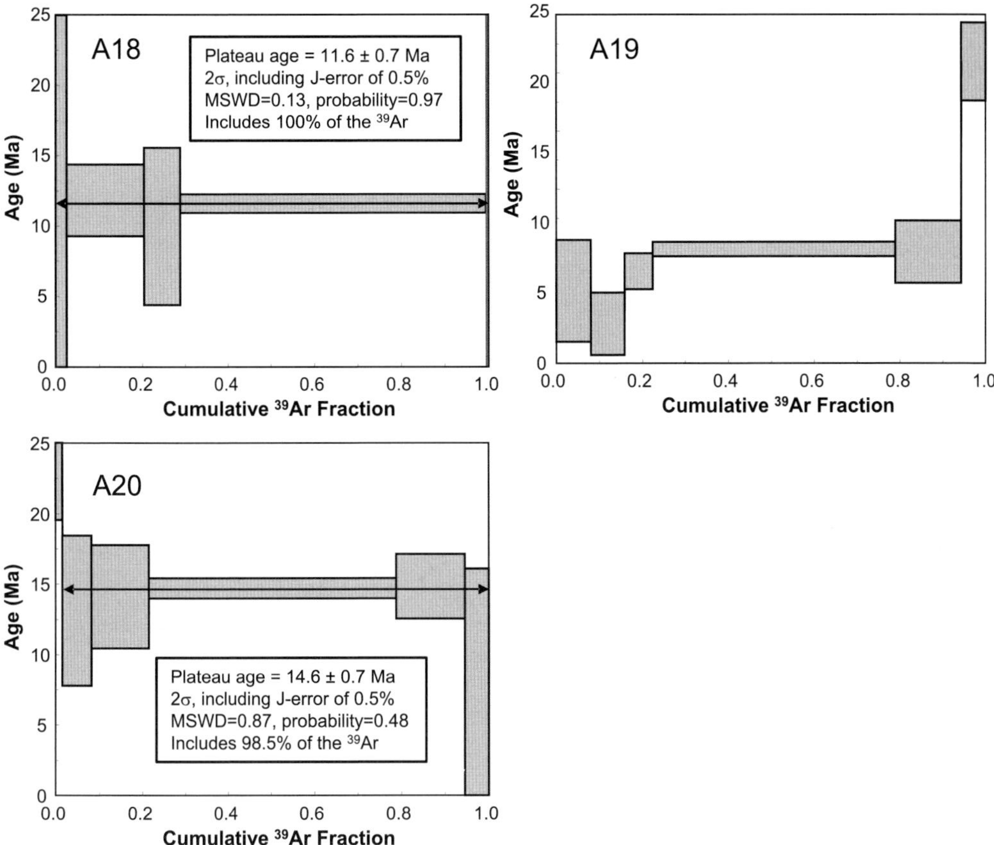

Fig. 6. Cumulative ^{39}Ar step heating plots for the nine new age determinations made on amphiboles in the Cabo de Gata volcanic rocks. Plateau ages are reported in Table 1 and also in the text boxes, with the exception of specimen A19, for which only a total age could be obtained. MSWD, mean square weighted deviation.

confined to the south side of the CFZ. Ages measured range between 12.9 and 10.5 Ma. The distribution of ages implies that the rock units overall dip gently towards the NW. A substantial area of about 2 km^2 lying to the south of El Saltador shows a high degree of hydrothermal alteration (Fig. 3) and hosts the hydrothermal gold deposits of El Palaín and La Islica (Morales Ruano *et al.* 2000; Carrillo Rosúa *et al.* 2002).

At GR 595450 4098160 (near specimen locations A7 and A8), on the western side of the Saltador valley, a few metres thickness of marly sediments make a vertical contact with phyllites of the Alpujárride tectonic unit to the north, which in turn to the north are truncated after a few metres by fault gouge of the southern strand of the CFZ. To the south the marls pass into volcaniclastic rocks via a vertical contact. Along strike, on the eastern side of the Saltador valley (which lies approximately along section line A–A′ on Fig. 3), specimen

A7 (Table 1) yielded an age of 10.7 ± 2.9 Ma. The marls were dated as Serravallian by Serrano (1990) from their microfauna. The age constraints confirm that the movements on the CFZ that led to upending of adjacent rocks did not begin before upper Serravallian time.

The younger volcanic rocks

These comprise red-weathering volcaniclastic rocks that lie above the older group, dominantly to the south of the southern strand of the CFZ, often with angular unconformity (e.g. GR 599670 4098495 and 599300 4100810) and that are not hydrothermally altered as are the underlying older rocks (e.g. GR 594720 4097195). The base generally dips gently to the SE, and whilst they overstep the upended unconformable contact between the older volcanic rocks and the Maláguide basement rocks, except for the very youngest components of the group they are still truncated against the southern strand of the CFZ. Their deposition therefore spans the main period of activity of the CFZ. $^{40}Ar-^{39}Ar$ ages from the younger group range from 10.5 to 6.5 Ma (Fig. 3), although there is overlap of the uncertainty ranges for ages obtained from rocks on either side of the unconformity.

The youngest volcaniclastic rocks are also found on the NW side of the main strands of the CFZ. Subsequent movements on the CFZ have not been of sufficient magnitude to remove the outcrop of the volcaniclastic rocks either through strike-slip or dip-slip displacements, and at GR 592620 4095260, where the southern strand of the CFZ is seen to cut the younger volcaniclastic rocks, the degree of cataclastic damage seems minor. Specimens A1 and A3 (Table 1) record ages of 6.5 $\pm$ 2.2 Ma and 8.9 $\pm$ 0.8 Ma for volcanciclastic rocks that outcrop to the NW of the CFZ trace. The anomalous age of 12.0 $\pm$ 1.9 Ma for specimen A13, taken from a volcanic clast lying just beneath Messinian marl is thought to be reworked older material, perhaps from the immediately underlying intrusive sheet (specimen A2, 12.5 Ma). From these ages and the stratigraphic constraints we concluded that despite the fact that minor movements have continued on at least the southern strand of the CFZ to the present day (e.g. Bell *et al.* 1997) the main movements were over by late Tortonian − early Messinian time.

Within this group are found volcanic breccias of the type Brèche Rouge de Carboneras (Uwe *et al.* 2003) (e.g. at GR 594720 4097195, 599650 4096885 and 597840 4095970), often immediately above an unconformable contact. The volcanic breccia fragments range in diameter up to about 100 mm, but possess an infiltrated pink carbonate matrix bearing a rich fauna of body fossils of Tortonian age (Uwe *et al.* 2003), although Serrano (1992) concludes that this distinctive facies can include rocks of Messinian age. Thus it seems likely that the age of the observed unconformity is not the same everywhere, that is, more than one unconformable contact has developed since the start of the period of CFZ movements.

Included within the younger volcanic group are dykes and larger minor andesitic igneous rocks up to 500 m wide injected along the CFZ, that are inferred to have used the fault zone as a route through the continental crust towards the surface (Figs 3 & 4). They range in age from 12.5 to 8.3 Ma (Table 1). The oldest specimen (A2, 12.5 $\pm$ 1.9 Ma) was taken from a 500 m wide intrusion (Fig. 3). It cuts the northern strand of the CFZ, that may point to this being the first-formed fault of the CFZ, or the intrusion may be a composite one with components dating from the pre-CFZ period of volcanic activity. The intrusive dykes may have fed

minor eruptive centres along the trace of the CFZ. Many of these igneous bodies are fresh and undamaged, whereas others show varying degrees of cataclastic damage, that might be expected of intrusion into an active fault zone. They are usually found cutting fault gouge or graphitic mica schist of the Nevado-Filábride basement, but also Serravallian and Tortonian marls and sands of the Saltador Fm (e.g. GR 594090 4097840). At GR 597610 4100100 several closely spaced vertical intrusive sheets cut tightly folded Saltador Fm (late Serravallian/Tortonian) marls and siltstones sub-parallel to their axial planes and have caused substantial baking and hardening of the marls near the contacts.

Some 10 km further SW from the study area, along the trace of the CFZ, lies La Serrata, a ridge of volcanic material some 7 km long and up to 1 km wide. This is a prominent uplift of intrusive and volcaniclastic rocks in the CFZ. At its northeastern extremity basement phyllites and dolomites are lifted and arched over the top of an intrusive body, forming its roof. This is geometrically similar to the antiformal arching of basement and sedimentary cover-rocks over the massive intrusive body that outcrops on the coast, at the northeastern extremity of the CFZ (Fig. 2).

Discussion

Dating constraints on CFZ movements

The dating of the volcanic rocks of the Carboneras region, coupled with stratigraphic relations observed in the field, help to constrain the main movement periods on the CFZ. The onset of transpressive displacements on the southern strand of the CFZ are quite tightly constrained at 11.5 $\pm$ 0.5 Ma. Older rocks are tilted to the vertical or even overturned adjacent to the fault zone as a result of transpressive left-lateral strike-slip movements, and the youngest uptilted rocks are about 11.5 Ma in age. Younger volcanic rocks (11 Ma and younger) overstep the base (18 Ma) of the older volcanic rocks on to basement Maláguide rocks but are then truncated against the southern strand of the CFZ. All of these older volcanic rocks that were probably originally deposited on basement rocks of the higher nappe units to the NW of the CFZ have now been transported to the SW by fault movements to be buried under Pliocene sediments of the Níjar basin fill, or removed by erosion as a result of a component of uplift to the north. Only volcanic rocks of upper Tortonian through mid-Messinian age are preserved to the NW of the CFZ, and whilst fault displacements have continued since that time (Bell *et al.* 1997), they seem to have been relatively small (tens of metres horizontally and vertically). Pliocene through recent history seems to have been dominated by tectonic uplift (Braga *et al.* 2003; Martín *et al.* 2003; Maher & Harvey 2008).

The main (upper Serravallian through Tortonian) period of fault activity was accompanied by injection of dykes and larger elongate intrusions of andesite into the fault zone, but the Serravallian age (12.5 Ma) of the south-westernmost of these intrusions suggests that fault activity may have begun on the northern strand of the CFZ by this time, and possibly also as pure strike-slip movements on the southern strand, such that it did not lead initially to uptilting of the adjacent rocks to the south. Welding of the northern strand of the CFZ as a result of igneous injection may have occurred, leading to the transfer of slip on to other fault strands.

Dating constraints on the unroofing of the metamorphic rocks of the Nevado-Filábride complex (Johnson *et al.* 1997) and the lithic content of derived sediments indicate they were not exposed until upper Serravallian/Tortonian time. The moderately to weakly consolidated upper Serravallian and Tortonian sediments rest unconformably upon basement schists that were never together buried deeper than about 1 km. The presently exposed fault rocks of the CFZ affecting metamorphic basement display evidence of near-perfect strike-slip shearing under only very shallow crustal conditions, with no relict higher-grade fault rocks that would have formed under deeper (higher pressure and temperature) crustal conditions prior to unroofing. These factors indicate that movement on any part of the CFZ prior to middle – upper Serravallian time (12.5 Ma) was unlikely.

Wider geodynamic implications

The Betic zone, lying to the north and west of the Trans-Alborán shear zone (that includes the CFZ), has undergone broadly NE–SW stretching since the lower Miocene (Aquitanian), culminating in the emergence and uplift of Alpine metamorphic basement rocks as metamorphic core complexes during Tortonian time (Johnson *et al.* 1997). The early history of tectonic unroofing led to rapid depressurization and cooling during 22–15 Ma (Platt & Whitehouse 1999; Platt *et al.* 2003*a*, *b*; 2005; Zeck 2004; Vissers 2012) with the displacement of the Alpujárride complex on top of higher metamorphic-grade rocks of the Nevado-Filábride complex (Jabaloy *et al.* 1993; Lonergan & Platt 1995; Platt *et al.* 2005; Vissers 2012; Fig. 2), potentially excising part of the metamorphic pile. The main flat-lying detachment zone (Betic Movement Zone, Platt & Vissers 1989; Jabaloy *et al.* 1993;

Vissers 2012) is marked by mylonitic rocks with internal structures that demonstrate roughly top-to-the-SW displacements under greenschist through amphibolite facies conditions (Vissers 2012). Progressive cooling and depressurization led to dominantly cataclastic deformation in the upper sheet and the formation of fault gouge in the phyllitic metapelitic rocks of the Alpujárride complex.

Post-orogenic sediments arising from the progressive unroofing during lower- and mid-Miocene times are found rarely (having been removed by erosion) and only sitting upon a substrate of rocks of the Maláguide and Alpujárride extensional nappes (Sanz de Galdeano & Vera 1992; Serrano *et al.* 2007). Absence of clasts derived from the Nevado-Filábride metamorphic basement rocks shows that they were not exposed by erosion until Tortonian time, from which point large amounts of post-orogenic sediment are preserved in inter-montane basins containing growth faults that developed as extension continued through the Tortonian (Rodríguez-Fernández & Sanz de Galdeano 2006). With the aid of fission track dating, Johnson *et al.* (1997) demonstrated the emergence of the Sierra Nevada – Sierra de Los Filabres metamorphic complex, the uplift and cooling migrating SW with time, through the latest Serravallian and Tortonian. Several studies have shown how NE–SW stretching through the Tortonian led to the development of the intermontane basins (Martínez-Díaz & Hernández Enrile 2004; Martínez-Martínez *et al.* 2006; Meijninger & Vissers 2006). NE–SW-directed stretching is regarded as responsible for the crustal and lithospheric mantle thinning that led to formation of the west Alborán basin through the middle and upper Miocene and its southwestward transport to its present position (Comas *et al.* 1999; Facenna *et al.* 2004). Díaz *et al.* (2010) showed that the direction of fast polarization of teleseismic S-waves is consistently NE–SW oriented throughout the Betic domain, and may be interpreted to indicate the long-term imprint of plastic stretching in the upper mantle.

Comas *et al.* (1999) also point out that evidence from basement metamorphic rocks for a heating event in the late stages of extension-related decompression can best be explained by an hypothesis, alternative to the slab rollback hypothesis, of lithospheric mantle delamination with upwelling of asthenospheric mantle. Variations on this model, including convective removal of subcontinental lithosphere had previously been proposed, for example by Platt & Vissers (1989) and García-Dueñas *et al.* (1992), Platt *et al.* (2003*a*), and some combination of processes may be involved in the geodynamic evolution of the region.

The initiation and development of the CFZ from mid to late Serravallian time coincides with the final emergence of the metamorphic core complexes and the formation of the Tortonian intramontane basins and their sedimentary fill. The earlier Miocene development of the calc-alkaline volcanism of the Cabo de Gata province, its arcuate trend across the Alborán Sea and extension eastwards on the south side of the Algero-Balearic basin, suggest it formed a continuous subduction-related arc (Lonergan & White 1997; Chalouan & Michard 2004; Duggen *et al.* 2005; Booth-Rea *et al.* 2007) before it became offset in its western part by the Tortonian activity of the Trans-Alborán shear zone. Several authors (Royden 1993; Lonergan & White 1997; Gutscher *et al.* 2002; Facenna *et al.* 2004; Gutscher 2012) inferred the western part of this arc migrated by subduction rollback towards the SW, until it arrived at its present position forming the Gibraltar arc by the end of the Miocene period (5 Ma), with a segment of accretionary complex lying offshore in the Atlantic ocean to the west (Fig. 7). The rollback is inferred to have caused ductile back-arc stretching of the lower crust and upper mantle rocks beneath the internal Betic zone. About 150 km of this stretching would have occurred since the beginning of the Tortonian stage, and hence been responsible for the final unroofing of the metamorphic core complexes and accumulation of Tortonian sediments in overlying extensional basins bounded by brittle faults. The initial crustal extension that occurred during lower and middle Miocene times would have developed under mid-crustal pressure and temperature conditions prior to the Tortonian emergence.

Gutscher (2012) explicitly suggests that the crustal segment in which the Tortonian–Messinian back-arc extension occurred became bounded to the north by the right-lateral Crevillente fault and internal-external zone boundary (Fig. 1) and to the south by the left-lateral Trans-Alborán shear system (Fig. 7). Viewed in this way the Trans-Alborán fault system can be regarded as a stretching transform fault system. The concept of a stretching fault was introduced by Means (1989) as a displacement discontinuity surface between materials stretched by different amounts on either side of the fault. Equivalently, Odé (1960) proposed that a fault surface could develop to accommodate a velocity discontinuity in an otherwise uniformly deforming plastic material. Such a fault can support large variations in the amount of displacement along its length, decreasing to zero at one end (in this case near Alicante), whilst the other end transforms into the migrating subduction arc. The strain within the back-arc region can be the same everywhere, however. The average displacement rate of the arc towards the SW would be 150 km in 6 Ma (relative to a fixed Alicante), or

25 mm a^{-1}, producing a lithospheric stretching strain of about 40%. The marked confinement of present-day seismicity to the back-arc stretching wedge (Martínez-Martínez *et al.* 2006; Martínez-Díaz *et al.* 2012) strongly suggests the Trans-Alborán shear zone separates the tectonically more active zone to the NW from the region to the SE. We do not know the stretching rate to the SE side of the Trans-Alboran shear zone, but the continuation of the upper mantle shear-wave splitting anisotropy and the existence of extensional basins and uplifted blocks of metamorphic basement (Sierra Almagrera and Sierra Almenara) on that side suggests it is non-zero, albeit less than to the NW. If it were assumed to be zero and displacements vary linearly along the fault system, the maximum expected displacement on the CFZ segment would be about 40 km. The minimum is the 15 km required to remove Tortonian volcanic deposits that would have lain to the NW of the CFZ.

For this hypothesis to be viable, the timing of the origin and main periods of activity on the Palomares and the Alhama de Murcia fault systems must be broadly the same (i.e. not older) as for the CFZ (upper Serravallian through Messinian). Much of the potential outcrop tract of the Palomares fault system (comprising the Palomares, Herrerías and El Arteal faults, (Booth-Rea *et al.* 2004)) is buried beneath Plio-Quaternary deposits, but Booth-Rea *et al.* (2003) report data confirming that it has been active at least between mid to upper Tortonian and Recent times. The Palomares fault cuts calc-alkaline volcanic rocks dated at 8.2 Ma (Bellon *et al.* 1983) that underlie upper Tortonian sediments, but it is unclear what movements might have occurred previously.

The Fortuna basin lies between the Alhama de Murcia and Crevillente faults at their eastern extremities. Garcés *et al.* (2001) link the development of the basin to movements on these faults. The earliest components of the sedimentary fill are lower Tortonian (Montenat *et al.* 1990). The faults (Crevillente fault and internal-external zone boundary, Fig. 1) that are inferred to have formed at least the southern edge of the northern side of the back-arc stretching wedge are less well exposed than those of the Trans-Alborán shear zone on the south side, and are partially buried beneath thrust sheets active between middle and uppermost Miocene time that push rocks towards the Iberian foreland. However, palaeomagnetic and structural evidence shows many of these have suffered large clockwise rotations during and after emplacement and hence are partially detached from the underlying basement that was undergoing right-lateral strike-slip displacements (Platt *et al.* 2003*b*). The latter authors estimate substantially more than 100 km of right-lateral displacement has occurred

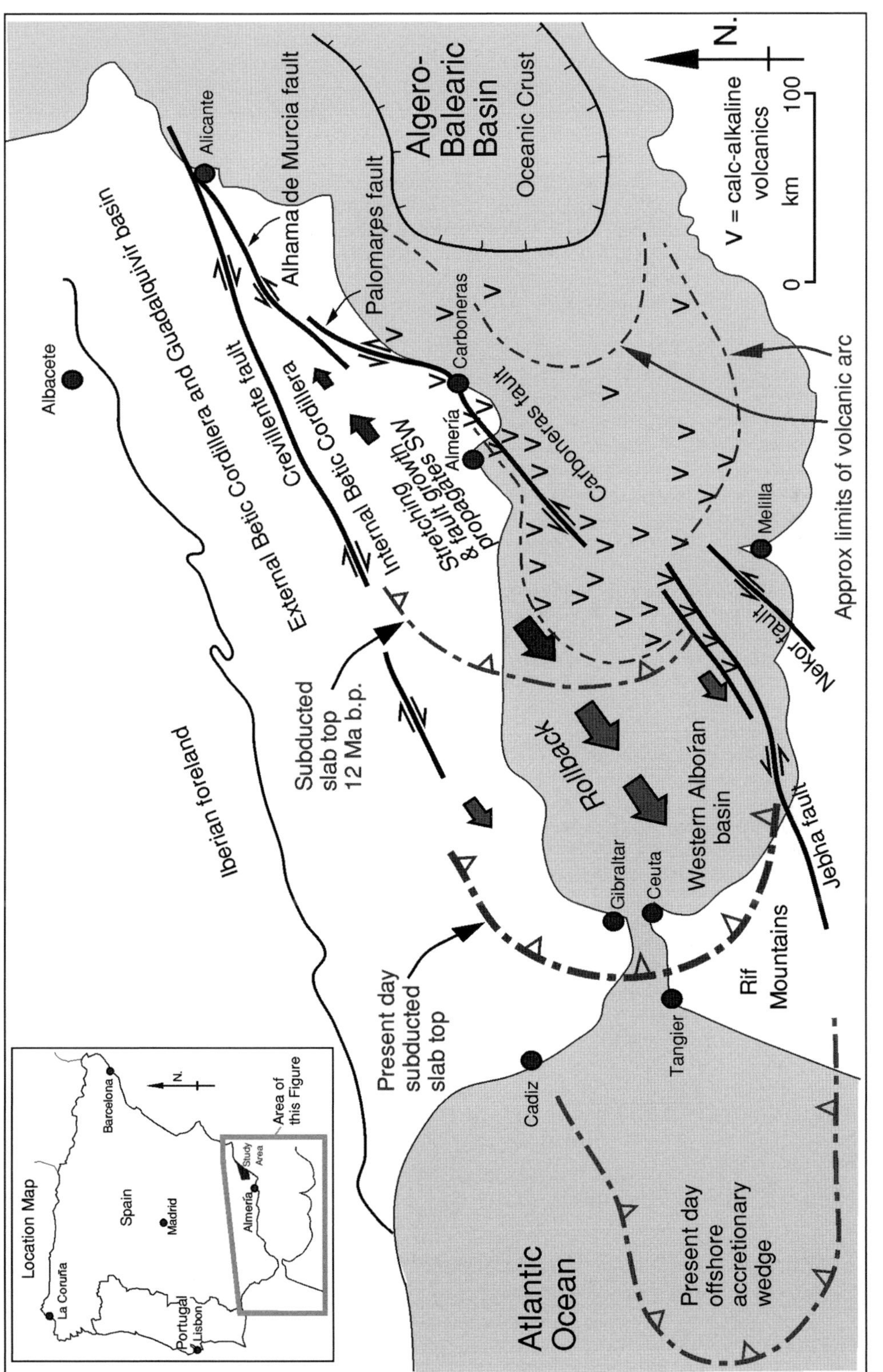

Fig. 7. Map of SE Spain and NW Africa, separated by the Alborán Sea basin, showing the geodynamic scheme (based on Lonergan & White 1997; Gutscher 2012) for the Serravallian through Recent evolution. Slab rollback is accommodated by stretching within the Betic-Alborán wedge that is bounded by right-lateral and left-lateral wrench-fault systems, and displacing part of the pre-Serravallian volcanic arc south-westwards, with Tortonian igneous rocks being injected into the active Carboneras fault segment. Offshore distributions of Neogene calc-alkaline rocks based on Comas *et al.* (1999) and Booth-Rea *et al.* (2007) shown offset by the Trans-Alborán shear zone, with intrusions into the fault zones as displacements occur.

along this boundary during Miocene time. Martín-Algarra *et al.* (2009) give the age of sediments that bury the outcrop of the internal-external zone boundary as upper Tortonian at the western end of its outcrop.

It should be noted that whilst we infer that these fault systems have accommodated major strike-slip displacements, it is clear that in the scenario of north−south tectonic convergence and contemporaneous NE–SW extension they can be reactivated with important dip-slip movements, for example, accommodating the uplift of basement blocks (e.g. Rutter *et al.* 1986; Meijninger & Vissers 2006).

Unlike other strike-slip faults bounding the wedge that contains the internal Betic zone, the CFZ movements (but possibly also early movements on the Palomares fault) were accompanied by intrusive and extrusive magmatic activity arising from the CFZ's situation transecting the calc-alkaline volcanic arc. The time span of the Tortonian−Messinian volcanic activity also defines the duration of the main movement period on the CFZ. Messinian-through-Recent faulting activity seems to be of lesser significance, although the present-day seismicity distribution demonstrates that movements still continue.

The uplift of the metamorphic core complexes, evidence for the transpressional nature of the CFZ, rotation and reactivation of basement block-bounding faults as high-angle reverse faults, thrusting in the external zones, formation of NE–SW-trending fold systems in basement rocks (folding schistosity) and in Tortonian and Messinian rocks within the internal and external orogenic zones, all testify to continuing NW–SE-directed shortening (Fig. 2) throughout the upper Miocene period (Galindo-Zaldívar *et al.* 2003), at the same time as NE–SW directed extension (Rutter *et al.* 1986; Vissers 2012). The shortening is generally inferred to be due to the continued convergence between North Africa and Iberia (Vissers 2012), and may have reduced the angle between the Trans-Alborán shear zone and the Crevillente fault by about 30% to its present 30° during the upper Miocene period.

Conclusions

The left-lateral strike-slip CFZ separates the Cabo de Gata volcanic terrain, lying to the SE, from the uplifted metamorphic basement block of the Sierra Cabrera lying to the NW, and is itself the focus of intrusion of andesitic dykes co-eval with the period of fault movement. $^{40}Ar-^{39}Ar$ dating of amphiboles in the volcanic rocks was used to determine the span of ages represented and, together

with stratigraphic constraints, to determine the period of movement of the fault zone.

The fault system is estimated to have become established between 12 and 11 Ma BP (upper Serravallian), mainly because all older volcanic rocks are rotated to the vertical adjacent to the southern strand of the fault zone. The earliest movements are inferred to be approximately co-eval with the earliest igneous intrusion into the CFZ northern strand (12.5 Ma). Younger Tortonian volcanic rocks overstep the base of the older volcanic rocks but are themselves truncated by the fault zone. The youngest volcanic rocks (6.5 Ma, Messinian) overstep the fault zone to the NW, but are affected by relatively minor, late movements on the fault zone. The fault zone has hosted the intrusion of a swarm of andesitic dykes and minor intrusives, and the duration of the main fault movements was intimately related to the duration of the Tortonian-Messinian period of volcanic activity.

The CFZ is inferred to form part of the Trans-Alborán shear zone, a system of linked left-lateral faults extending from Alicante south-westwards to Morocco. We infer that the fault system forms a bounding velocity discontinuity in the NE–SW-directed upper Miocene stretching of the internal Betic zone, in response to subduction rollback that led to the formation of the present-day Gibraltar arc. The Trans-Alborán shear zone is therefore an example of a stretching transform fault system, with an estimated 40 km of offset in the vicinity of Carboneras.

This work was carried out with support from UK NERC, grant NE/F019475/1. Fellow members of the research group and those who provided support during field-work are: Andreas Rietbrock, Stuart Nippress, Rochelle Taylor, Oshaine Blake, John Gates, Joe Tant, Lindy Walsh and Antonio Ruiz-Zamora. The Consejería del Medio-Ambiente, Junta de Andalucía gave permission for geological work to be carried out in the Gabo de Gata-Níjar Natural Park area. We appreciate the helpful and constructive reviews received from Hans de Bresser and an anonymous referee.

References

ALONSO-CHAVES, F. M., SOTO, J. I., OROZCO, M., KILIAS, A. A. & TRANOS, M. D. 2004. Tectonic evolution of the Betic Cordillera: an overview. *Bulletin of the Geological Society of Greece*, **36**, 1598–1607.

AUGIER, R., JOLIVET, L. & ROBIN, C. 2005. Late Orogenic doming in the eastern Betic Cordilleras: final exhumation of the Nevado-Filábride complex and its relation to basin genesis. *Tectonics*, **24**, TC4003, http://dx.doi.org/10.1029/2004TC001687

BELL, J. W., AMELUNG, F. & KING, G. C. P. 1997. Preliminary late quaternary slip history of the Carboneras fault, southeastern Spain. *Geodynamics*, **24**, 5–14.

BELLON, H., BORDET, P. & MONTENAT, C. 1983. Chronologie du magmatisme Néogène des Cordillères Bétiques (Espagne méridionale). *Bulletin de la Societe Géologique du France*, **7**, 205–217.

BOOTH-REA, G., AZAÑÓN, J. M., VÍCTOR GARCÍA-DUEÑAS, V. & AUGIER, R. 2003. Uppermost Tortonian to Quaternary depocentre migration related with segmentation of the strike-slip Palomares Fault Zone, Vera Basin (SE Spain). *Comptes Rendus Geoscience*, **335**, 751–761.

BOOTH-REA, G., AZAÑÓN, J. M., AZOR, A. & GARCÍA-DUEÑAS, V. 2004. Influence of strike-slip fault segmentation on drainage evolution and topography. A case study: the Palomares Fault Zone (southeastern Betics, Spain). *Journal of Structural Geology*, **26**, 1615–1632.

BOOTH-REA, G., RANERO, C. R., MARTÍNEZ-MARTÍNEZ, J. M. & GREVEMEYER, I. 2007. Crustal types and Tertiary tectonic evolution of the Alborán sea, western Mediterranean. *Geochemistry, Geophysics, Geosystems*, **8**, 1–25, Q10005, http://dx.doi.org/10.1029/2007GC001639

BRAGA, J. C., MARTÍN, J. M. & QUESADA, C. 2003. Patterns and average rates of late Neogene–recent uplift of the Betic Cordillera, S. E. Spain. *Geomorphology*, **50**, 3–26.

CARRILLO ROSÚA, F. J., MORALES RUANO, S. & FENOLL HACH-ALÍ, P. 2002. The three generations of gold in the Palaí–Islica epithermal deposit, southeastern Spain. *The Canadian Mineralogist*, **40**, 1465–1481.

CHALOUAN, A. & MICHARD, A. 2004. The Alpine Rif Belt (Morocco): a case of mountain building in a subduction-subduction-transform fault triple junction. *Pure and Applied Geophysics*, **161**, 489–519, http://dx.doi.org/10.1007/s00024-003-2460-7

COMAS, M. C., PLATT, J. P., SOTO, J. I. & WATTS, A. B. 1999. The origin and tectonic history of the Alborán basin: insights from leg 161 results. *Proceedings of the Ocean Drilling Program, Scientific Results*, **161**, 555–580.

DABRIO, C. J. & POLO, M. D. 1991. Fan-delta slope deposits and sequences in the Murcia-Carrascoy Basin (Late Neogene, S.E. Spain). *Cuadernos de Geología Ibérica*, **15**, 49–71.

DE LAROUZIÈRE, F. D., BOLZE, J., BORDET, I. P., HERNANDEZ, J., MONTENAT, C. & OTT D'ESTEVOU, P. 1988. The Betic segment of the lithospheric Trans-Alborán shear zone during the Late Miocene. *Tectonophysics*, **152**, 41–52.

DÍAZ, J., GALLART, J. *ET AL.* 2010. Mantle dynamics beneath the Gibraltar Arc (western Mediterranean) from shear-wave splitting measurements on a dense seismic array. *Geophysical Research Letters*, **37**, L18304, http://dx.doi.org/10.1029/2010GL044201

DUGGEN, S., HOERNLE, K., VAN DEN BOGAARD, P. & GARBE-SCHÖNBERG, D. 2005. Post-collisional transition from subduction to intraplate-type magmatism in the westernmost Mediterranean: evidence for continental-edge delamination of subcontinental lithosphere. *Journal of Petrology*, **46**, 1155–1201.

FACENNA, C., PIROMALLO, C., CRESPO-BLANC, A., JOLIVET, L. & ROSSETTI, F. 2004. Lateral slab deformation and the origin of the western Mediterranean arcs. *Tectonics*, **23**, TC1012.

FAULKNER, D. R., LEWIS, A. C. & RUTTER, E. H. 2003. On the internal structure and mechanics of large strike-slip fault zones: field observations of the Carboneras fault zone, S.E. Spain. *Tectonophysics*, **367**, 235–251.

FORTUIN, A. R. & KRIJGSMAN, W. 2003. The Messinian of the Nijar Basin (SE Spain): sedimentation, depositional environments and paleogeographic evolution. *Sedimentary Geology*, **160**, 213–242.

GALINDO-ZALDÍVAR, J., GIL, A. J. *ET AL.* 2003. Active faulting in the internal zones of the central Betic Cordilleras (SE, Spain). *Journal of Geodynamics*, **36**, 239–250.

GARCÉS, M., KRIJGSMAN, W. & AGUSTI, J. 2001. Chronostratigraphic framework and evolution of the Fortuna basin (Eastern Betics) since the Late Miocene. *Basin Research*, **13**, 199–216.

GARCÍA-DUEÑAS, V., BALANYÁ, J. C. & MARTÍNEZ-MARTÍNEZ, J. M. 1992. Miocene extensional detachments in the outcropping basement of the northern Alborán basin (Betics) and their tectonic implications. *Geo-Marine Letters*, **12**, 88–95.

GIACONIA, F., BOOTH-REA, G., MARTÍNEZ-MARTÍNEZ, J. M. & AZAÑÓN, J. M. 2011. Extensional transfer faults versus transcurrent faults in the eastern Betics, an example from the Sorbas basin (SE Spain). *Geophysical Research Abstracts*, **13**, 5–15. EGU2011-4011-1.

GRÀCIA, E., PALLÁS, R. *ET AL.* 2006. Active faulting offshore SE Spain (Alborán Sea): implications for earthquake hazard assessment in the Southern Iberian Margin. *Earth and Planetary Science Letters*, **241**, 734–749.

GUTSCHER, M. A. 2012. Subduction beneath Gibraltar? Recent studies provide answers. *Eos*, **93**, 133–134.

GUTSCHER, M. A., MALOD, J., REHAULT, J. P., CONTRUCCI, I., KLINGELHOEFER, F., MENDES-VICTOR, L. & SPAKMAN, W. 2002. Evidence for active subduction beneath Gibraltar. *Geology*, **30**, 1071–1074.

JABALOY, A., GALINDO-ZALDÍVAR, J. & GONZÁLEZ-LODEIRO, F. 1993. The Alpujárride Nevado-Filábride extensional shear zone, Betic-Cordillera, SE Spain. *Journal of Structural Geology*, **15**, 555–569.

JOHNSON, C., HARBURY, N. & HURFORD, A. J. 1997. The role of extension in the Miocene denudation of the Nevado-Filábride Complex, Betic Cordillera (SE Spain). *Tectonics*, **16**, 189–204.

JOURDAN, F., VERATI, C. & FERAUD, G. 2007. Intercalibration of the Hb3gr ^{40}Ar/^{39}Ar dating standard. *Chemical Geology*, **231**, 177–189.

LEBLANC, D. & OLIVIER, P. 1984. Role of strike-slip faults in the Betic-Rifian orogeny. *Tectonophysics*, **101**, 345–355.

LONERGAN, L. & PLATT, J. P. 1995. The Maláguide–Alpujárride boundary: a major extensional contact in the Internal Zone of the eastern Betic Cordillera, S. E. Spain. *Journal of Structural Geology*, **17**, 1655–1671.

LONERGAN, L. & WHITE, N. 1997. Origin of the Betic-Rif mountain belt. *Tectonics*, **16**, 504–522.

LUDWIG, K. 2003. *Isoplot/Ex, Version 3: A Geochronological Toolkit for Microsoft Excel*. Geochronology Center, Berkeley, USA.

MAHER, E. & HARVEY, A. M. 2008. Fluvial system response to tectonically induced base-level change during the late-Quaternary: the Rio Alias southeast Spain. *Geomorphology*, **100**, 180–192.

MARTÍN, J. M., BRAGA, J. C. & BETZLER, C. 2003. Late Neogene-Recent uplift of the Cabo de Gata volcanic province, Almeria, SE Spain. *Geomorphology*, **50**, 27–42.

MARTÍN-ALGARRA, A., MAZZOLI, S., PERRONE, V. & RODRÍGUEZ-CAÑERO, R. 2009. Variscan tectonics in the Malaguide Complex (Betic Cordillera, Southern Spain): stratigraphic and structural Alpine versus Pre-Alpine constraints from the Ardales area (Province of Malaga). II. Structure. *Journal of Geology*, **117**, 263–284.

MARTÍNEZ-DÍAZ, J. & HERNÁNDEZ-ENRILE, J. L. 2004. Neotectonics and morphotectonics of the southern Almería region (Betic Cordillera-Spain) kinematic implications. *International Journal of Earth Sciences (Geologisches Rundschau)*, **93**, 189–206, http://dx.doi.org/10.1007/s00531–003–0379–y

MARTÍNEZ-DÍAZ, J. J., MASANA, E. & ORTUÑO, M. 2012. Active tectonics of the Alhama de Murcia fault, Betic Cordillera, Spain. *Journal of Iberian Geology*, **38**, 269–286.

MARTÍNEZ-MARTÍNEZ, J. M., BOOTH-REA, G., AZAÑÓN, J. M. & TORCAL, F. 2006. Active transfer fault zone linking a segmented extensional system (Betics, southern Spain): Insight into heterogeneous extension driven by edge delamination. *Tectonophysics*, **422**, 159–173.

MEANS, W. D. 1989. Stretching faults. *Geology*, **17**, 893–896, http://dx.doi.org/10.1130/0091-7613(1989) 017<0893:SF>2.3.CO;2

MEIJNINGER, B. M. L. & VISSERS, R. L. M. 2006. Miocene extensional basin development in the Betic Cordillera, SE Spain revealed through analysis of the Alhama de Murcia and Crevillente Faults. *Basin Research*, **18**, 547–571.

MONTENAT, C., OTT D'ESTEVOU, P. H. & COPPIER, G. 1990. Les bassins Néogènes entre Alicante et Cartagena. *In*: MONTENAT, C. (ed.) *Les Bassins Néogènes du Domaine Bétique Oriental (Espagne)*. Documents et Travaux, Institut Géologique Albert de Lapparent, Paris, **12/13**, 313–368.

MORALES RUANO, S., CARRILLO ROSÚA, F. J., HACH-ALÍ, P. F., DE LA FUENTE CHACÓN, F. & CONTRERAS LÓPEZ, E. 2000. Epithermal Cu–Au mineralization in the Palai–Islica deposit, Almería, southeastern Spain: fluid-inclusion evidence for mixing of fluids as a guide to gold mineralization. *The Canadian Mineralogist*, **38**, 553–565.

MORENO MOTA, X. 2010. *Neogene and paleoseismic onshore-offshore integrated study of the Carboneras fault (Eastern Betics, SE Iberia)*. PhD thesis, University of Barcelona.

ODÉ, H. 1960. Faulting as a velocity discontinuity in plastic deformation. *In*: GRIGGS, D. & HANDIN, J. (eds) *Rock Deformation*. Geological Society of America Memoir, Waverly Press, Baltimore, **79**, 293–321.

PLATT, J. P. & VISSERS, R. L. M. 1989. Extensional collapse of thickened continental lithosphere: a working hypothesis for the Alborán Sea and Gibraltar arc. *Geology*, **17**, 540–543.

PLATT, J. P. & WHITEHOUSE, M. J. 1999. Early Miocene high-temperature metamorphism and rapid exhumation in the Betic Cordillera (Spain): evidence from U–Pb zircon ages. *Earth and Planetary Science Letters*, **171**, 591–605.

PLATT, J. P., KELLEY, S. P., CARTER, A. & OROZCO, M. 2005. Timing of tectonic events in the Alpujárride Complex, Betic Cordillera, southern Spain. *Journal of the Geological Society, London*, **162**, 451–462.

PLATT, J. P., WHITEHOUSE, M. J., KELLEY, S. P., CARTER, A., HOLLICK, L., KELLEY, S. P. & CARTER, A. 2003a. Simultaneous extensional exhumation across the Alboran Basin: implications for the causes of late orogenic extension. *Geology*, **31**, 251–254.

PLATT, J. P., ALLERTON, S., KIRKER, A., MANDEVILLE, C., MAYFIELD, A., PLATZMAN, E. S. & RIMI, A. 2003b. The ultimate arc: differential displacement, oroclinal bending, and vertical axis rotation in the External Betic-Rif arc. *Tectonics*, **22**, 1017, http://dx.doi.org/10.1029/2001TC001321

RODRÍGUEZ-FERNÁNDEZ, J. & SANZ DE GALDEANO, C. 2006. Late orogenic intramontane basin development: the Granada basin, Betics (southern Spain). *Basin Research*, **18**, 85–102, http://dx.doi.org/10.1111/j.1365-2117.2006.00284.x

ROYDEN, L. H. 1993. Evolution of retreating subduction boundaries formed during continental collision. *Tectonics*, **12**, 629–638.

RUTTER, E. H. & WHITE, S. H. 1979. The microstructure and rheology of fault gouges produced experimentally at temperatures up to 400 °C. *Bulletin Minéralogie*, **102**, 101–109.

RUTTER, E. H., MADDOCK, R. H., HALL, S. H. & WHITE, S. H. 1986. Comparative microstructures of naturally and experimentally produced clay-bearing fault gouge. *Pure and Applied Geophysics*, **124**, 3–30.

RUTTER, E. H., FAULKNER, D. R. & BURGESS, R. 2012. Structure and geological history of the Carboneras Fault Zone, SE Spain: part of a stretching transform fault system. *Journal of Structural Geology*, **45**, 68–86.

SANZ DE GALDEANO, C. & VERA, J. A. 1992. Stratigraphic record and palaeogeographical context of the Neogene basins in the Betic Cordillera, Spain. *Basin Research*, **4**, 21–36.

SANZ DE GALDEANO, C., GARCÍA-TORTOSA, F. J. *ET AL.* 2012. Main active faults in the Granada and Guadix Baza basins, (Betic cordillera). *Journal of Iberian Geology*, **38**, 223–238.

SCOTNEY, P., BURGESS, R. & RUTTER, E. H. 2000. $^{40}Ar–^{39}Ar$ dating of the Cabo de Gata volcanic series and displacements on the Carboneras fault, S. E. Spain. *Journal of the Geological Society, London*, **157**, 1003–1008.

SERRANO, F. 1990. El Mioceno Medio en el area de Nijar (Almeria, España). *Reviews Geological Society España*, **3**, 65–77.

SERRANO, F. 1992. Biostratigraphic control of Neogene volcanism in Sierra de Gata (SE Spain). *Geologie en Mijnbouw*, **60**, 209–214.

SERRANO, F. & GONZÁLES DONOSO, J. M. 1989. Cronoestratigrafía de la sucesión volcano-sedimentaria del área de Carboneras (Sierra de Gata, Almería). *Reviews Geological Society España*, **2**, 143–151.

SERRANO, F., GUERRA-MERCHÁN, A. *ET AL.* 2007. Tectono-sedimentary setting of the Oligocene-early Miocene deposits on the Betic-Rifian Internal Zone (Spain and Morocco). *Geobios*, **40**, 191–205.

SOLUM, J. G. & VAN DER PLUIJM, B. A. 2009. Quantification of fabrics in clay gouge from the Carboneras fault, Spain and implications for fault behaviour. *Tectonophysics*, **475**, 554–562.

TURNER, S. P., PLATT, J. P., GEORGE, R. M. M., KELLEY, S. P., PEARSON, D. G. & NOWELL, G. M. 1999. Magmatism associated with orogenic collapse of the Betic–Alboran Domain, SE Spain. *Journal of Petrology*, **40**, 1011–1036.

UWE, M. R., KRAUTWORST, T. & BRACHERT, C. 2003. Sedimentary facies during early stages of flooding in an extensional basin: the Brèche Rouge de Carboneras (Late Miocene, Almería, S. E. Spain). *International Journal of Earth Sciences (Geologisches Rundschau)*, **92**, 610–623, http://dx.doi.org/10.1007/s00531-003-0337-8

VAN DER POEL, H. M. 1992. Foraminiferal biostratigraphy and palaeoenvironments of the Miocene-Pliocene Carboneras-Nijar basin (S. E. Spain). *Scripta Geologica*, **102**, 1–31.

VISSERS, R. L. M. 2012. Extension in a convergent tectonic setting: a lithospheric view on the Alboran system of SW Europe. *Geologica Belgica*, **15**, 53–72.

WESTRA, G. 1969. *Petrogenesis of a Composite Metamorphic Facies Series in an Intricate Fault-Zone in the South-Eastern Sierra Cabrera, S. E. Spain*. GUA Series in Geology, Geological Institute, University of Amsterdam.

ZECK, H. P. 2004. Rapid exhumation in the alpine belt of the Betic-Rif (W Mediterranean): Tectonic extrusion. *Pure and Applied Geophysics*, **161**, 477–487, http://dx.doi.org/10.1007/s00024–003–2459–0

Age of the magmatism related to the inverted Stephanian–Permian basin of the Sallent area (Pyrenees)

LIDIA RODRÍGUEZ-MÉNDEZ[1]*, JULIA CUEVAS[1], JOSÉ JULIÁN ESTEBAN[1], JOSÉ M. TUBÍA[1], SERGEI SERGEEV[2] & ALEXANDER LARIONOV[2]

[1]*Departamento de Geodinámica, Facultad de Ciencia y Tecnología, Universidad del País Vasco UPV/EHU, Apartado 644, 48080 Bilbao, Spain*

[2]*Centre of Isotopic Research, VSEGEI, 199106 St Petersburg, Russia*

**Corresponding author (e-mail: lidia.rodriguez@ehu.es)*

Abstract: We have dated an alkaline diabase dyke from the Axial Zone of the Pyrenees near the village of Sallent (Spain), using a sensitive high-resolution ion microprobe (SHRIMP). The diabase dyke intrudes into Devonian rocks that constitute the basement of an inverted Stephanian–Permian basin. Five U–Pb analyses of zircon yield a concordant age of 259.2 ± 3.2 Ma (2σ) for the intrusion of the dyke. This age links diabase dykes with the magmatism associated with the opening of the Stephanian–Permian basins scattered along the Pyrenees. Previous data from the largest Anayet and Midi d'Ossau volcanic massifs, located to the NW of the studied dyke, indicate that the Permian magmatism develops from late Autunian/early Saxonian to Wordian ages. Consequently, this new age broadens the time span of the Permian magmatism of the Pyrenees. The work relates the structure of the Stephanian–Permian basin and the magmatism in order to explain the transition to the new geodynamic setting prevailing in the Pyrenees during the Mesozoic.

The formation of the Pangaea supercontinent and the development of the European Variscan belt ended by Late Carboniferous times, when the amalgamation of Gondwana and Laurentia was almost completed (Stampfli *et al.* 2002). The Jurassic break-up of Pangaea was preceded by significant tectonic instability, leading to the opening of many intracontinental basins in Stephanian–Permian times. In Europe, the largest Permian basin is that of the Southern Permian that extends over more than 1700 km from England to the Baltic States (van Wees *et al.* 2000). Isolated and smaller Permian troughs are also found in more southern regions, such as in France and Spain (McCann *et al.* 2006). The frequent presence of Permian volcanic rocks with mantle signature attests that the asthenosphere was actively involved in the formation of these basins.

The Pyrenees are marked out by a number of Stephanian–Permian basins that are scattered along more than 500 km. These basins frequently trend parallel to the east–west elongation of the mountain belt. The basins were filled mostly with clastic rocks, red sandstones and conglomerates, but also contain co-eval volcanic and shallow intrusive rocks (Bixel & Lucas 1983). As a consequence of the local provenance of the sediments, the correlation between one basin and another remains a major question, making it difficult to unravel the post-Variscan readjustment of the lithosphere.

Precise isotopic dating of the volcanic rocks may provide some answers to these questions. However, reliable ages for the Permian volcanic events in the Pyrenees are limited and are not yet well constrained. As pointed out by Gilbert *et al.* (1995) only four U–Pb zircon ages were available in the 1990s, a condition that has remained virtually unchanged until now. The scarcity of geochronological data can be partly attributed to the basic composition of most of these volcanic rocks, which usually involves a deficiency of zircon crystals. Moreover, a late or post-magmatic hydrothermalism has been detected, resetting the primary age of peraluminous felsic volcanic rocks (rhyolites) from the Midi d'Ossau dome (Innocent *et al.* 1994). Finally, the fact that the most abundant country rocks are azoic sandstones and conglomerates also makes it difficult to unravel the relative ages of the co-eval volcanic rocks.

In this contribution, we report structural evidence reflecting the positive inversion tectonics of the Anayet Stephanian–Permian basin, located near Sallent, in the Axial Zone of the Pyrenees (Fig. 1). We also aim to put a constraint on how long the extensional conditions lasted in the Stephanian–Permian basin. To that end we present new U–Pb zircon analyses, using a sensitive high-resolution ion microprobe (SHRIMP), from a dyke intruded into the Devonian slates and limestones that form the basement of the Stephanian–Permian basin.

From: LLANA-FÚNEZ, S., MARCOS, A. & BASTIDA, F. (eds) 2014. *Deformation Structures and Processes within the Continental Crust*. Geological Society, London, Special Publications, **394**, 101–111.
First published online November 21, 2013, http://dx.doi.org/10.1144/SP394.2

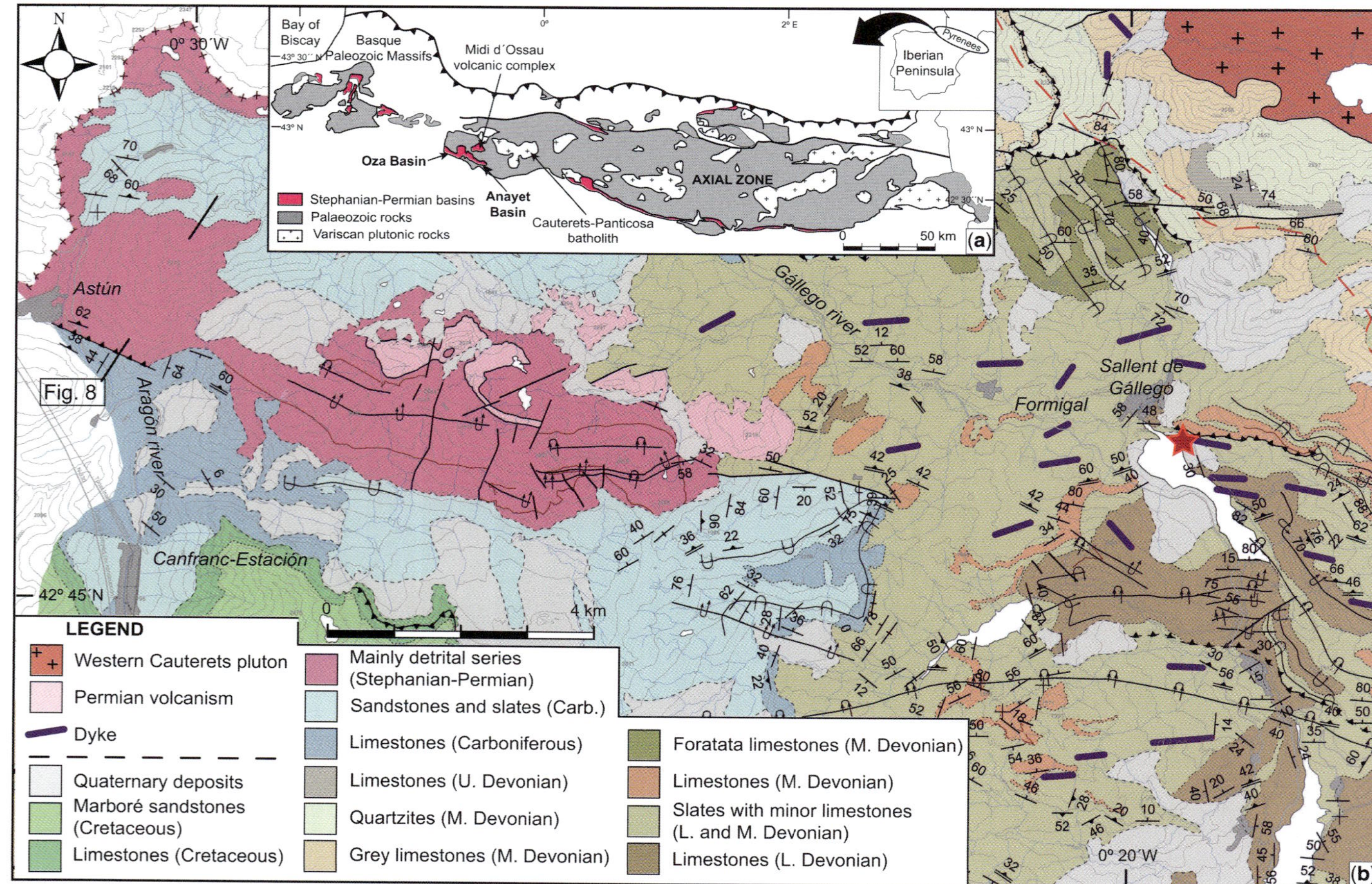

Fig. 1. (**a**) Simplified map of the Pyrenees showing the main Stephano-Permian basins (red). The Cauterets–Panticosa batholith is indicated as a reference for the location of the studied region. (**b**) Geological map of the Sallent region, in the Pyrenean Axial Zone (modified from Rodríguez 2012). The dashed red line marks the thermal metamorphic aureole of the Western Cauterets pluton and the red star marks the location of the analysed dyke (Li-26).

In an attempt to obtain zircon crystals from these subvolcanic rocks, we have processed a sample mass of over 50 kg. This procedure has proven to be successful and the extracted zircon grains give a concordant age of 259.2 ± 3.2 Ma, which we interpret as the intrusion age of the dyke.

Geological setting

The Pyrenees were raised in response to the convergence of the Iberian and European plates in Alpine times and consist of a Palaeozoic domain, the Axial Zone, bounded by the North Pyrenean and South Pyrenean zones, where sedimentary rocks of Mesozoic or Cenozoic age are dominant. The region around Sallent (Spain) belongs to the Axial Zone and is composed mainly of slates and limestones of Devonian and Carboniferous ages, intruded by the Cauterets–Panticosa granite batholith (Fig. 1). The emplacement of this batholith is dated by the U–Pb age of 301 ± 7 Ma obtained using zircons from granite samples of the Eastern Cauterets pluton (Majesté-Menjoulas et al. 1999). The emplacement of the Cauterets–Panticosa batholith is followed by the late – but nearly co-eval – intrusion of basic dykes in the batholith and its metamorphic aureole (Debon & Zimmermann 1993). A late tectonic emplacement has been proposed for the Cauterets–Panticosa batholith, based on field structural data and detailed studies of its internal structure by applying the method of the anisotropy of the magnetic susceptibility (Gleizes et al. 1998; Santana 2001, 2002). Consequently, the age of 301 ± 7 Ma also constrains the timing of the Variscan deformation (as the Panticosa pluton transects the structures, major folds and associated foliation) of the Devonian country rocks.

From the structural point of view, the Sallent region has been integrated within the so-called 'suprastructure' of the Variscan Pyrenees (Zwart 1963; Carreras & Capella 1994), which means that it should be characterized by upright or south-verging folds with steep-dipping foliation. However, this region preserves type 2 fold interference patterns between older recumbent folds and younger upright folds, which lead to the common presence of overturned folds of kilometre-scale (Fig. 2a) with north-trending axes and flat-dipping foliations. At the outcrop scale, these superposed deformations are represented by the frequent presence of a slaty cleavage deformed by a crenulation cleavage in the metapelitic rocks.

In the Pyrenees, the Variscan Orogeny was replaced by the opening of narrow intracontinental basins filled with detrital deposits during Stephano–Permian times. In the Anayet basin, located towards the west of Sallent, Stephanian to Triassic rocks unconformably rest over the older Palaeozoic rocks. The predominance of breccias, conglomerates and red sandstones supports a continental, locally derived, origin of the sediments (Bixel & Lucas 1983). The formation of these basins also marked a period of thermal instability in the subcontinental mantle, as evidenced by the co-eval extrusion of calc-alkaline to alkaline volcanic rocks (Bixel 1988) with mantle signature and variable degrees of crustal contamination (Innocent et al. 1994). Five Permian volcanic events are recognized in the Pyrenees (Bixel 1988) but only three of them are represented in the Sallent region (Bixel & Lucas 1983; Briqueu & Innocent 1993; Ternet et al. 2004).

The Anayet Stephano-Permian basin is now preserved as a narrow trough of several kilometres width and displaying a WNW–ESE elongation (Fig. 1). From bottom to top we have distinguished four lithological units in this basin, following the subdivision proposed by Gisbert (1984) for the Stephanian–Permian deposits of the Pyrenees: Grey Unit, Transition Unit, Lower Red Unit and Upper Red Unit. Due to the common lack of fossils, the age of the Permian deposits in the Anayet Stephanian–Permian basin remains imprecise.

The Grey Unit (50 to 120 m thick) is in structural unconformity over the Devonian and Carboniferous rocks and is made of volcanic flows interlayered in shales, coal and brecciated conglomerates with a diachronic base and a lateritic roof. The volcanic rocks interlayered in this unit define the first volcanic event recognized in the Permian magmatism of this region, the so-called Midi d'Ossau episode. The Midi d'Ossau volcanism is represented mainly by peraluminous ignimbrites of rhyolitic to dacitic composition at the base, and calc-alkaline basalts, andesites and dacites on top. Ages of 278 ± 5 Ma and 272 ± 3 Ma have been reported for the basal peraluminous rhyolites and for the overlying dacites, respectively (Briqueu & Innocent 1993). The Transition Unit, reaching a maximum thickness of 50 m to the west of the studied area, is composed of alternating layers of shale, sandstone and oolitic limestone of grey colours. The Lower Red Unit consists of 250 m of thick red sandstones and incorporates the second magmatic episode, Anayet, in its upper part. The Anayet magmatism includes lava flows, dykes and laccoliths of andesitic composition (Ternet et al. 2004). In addition, we have recognized that the bigger volume of intrusions is encountered along the northern and NE border of the basin (see Fig. 1b). The age of this second volcanic event is not well constrained, although late Autunian to Saxonian ages are postulated from the stratigraphic data of the Permian country rocks (Ternet et al. 2004). The Upper Red Unit is made of three

Fig. 2. (a) Variscan overturned folds in Devonian limestones near Sallent. **(b)** Dyke intruded into Carboniferous rocks showing small branches that emanate from the main dyke. **(c)** Corrugated boundaries in a dyke that contains an enclave of the Carboniferous country rock. **(d)** Internal chilled bands in the analysed dyke (Li-26) showing the inhomogeneity of the intrusion. **(e)** Field view of the studied dyke (D, Li-26), intruded into Devonian foliated slates and limestones.

megasequences of red conglomerates, sandstones and minor shales that progressively thin upwards. The alkaline basalts that compose the third magmatic episode are interlayered in these megasequences. The overall thickness of this unit is more than 1500 m.

The Devonian and Carboniferous sequences located around Sallent contain a significant number of basic dykes lacking an accurate adscription to either the Carboniferous or the Permian magmatism. The studied dyke belongs to these poorly constrained magmatic events. The contacts of such dykes with the country rocks are sharp and the dyke walls are planar surfaces except in a few dykes showing small branches that emanate from the main dyke (Fig. 2b) or corrugated boundaries (Fig. 2c). Enclaves of country rocks are scarce and usually come from dykes with corrugated contacts (Fig. 2c). The observation of internal chilled bands in several thick dykes (Fig. 2d) attests that they are composite bodies resulting from repeated intrusions of magma along the same conduit.

Figure 3 shows that many dykes are nearly parallel and concentrate around an east–west direction, which also marks the mean strike of the main foliation of the country rocks (Fig. 1b). These east-trending dykes allow two groups to be distinguished depending on whether they are vertical or parallel to the foliation of the country rocks, which suggests that the dykes could have a saw-like geometry defined by alternating vertical and north-dipping segments. These observations are consistent with the effects on fracturing in anisotropic media (Donath 1961) and suggest that the opening of the dilatant fractures where the dykes are lodged was partly controlled by the orientation of the Variscan structures.

Age of the dykes

Deciphering whether the diabase dykes are associated with the Permian volcanism or with the Carboniferous Cauterets–Panticosa pluton is not straightforward from arguments based on field data. The difficulty of interpreting the field data is enhanced because several groups of intrusive sheets have been distinguished in the Sallent area, ranging from dacites to diabases (Wensink 1962), although no reliable intrusion ages are available yet.

Table 1. *Whole-rock chemical composition*

Samples	Li-26 A	Li-26 B
Major elements (wt%)		
SiO_2	48.90	48.12
Al_2O_3	17.53	16.82
$Fe_2O_3(t)$	9.31	9.90
MnO	0.18	0.18
MgO	6.02	6.19
CaO	7.60	7.79
Na_2O	4.07	4.10
K_2O	0.04	0.00
TiO_2	1.88	1.83
P_2O_5	0.60	0.58
LOI	4.50	4.79
Total	100.63	100.29
Mg#	0.56	0.55
Trace elements (ppm):		
Ba	42.60	31.71
Co	57.75	34.04
Cr	0.00	126.90
Cu	39.51	35.45
Hf	5.21	4.70
Nb	30.15	24.85
Ni	84.69	91.52
Rb	0.00	1.01
Sc	19.53	–
Sr	661.60	502.90
Ta	1.91	1.86
Th	3.86	3.28
U	1.01	0.90
V	144.80	151.90
Y	32.26	29.28
Zn	74.03	79.97
Zr	257.70	228.30
Rare earth elements (ppm):		
La	29.88	27.39
Ce	62.25	55.84
Pr	7.43	6.63
Nd	29.50	26.79
Sm	6.19	5.71
Eu	1.69	1.67
Gd	6.59	5.42
Tb	1.00	0.86
Dy	5.88	5.25
Ho	1.19	1.04
Er	3.39	2.95
Tm	0.52	0.43
Yb	3.26	2.89
Lu	0.50	0.46
ΣREE	159.27	143.33
(La/Lu)N C1	6.41	5.87
(La/Sm)N C1	3.11	3.10

LOI, loss on ignition.
N C1: normalization to Sun & McDonough's (1989) C1 chondrite.

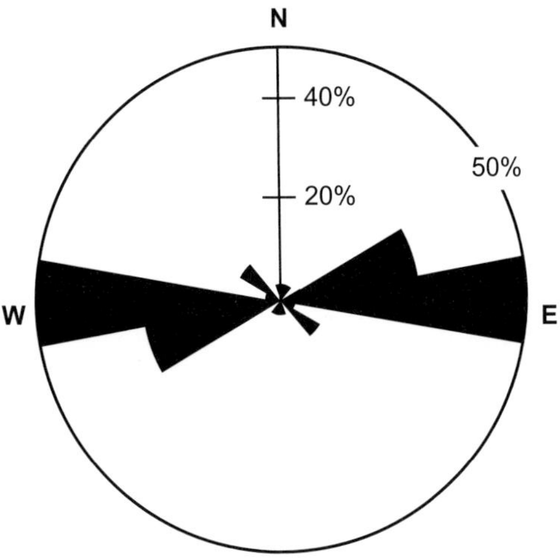

Fig. 3. Rose diagram for the strike of Permian dykes in the study area (Fig. 1b). $n = 32$; interval 20°; maximum 50%.

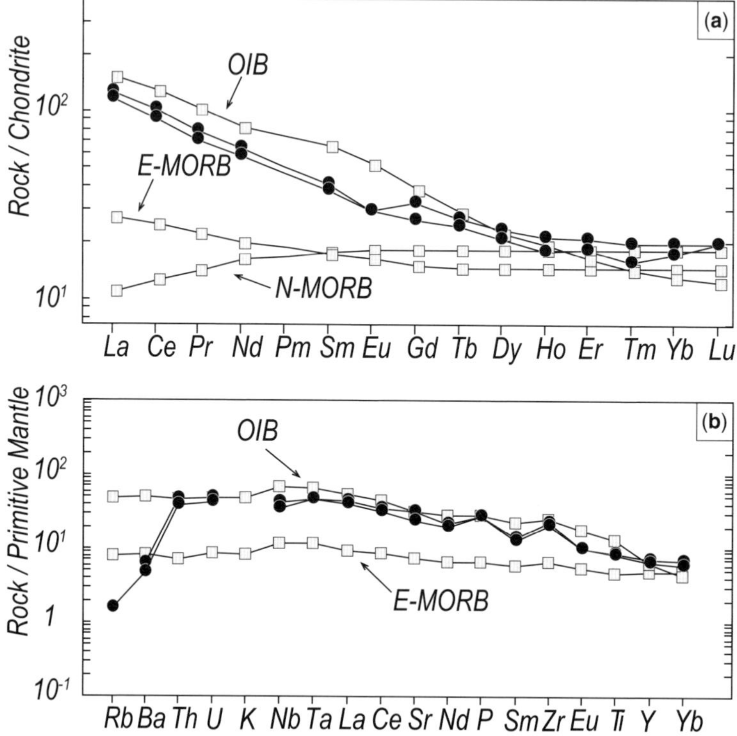

Fig. 4. Geochemical trends of the analysed samples (black dots) (**a**) Chondrite-normalized rare earth elements patterns (Sun & McDonough 1989) and (**b**) primitive-mantle-normalized multi-elemental plot (Sun & McDonough 1989). E-MORB, enriched mid-ocean ridge basalt; N-MORB, normal mid-ocean ridge basalt; OIB, ocean island basalt.

As an alternative, this work tries to constrain the emplacement age of the diabase dykes using U–Pb SHRIMP analyses on zircons extracted from the largest and apparently most pristine diabase dyke we have found around Sallent, emplaced in slates and limestones of Middle Devonian age (see Fig. 1b for location). The studied dyke (Li-26), which spreads along 200 m and reaches a thickness of around 3 m (Fig. 2e), is a medium-grained rock, that shows internal chilled bands (Fig. 2d) with an isotropic microstructure defined by plagioclase and pyroxene crystals with random orientation. It contains calcite-bearing vacuoles, shows a greenish colour and, in spite of its fresh appearance, is affected by a moderate hydrothermal alteration (loss on ignition, LOI 4.5–4.8%). Microscopic

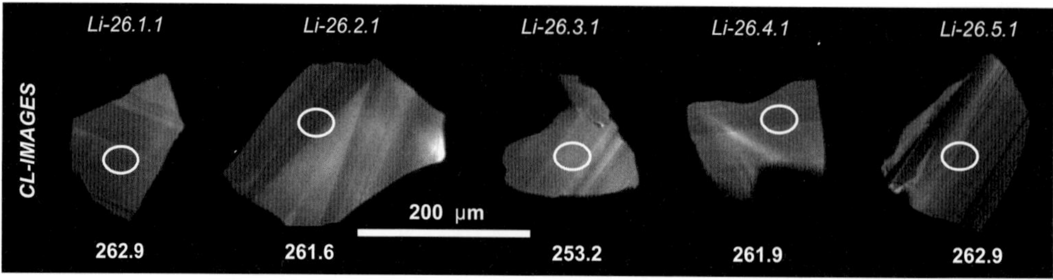

Fig. 5. Cathodoluminescense images of analysed zircons. The ellipses show the spot location. Attached numbers yield the obtained $^{206}Pb/^{238}U$ age.

Table 2. *SHRIMP U–Pb isotopic data for analysed zircons*

Spot name	$^{206}Pb_c$ (%)	U (ppm)	Th (ppm)	$^{232}Th/^{238}U$	$^{206}Pb*$ (ppm)	[†]Age (Ma) $^{206}Pb/^{238}U$	[†]Age (Ma) $^{207}Pb/^{206}Pb$	Dis. (%)	$^{238}U/^{206}Pb*$ $\pm(\%)^†$	$^{207}Pb*/^{206}Pb*$ $\pm(\%)^†$
Li-26.1.1	0.85	180	175	1.00	6.5	262.9 ± 4.1	276 ± 210	5	24.01 ± 1.6	0.0518 ± 9.1
Li-26.2.1	–	218	206	0.98	7.68	261.6 ± 3.5	534 ± 160	104	24.15 ± 1.4	0.0581 ± 7.5
Li-26.3.1	0.42	156	127	0.84	5.39	253.2 ± 3.6	410 ± 180	62	24.96 ± 1.4	0.0549 ± 7.9
Li-26.4.1	1.15	112	79	0.73	4.04	261.9 ± 4.5	135 ± 290	−49	24.11 ± 1.8	0.0487 ± 12
Li-26.5.1	0.00	274	157	0.59	9.62	258.4 ± 2.9	332 ± 72	28	24.45 ± 1.1	0.0531 ± 3.2

Errors are 1σ; Pb_c and $Pb*$ indicate the common and radiogenic portions, respectively; Dis., discordance. Error in Temora Standard calibration was 0.34%.
[†]Common Pb corrected using measured ^{204}Pb.

study shows a diabasic texture made of euhedral to subhedral phenocrysts of plagioclase and altered interstitial pyroxene.

Geochemistry of the dykes

In order to determine the mineral chemistry, quantitative electron microprobe mineral analyses were performed using a CAMECA SX100 at Oviedo University (Spain). Measurement conditions were 15 kV accelerating voltage, 10 s counting time and 15 nA beam current. Primary labradorite is largely transformed to albite (>70%). Primary pyroxene (augite) is almost chloritized. Apatite, pyrite, rutile and titanite are minor constituents of the dykes.

Whole-rock geochemistry is based on two dyke-sample analyses (Table 1). Major-element compositions were determined by inductively coupled plasma-atomic emission spectrometry (ICP-AES), and the rare earth element (REE) and trace-element compositions by inductively coupled plasma-mass spectrometry (ICP-MS) at the 'Centre de Recherches Pétrographiques et Géochimiques' (Nancy, France). The major, rare and trace elements were also determined by Quadrupole-ICP-MS (Q-ICP-MS) at the 'Servicio de Geocronología y Geoquímica Isotopica' (Bilbao, Spain). The whole-rock composition of the samples shows low SiO_2 (48.1–48.9%) and alkali (4.1%) contents, with 4.50–4.80% LOI due to their degree of alteration. On discrimination diagrams like Zr/TiO_2 v. SiO_2, Zr/P_2O_5 v. TiO_2 and Zr/TiO_2 v. Nb/Y (Winchester & Floyd 1977) the two samples plot on the 'alkaline basalt' field. High Ti/V >50 values (Shervais 1982) also support their alkali nature.

REE abundances (Fig. 4a) are up to one hundred times the C1 chondrite values and they display highly heavy rare earth element (HREE)-enriched patterns ($La_N/Lu_N = 5.87–6.41$; $La_N/Sm_N = 3.10–3.11$) that exclude a normal mid-ocean ridge basalt (N-MORB) for the source of the magma. Slightly negative europium anomalies attest plagioclase fractionation. In a primitive-mantle-normalized multi-elemental plot (Fig. 4b), the samples exhibit upward-convex patterns and positive anomalies in P and Zr, according to apatite and zircon modal contents, and negative in K. The comparison with reference patterns of enriched mid-ocean ridge basalt (E-MORB), N-MORB and ocean island basalt (OIB) normalized to the primitive mantle reveals a clear similarity of these rocks with the ones formed in OIB settings (Fig. 4b), in particular for the more immobile elements (Th, Yb). In commonly used tectono-discrimination diagrams (not shown in this work) samples are plotted over within-plate or continental basalt fields (Rollinson 1993), an approach which accords with a post-orogenic extensional setting.

U–Pb SHRIMP dating

Zircons were extracted from blocks with a mass of over 50 kg, according to conventional mineral separation routines. Zircons from the heaviest fractions were extracted using HNO_3 in order to

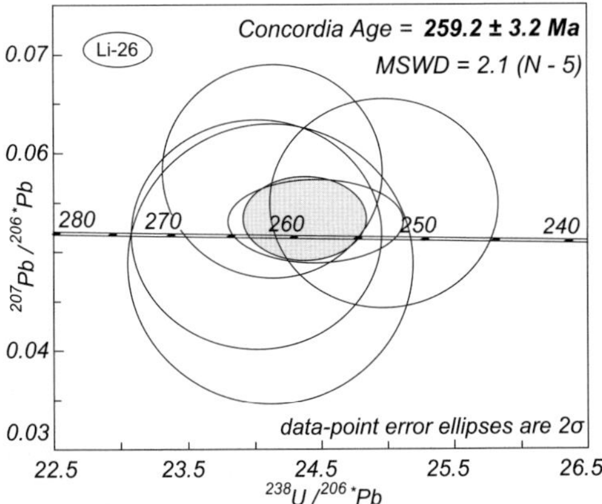

Fig. 6. Tera-Wasserburg diagram for zircon data according to radiogenic Pb ratios. MSWD, mean square weighted deviation.

dissolve the huge proportion of pyrite. Zircon grains were mounted in epoxy resin together with the TEMORA 1 reference zircons (Black *et al.* 2003) and analysed on a SHRIMP-II SIMS in the Centre of Isotopic Research (CIR) at VSEGEI (Saint Petersburg). The grains were polished to half-thickness. Electron back-scattered and cathodoluminescence images (Fig. 5) were used to reveal the internal structures of the zircon grains and as a means of selecting target areas for analysis. The results were obtained with a secondary electron multiplier in peak-jumping mode and following the procedure described in Larionov *et al.* (2004). Seven cycles for each analysed spot were acquired. Apart from 'unknown' zircons, each fourth measurement was carried out on the standard TEMORA 1 zircon (Black *et al.* 2003).

The studied zircons are large fragments (>200 μm) of euhedral grains, possibly crushed during separation, with oscillatory zoning (Fig. 5). Elevated Th/U ratios obtained (0.59–1.00) are often observed in zircon from igneous rocks suggesting the authigenic nature of the zircon. The U–Pb ion microprobe data were processed with the SQUID 1.02 (Ludwig 2001) and Isoplot/Ex 3.00 (Ludwig 2003) software, using the decay constants of Steiger & Jäger (1977). The common lead data were corrected using measured $^{204}Pb/^{206}Pb$ according to the model of Stacey & Kramers (1975). The data of five analyses from the central areas of these zircons (Fig. 5, Table 2) are plotted on a Tera-Wasserburg diagram (Fig. 6). The analyses define a Concordia age of 259.2 ± 3.2 Ma (2σ; mean square weighted deviation (MSWD) = 2.1)

Fig. 7. (**a**) General view of asymmetrical folds in the easternmost sector of the inverted Permian basin. S, sill. The NE is on the right. (**b**) Gently folded unconformity between Permian beds above and Carboniferous rocks (Culm facies) below. The NE is on the right. Note the NE-dipping slate cleavage in the Permian sequence and the presence of recumbent Variscan folds in the Carboniferous materials.

(Fig. 6), which we interpret as the time of the dyke intrusion into the Devonian slates and limestones.

Structures in the inverted Stephanian–Permian basin

The Stephanian–Permian materials in the Sallent region display chevron folds with a consistent southward vergence. These asymmetrical folds show vertical to reversed short limbs and long limbs dipping 20–30° to the north, and have an average wavelength of 400 m (Fig. 7a). Fold axes are subhorizontal and N110° east-trending. An axial planar cleavage (S_1) dipping 40–60° to the north is recognized everywhere and minor folds are common.

Previous studies in westernmost areas have interpreted the southern contact between the Permian and older Palaeozoic rocks in two different ways: (1) as a normal fault reactivated as a thrust during the Pyrenean orogeny (van der Lingen 1960; Ríos *et al.* 1987), and (2) as an unconformity with no major detachments between the Stephanian–Permian sequence and the underlying Carboniferous and Devonian basement (Teixell 1992). Our results suggest that these apparently contradictory interpretations probably reflect partial observations on different locations of the Stephanian–Permian basin. In fact, in the Aragón valley the contact is a fault that dips 60°N and juxtaposes Permian shales on the top of Carboniferous limestones, while to the east, the contact corresponds to a gently folded unconformity (Figs 1 & 7b). Close to the fault, the Permian shales present a pervasive cleavage

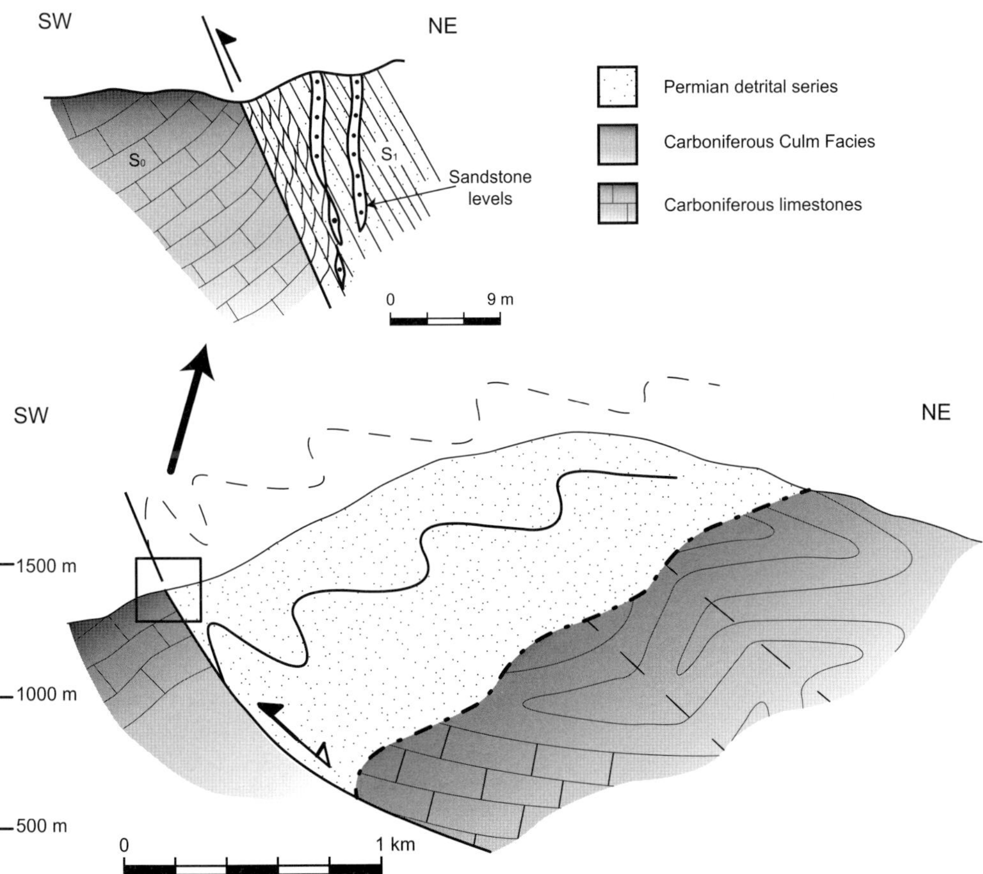

Fig. 8. Schematic cross-section showing the inversion tectonics of the Stephanian–Permian basin (see Fig. 1b for location). The southern boundary fault of the basin was reactivated as a southward thrust during the Pyrenean compression. Inside the box: Detail of the fault zone, showing minor shear bands with kinematic criteria supporting a southward, reverse motion of the Permian hanging-wall rocks.

parallel to the fault contact. The obliquity between the cleavage and bedding is consistent with the reverse limb of one of the previously described south-verging folds. The fault zone contains minor shear bands and S–C tectonites linked to north–south-trending slickensides. The asymmetry of these kinematic markers provides a systematic top-to-the-south motion, in agreement with a reverse fault. Such shear sense, coupled with the existence of rocks in the hanging-wall block that are younger (Permian) than those in the footwall block (Carboniferous), imply that the Aragón valley fault is evidence of positive inversion tectonics (Fig. 8), the fault contact acting as a buttress which promoted the localized deformation.

Discussion and conclusions

The late Permian age of 259.2 ± 3.2 Ma allows us to discard any possible link of the diabase dykes with the quite discrete Devonian to Dinantian intermittent magmatism identified in easternmost sectors of the Pyrenees, as was proposed by Barnolas & Chiron (1996). In contrast, the age of 259.2 ± 3.2 Ma demonstrates the link between the intrusion of the diabase dykes and the Permian volcanism of the Pyrenees. However, it should be stressed that the nearby Midi d'Ossau and Anayet volcanic events yield older ages. Autunian ages, ranging between 278 ± 5 Ma and 272 ± 3 Ma, have been obtained in samples from the Midi d'Ossau (Briqueu & Innocent 1993) and late Autunian to Saxonian ages are suggested for the Anayet volcanic rocks (Ternet *et al.* 2004). The Ossau massif experienced a post-magmatic hydrothermalism leading to the generation of secondary muscovite about 260 myr ago (Innocent *et al.* 1994).

The igneous rocks with ages closer to that of the diabase dykes are some lamprophyres that intruded into the Panticosa pluton. These rocks have been correlated with the fifth Permian volcanic event defined by Bixel (1988) and yielded K–Ar ages in kaersutite between 271 ± 9 Ma and 266 ± 10 Ma (Debon & Zimmermann 1993).

From the obtained age, 259.2 ± 3.2 Ma, these diabases emerge as the youngest expression reported so far of the Permian magmatism in the Pyrenees. On the other hand, we propose that this age does mark the fading of the Pyrenean Permian magmatism, which is consistent with the small volume of these magmatic rocks and with their occurrence as dykes and sills at deeper structural levels than in the older Anayet and Midi d'Ossau massifs. Finally, although it is beyond the scope of this work, the age and the geochemistry of the Sallent dyke could mark the transition to the new geodynamic setting prevailing in the Pyrenees

during the Mesozoic, responsible for the generation of the Triassic 'ophites' and the 'Buntsandstein' facies. In this regard, it is worth noting that a nearly north–south crustal extension is deduced from the dominant east–west strike of the dykes, assuming they are emplaced in dilatant fractures of Andersonian type. In conclusion, the results obtained here contribute towards a more accurate timing for the transition to the Early Alpine geodynamic scenario in the central Pyrenees.

This work has been supported by grants CGL2010-14869, CGL2011-23755 and IT-364-10. E. Lepekhina and S. Presnyakov are also thanked for their supporting analytical work. This work is part of the PhD thesis of L. Rodríguez (FPU-AP2006-00648). We are grateful to A. Casas and a second referee for their constructive criticism of the original manuscript.

References

Barnolas, A. & Chiron, J. C. 1996. *Synthèse géologique et géophysique des Pyrénées. Vol. 1: Introduction. Géophysique. Cycle hercynien.* Édition BRGM-ITGE, Orléans-Madrid.

Bixel, F. 1988. Le volcanisme stéphano-pérmian des Pyrénées. *Bulletin des Centres de Recherche Exploration-Production Elf-Aquitaine*, **12**, 661–706.

Bixel, F. & Lucas, C. 1983. Magmatisme, tectonique et sédimentation dans le fossés stéphano-permiens des Pyrénées occidentales. *Revue de Géographie Physique et Géologie Dynamique*, **24**, 329–342.

Black, L. P., Kamo, S. L., Allen, C. M., Aleinikoff, J. N., Davis, D. W., Korsch, R. J. & Foudoulis, C. 2003. TEMORA 1: a new zircon standard for Phanerozoic U–Pb geochronology. *Chemical Geology*, **200**, 155–170.

Briqueu, L. & Innocent, C. 1993. Datation U/Pb sur zircon et géochimie isotopique Sr et Nd du volcanisme permien des Pyrénées occidentales (Ossau et Anayet). *Comptes Rendus de l'Académie des Sciences, Paris*, **316**, 623–628.

Carreras, J. & Capella, I. 1994. Tectonic levels in the Paleozoic basement of the Pyrenees: a review and a new interpretation. *Journal of Structural Geology*, **16**, 1509–1524.

Debon, F. & Zimmermann, J. L. 1993. Mafic dykes from some plutons of the Western Pyrenean Axial Zone (France, Spain): markers of the transition from Late-Hercynian to early Alpine events. *Schweizerische Mineralogische und Petrographische Mitteilungen*, **73**, 421–433.

Donath, F. 1961. Experimental study of shear failure in anisotropic rocks. *Geological Society of America Bulletin*, **72**, 985–990.

Gilbert, J. S., Bickle, M. J. & Chapman, H. J. 1995. The origin of the Pyrenean Hercynian volcanic rocks (France–Spain): REE and Sm–Nd isotope constraints – reply. *Chemical Geology*, **122**, 298–300.

Gisbert, J. 1984. Las molasas tardihercínicas del Pirineo. *In*: Fontboté, J. M. (ed.) *Libro Jubilar J.M. Rios. Geología de España, tomo II*. IGME, Madrid, 168–185.

GLEIZES, G., LEBLANC, D., SANTANA, V., OLIVIER, PH. & BOUCHEZ, J. L. 1998. Sigmoidal structures featuring dextral shear during emplacement of the Hercynian granite complex of Cauterets–Panticosa (Pyrenees). *Journal of Structural Geology*, **20**, 1229–1245.

INNOCENT, C., BRIQUEU, L. & CABANIS, B. 1994. Sr–Nd isotope and trace-element geochemistry of late Variscan volcanism in the Pyrenees: magmatism in post-orogenic extension? *Tectonophysics*, **238**, 161–181.

LARIONOV, A. N., ANDREICHEV, V. A. & GEE, D. G. 2004. The Vendian alkaline igneous suite of northern Timan: ion microprobe U–Pb zircon ages of gabbros and syenite. *In*: GEE, D. G. & PEASE, V. L. (eds) *The Neoproterozoic Timanide Orogen of Eastern Baltica*. Geological Society, London, Special Publications, **30**, 69–74.

LUDWIG, K. R. 2001. *SQUID 1.02, a user Manual*. Berkeley Geochronology Center Special Publication, Berkeley, CA.

LUDWIG, K. R. 2003. *User's Manual for Isoplot/Ex, Version 3.00, A Geochronological Toolkit for Microsoft Excel*. Berkeley Geochronology Center Special Publication, Berkeley, CA.

MAJESTÉ-MENJOULÀS, C., DEBON, F. & BARRÈRE, P. 1999. *Notice explicative, Carte géologique de la France (1/50 000)*. Feuille Gavarnie (1082). BRGM, Orléans.

MCCANN, T., PASCAL, C. ET AL. 2006. Post-Variscan (end Carboniferous–Early Permian) basin evolution in Western and Central Europe. *In*: GEE, D. G. & STEPHENSON, R. A. (eds) *European Lithosphere Dynamics*. Geological Society, London, Special Publications, **32**, 355–388.

RÍOS, J. M., GALERA, J. M., BARETTINO, D. & LANAJA, J. M. 1987. *Hoja y memoria de Sallent (145). Mapa geológico de España a escala 1:50 000*. IGME, Madrid.

RODRÍGUEZ, L. 2012. *Análisis de la estructura varisca y alpina en la transversal Sallent-Biescas (Pirineos Centrales, Huesca)*. Laboratorio Xeolóxico de Laxe, Serie Nova Terra, A Coruña, **41**.

ROLLINSON, H. R. 1993. *Using Geochemical Data: Evaluation, Presentation, Interpretation*. Longman Scientific and Technical, Harlow.

SANTANA, V. 2001. *El plutón de Panticosa (Huesca, Pirineos): Estructura y modelo de emplazamiento a partir del análisis de la Anisotropía de la Susceptibilidad Magnética*. PhD thesis, University of the Basque Country UPV/EHU.

SANTANA, V. 2002. El plutón de Panticosa (Pirineos Occidentales, Huesca): fábrica magnética y modelo de emplazamiento. *Revista de la Sociedad Geológica de España*, **15**, 175–191.

SHERVAIS, J. W. 1982. Ti–V Plots and the petrogenesis of modern and ophiolitic lavas. *Earth and Planetary Science Letters*, **59**, 101–118.

STACEY, S. & KRAMERS, J. D. 1975. Approximation of terrestrial lead isotope evolution by a two-stage model. *Earth and Planetary Science Letters*, **26**, 207–221.

STAMPFLI, G. M., BOREL, G. D., MARCHANT, R. & MOSAR, J. 2002. Western Alps geological constraints on western Tethyan reconstructions. *In*: ROSENBAUM, G. & LISTER, G. S. (eds) *Reconstruction of the Evolution of the Alpine–Himalayan Orogen, Journal Virtual Explorer*, **8**, 77–106.

STEIGER, R. H. & JÄGER, E. 1977. Subcommission on geochronology: convention on the use of decay constants in geo- and cosmochronology. *Earth and Planetary Science Letters*, **36**, 359–362.

SUN, S. S. & MCDONOUGH, W. F. 1989. Chemical and isotopic systematics of oceanic basalts: implications for mantle composition and processes. *In*: SUN, S. S. & MCDONOUGH, W. F. (eds) *Magmatism in the Ocean Basins*. Geological Society, London, Special Publications, **42**, 313–345.

TEIXELL, A. 1992. *Estructura alpina en la terminación occidental de la Zona Axial pirenaica*. PhD thesis, Barcelona University.

TERNET, Y., MAJESTÉ-MENJOULÀS, C., CANÉROT, J., BAUDIN, T., COCHÉRIE, A., GUERROT, C. & ROSSI, P. 2004. *Notice Explicative. Carte géologique de la France (1/50 000). Feuille Laruns-Somport (1069)*. BRGM, Orléans.

VAN DER LINGEN, G. J. 1960. Geology of the Spanish Pyrenees, North of Canfranc, Huesca Province. *Estudios Geológicos*, **16**, 205–242.

VAN WEES, J. D., STEPHENSON, R. A. ET AL. 2000. On the origin of the Southern Permian Basin, Central Europe. *Marine and Petroleum Geology*, **17**, 43–59.

WENSINK, H. 1962. Paleozoic of the upper Gállego and Ara valleys, Huesca province, Spanish Pyrenees. *Estudios Geológicos*, **18**, 1–74.

WINCHESTER, J. A. & FLOYD, P. A. 1977. Geochemical discrimination of different magma series and their differentiation products using immobile elements. *Chemical Geology*, **20**, 325–343.

ZWART, H. J. 1963. The structural evolution of the Paleozoic of the Pyrenees. *Geologische Rundschau*, **53**, 170–205.

Cross-strike structures controlling magmatism emplacement in a flat-slab setting (Precordillera, Central Andes of Argentina)

S. ORIOLO[1]*, M. S. JAPAS[2], E. O. CRISTALLINI[3] & M. GIMÉNEZ[4]

[1]*Departamento de Ciencias Geológicas, Universidad de Buenos Aires, Intendente Güiraldes 2160, Buenos Aires, Argentina*

[2]*Instituto de Geociencias Básicas, Aplicadas y Ambientales (CONICET-UBA), Intendente Güiraldes 2160, Buenos Aires, Argentina*

[3]*Laboratorio de Modelado Geológico, IDEAN (UBA-CONICET), Intendente Güiraldes 2160, Buenos Aires, Argentina*

[4]*Instituto Geofísico Sismológico Volponi, FCEN, Universidad Nacional de San Juan, Av. José Ignacio de la Roza y Meglioli S/N, San Juan, Argentina*

**Corresponding author (e-mail: seba.oriolo@gmail.com)*

Abstract: Detailed structural, kinematic and geophysical data on the foreland of the Pampean flat-slab segment (Hualilán area, Andean Precordillera of Argentina) have shown that cross-strike structures have had an important role in the evolution of this Andean segment since Miocene times. These structures represent pre-existing crustal fabrics reactivated during the Andean orogeny and could have controlled the emplacement of the Miocene arc-magmatism migrating into foreland due to the flattening of the slab. Likewise, kinematic results obtained for these structures support a similar stress frame to that obtained elsewhere in the Precordillera but showing different motions as a consequence of their high obliquity to the orogen trend. Moreover, they record a reorientation of kinematic axes during Late Miocene–Pliocene times.

Cross-strike structures are steeply dipping cross-strike structural discontinuities represented by wide zones of brittle–ductile deformation (Wheeler 1980; Twiss & Moore 1992; Mueller & Talling 1997). Usually, they represent reactivated pre-existing structures inherited from earlier tectonic phases. They are considered important structures since they can behave either as barriers for seismic ruptures of major faults or as transfer structures linking two adjacent segments of the same fault system (Wheeler 1980; Hudnut *et al.* 1989; Pizzi & Galadini 2009). Likewise, cross-faults cutting across main faults resulted in the formation of fault blocks and therefore could have relevant economic significance (Johnson *et al.* 2002; Saha *et al.* 2006). In some orogenic systems, cross-faults act as channels for crustal fluids and highlight basement-involved deformation. Studies of these faults have renewed discussion about thin-skinned v. thick-skinned tectonics. Hence, the true amount of orogenic shortening when cross-faults are present has to be carefully considered (Tavarnelli *et al.* 2004).

The Central Andes display two flat-slab sectors, the Peruvian (5–15°S) and the Pampean (28–33°S) segments (Ramos *et al.* 2002), which resulted from the subduction of the Nazca and Juan Fernández aseismic ridges, respectively (Pilger 1981; Gutscher *et al.* 2000). Linked to the subduction of the Nazca Plate under the overriding South American Plate, the Central Andes (5–47°S; Gansser 1973) show oblique structures compartmentalizing the orogen, and controlling its architecture (Baldis & Vaca 1985; Salfity 1985) as well as its kinematic evolution (Rossello *et al.* 1996; Urreiztieta 1996; Ré *et al.* 2001; Japas & Ré 2005, 2012). Both flat-slab regions show similar tectonic features such as arc-related magmatism and compressional/transpressional deformation migrating more than 600 km inland from the trench. Although inland migration of arc-magmatism and deformation has been much described along these segments, studies concerning the structural control of the former are scarce. Hence, the Central Andes between 28° and 33°S provide an excellent opportunity to study the evolution of an active margin linked to flat-slab subduction and, particularly, the Precordillera offers the chance to analyse the relationship between inland migration of deformation and magmatism.

References about cross-strike structures in the Pampean flat-slab segment are still scarce. Within this framework, contributions from Baldis & Vaca (1985), Japas (1998), Ré *et al.* (2000, 2001), Japas *et al.* (2002*a*, *b*) and Ré & Japas (2004) stand out. These latter authors (Japas 1998; Ré *et al.* 2000,

From: LLANA-FÚNEZ, S., MARCOS, A. & BASTIDA, F. (eds) 2014. *Deformation Structures and Processes within the Continental Crust.* Geolcical Society, London, Special Publications, **394**, 113–127.
First published online November 21, 2013, http://dx.doi.org/10.1144/SP394.6

2001; Japas *et al.* 2000*a*, *b*; Ré & Japas 2004) have recognized two systems of conjugated brittle–ductile megashear zones in the Precordillera: NNW left-lateral and NNE right-lateral transpressional structures, and WNW left-lateral and ENE right-lateral transtensional ones. The former consist of gently oblique zones whereas the latter comprise cross-strike structures. Contributions on the relationship between cross-strike structures and magmatism in the flat-slab segment are restricted to the Sierras Pampeanas, particularly to the Sierra de San Luis (Urbina *et al.* 1995, 1997; Sruoga *et al.* 1996; Sruoga & Urbina 2008; Urbina & Sruoga 2009; Japas *et al.* 2010, 2011*a*, *b*). However, many authors have outlined the importance of the structural control of magmatism. De Saint Blanquat & Tikoff (1997), de Saint Blanquat *et al.* (1998), Brown & Solar (1999), Žák *et al.* (2005), Lara *et al.* (2006) and Romeo *et al.* (2006) have remarked on the interaction between transpression/transtension and magmatism. Recently, Acocella & Funiciello (2010) have provided a detailed study of many arcs and their kinematic and structural settings, concluding that regional or local extension is always required for volcanic output.

The aim of this paper is to contribute to the understanding of the development and kinematics of cross-strike structures as well as their interaction with magmatism in a flat-slab setting. The integration of tectonic fabric analysis and structural, kinematic and magnetometric data have provided information that helps to constrain temporal and spatial relationships between magmatism and deformation.

Geological setting

The main morphostructural units in the Pampean flat-slab segment in Argentina are the Frontal Cordillera, the Precordillera and the Sierras Pampeanas (Fig. 1a). The Frontal Cordillera represents a basement block made up mostly of Palaeozoic sedimentary rocks intruded by Permo-Triassic and Jurassic granitoids uplifted during the Andean orogeny (Heredia *et al.* 2002). The Precordillera comprises a NNE-trending range that represents a foreland composite fold-and-thrust belt and can be divided into three main morphostructural units: Western, Central and Eastern Precordillera (Fig. 1b). Western and Central Precordillera show east-verging epidermic deformation while the Eastern Precordillera represents a west-verging thick-skinned fold-and-thrust belt (Allmendinger *et al.* 1990; Cristallini & Ramos 2000). The Sierras Pampeanas represent a series of uplifted blocks constituted mostly of metamorphic and igneous rocks of Neoproterozoic to Palaeozoic ages.

The study area is located in Hualilán in the Central Precordillera of San Juan, Argentina (Fig. 1b). This region includes an intermontane basin known as the Pampa de Hualilán that is bounded by several ranges (Fig. 2). The locality is of special interest because of the presence of outcrops of Miocene dacites that represent one of the very few manifestations of the Miocene arc migration into the foreland due to the flat-slab process in the Precordillera (Kay & Abbruzzi 1996). These volcanic and subvolcanic rocks are associated with hydrothermal veins related to the late magmatic stages (Logan 2000). The stratigraphic column also comprises Ordovician siliciclastic rocks (Sierra de la Invernada Formation) and limestones (San Juan Formation), Siluro-Devonian metasedimentary rocks (Tucunuco and Gualilán Groups), Miocene continental sediments (Cuculí Formation), and Quaternary alluvial, aeolian and lacustrine sediments (Fig. 2).

Tectonic fabric analysis

Tectonic fabric of the Precordillera

Tectonic fabric analysis was carried out in the Precordillera and the Hualilán area. In the first case, a chromatic scale was used to discriminate between fabric elements (structural lineaments, traces of fold axes and faults) with different orientations (van Gool & Piazolo 2006; Japas *et al.* 2013) using data already published by Ragona *et al.* (1995). Fabric elements in Hualilán were recognized on the basis of satellite imagery, aerial photographs at 1:25 000 and 1:50 000 scales, and field work.

On a regional scale, the Precordillera fold-and-thrust belt shows a north − NNE trend of major faults and fold axes (Fig. 3a). However, detailed fabric analysis revealed the presence of three domains according to the orientation of their constituent elements (Fig. 3b).

Domain I displays a WNW orientation and is distinguishable in three areas (IA, IB, IC; Fig. 3b). Tectonic fabric elements have mostly a NW–NNW strike but ENE to WNW structural features are also common. It must be emphasized that all Miocene volcanic and subvolcanic rocks crop out within this domain. On the other hand, domain II exhibits a NNE trend and fabric elements with the same orientation, whereas domain III has a NNE orientation and NNW structural features.

Sector IA (Fig. 3b) shows mostly NW to NNW fabric elements and matches the Miocene Northern Precordillera volcanic belt defined by Limarino *et al.* (2002). Sector IB includes many structures striking WNW−ENE and shows a regional WNW distribution of Miocene igneous rocks.

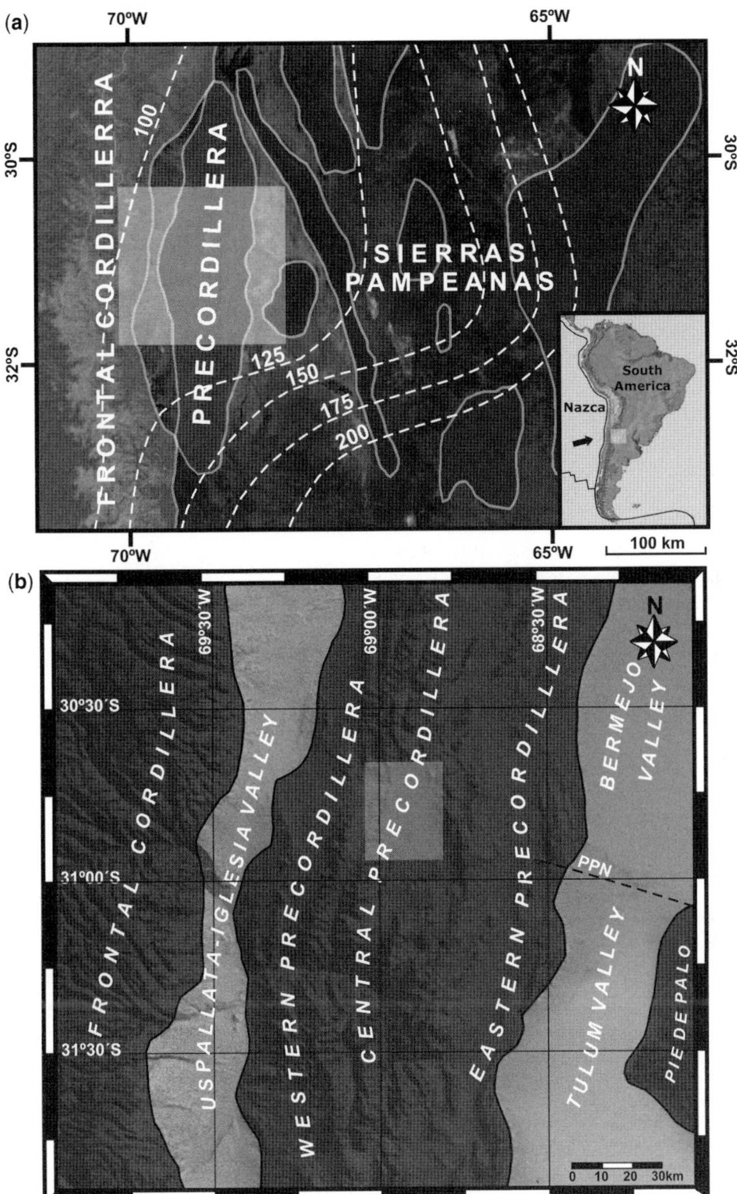

Fig. 1. (**a**) Location of the Andean flat-slab segment in the Central Andes. Contours of depth (km) of oceanic slab after Cahill & Isacks (1992) and the main morphostructural units are shown. Inset indicates the regional Andean subduction and the Pampean flat-slab segment in Argentina. The shaded area is shown in (b). (**b**) Main morphostructural units between 30° and 32°S. The study area is shown by the light-grey rectangle. PPN, Pie de Palo Norte lineament.

These outcrops are mostly located in the Hualilán area and in the Cerro Negro de Iglesia which is placed in the Western Precordillera (Leveratto 1976; Gómez Rivarola 2007). A more complex pattern of deformation is observed in sector IC where NNW structures dominate.

Structure of the Hualilán area

Detailed mapping in the eastern area of sector IB shows the presence of two main structural systems (Fig. 3a). The first system includes NNW to NNE west-dipping thrusts, while the second one

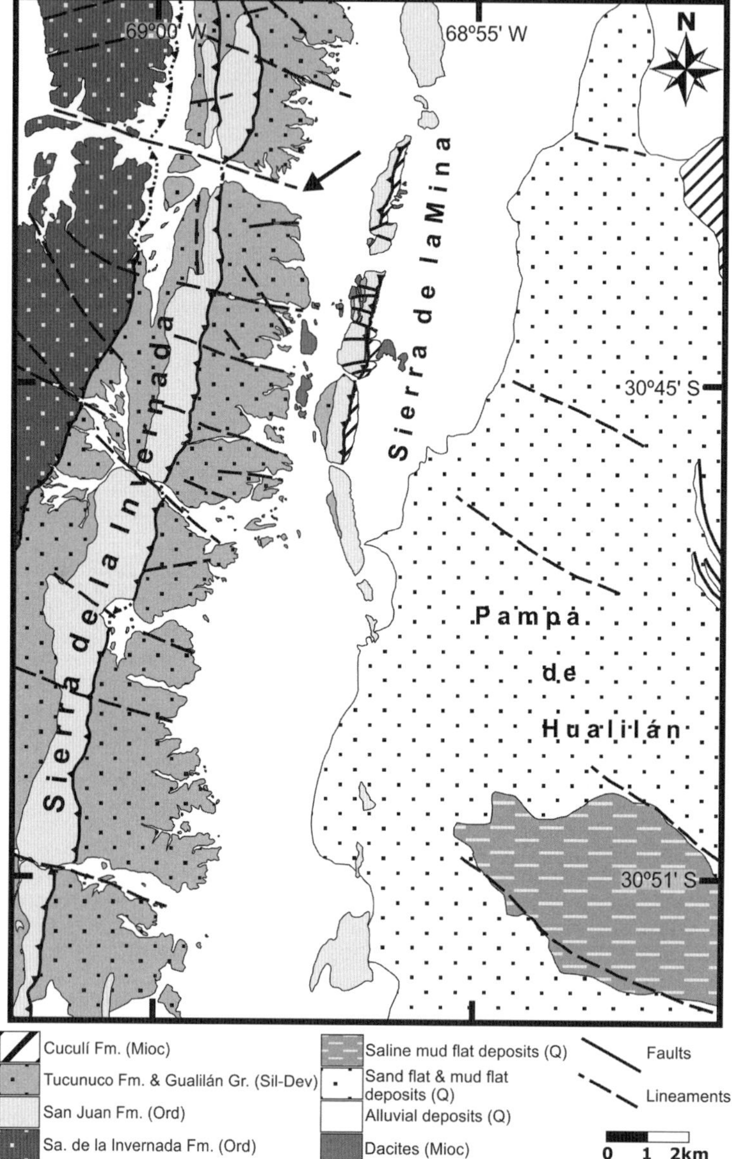

Fig. 2. Geological map of the study area. The black arrow indicates the WNW lineament that cross-cuts the Sierra de la Invernada.

contains WNW to ENE structures with strike-slip displacement.

The Sierra de la Invernada is located between the Western and Central Precordillera and shows a NNE regional trend (Fig. 2). It is made up of Early Palaeozoic units thrusted and folded with NNE and NNW strikes. Many WNW to ENE lineaments are also present, one of which exhibits a WNW strike, and cross-cuts the whole range (Fig. 2).

East of the Sierra de la Invernada (Fig. 2), the Sierra de la Mina is comprised of a west-dipping homoclinal structure with a basal detachment thrust which strikes NNE in the central area, to NNW in the southern and northern ones. Despite its short length, the Sierra de la Mina is of special interest because of the presence of several oblique structures associated with Miocene volcanic rocks. WNW to ENE cross-faults controlled

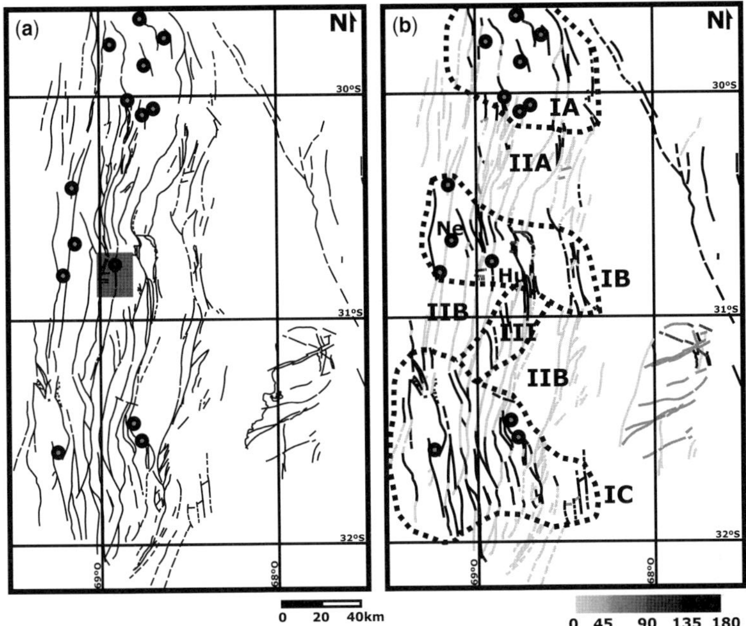

Fig. 3. (**a**) Map of tectonic fabric elements (faults, lineaments and fold axes traces) in the Precordillera and Western Sierras Pampeanas (modified after Ragona *et al.* 1995). Main outcrops of Miocene igneous rocks are represented by grey dots. The shaded area indicates the study area from Figure 2. (**b**) Map from (a) where tectonic fabric elements have been discriminated using a chromatic scale according to their orientation. Different domains (I, II, III) and sectors (A, B, C) recognized from the chromatic analysis are shown, as well as the two main localities with Miocene igneous rocks mentioned in this work (Hu, Hualilán; Ne, Cerro Negro de Iglesia). It is observed that Miocene igneous rock outcrops are restricted to domain I.

the development of the gullies that cross-cut the range. Likewise, neotectonic activity reported by Bastías *et al.* (1984) and Bastías (1985) in the central block of the Sierra de la Mina as well as along-strike changes in geomorphological and structural features suggest strong segmentation of drainage patterns controlled by these cross-strike structures (Fig. 2). They not only separate areas with differences in their drainage networks but they also represent the watershed in the southern margin of the basin (Oriolo 2012 and references therein).

Surveying of brittle–ductile shear zones of centimetre to metre widths affecting the Miocene dacites reveals the presence of a great number of steeply dipping structures (Fig. 4a). A three-modal distribution includes a WNW mean mode with two subordinated NNE and ENE modes (Fig. 4b). Kinematic analysis of these shear zones is presented in the following section.

The cross-strike structures that are frequent in the Precordillera (Fig. 3) are well represented in the Hualilán area, particularly in the Sierra de la Mina (Fig. 4). Mesoscopic brittle–ductile shear zones also support these observations, though the

presence of many oblique WNW to ENE cross-structures seems to be verified at regional to local scale in this region.

Kinematic analysis

Measurements of 3D kinematic indicators were made in brittle–ductile shear zones (Fig. 5) (these zones defined in the sense of Ramsay & Huber 1987)) developed in Miocene dacites and their wall-rock following the method described by Japas *et al.* (2008). These structures consist of planar or curviplanar zones of widely distributed deformation that have centimetre to metre width where mostly *R* (*Riedel*) shears were measured. Scarce tensional fractures have also been measured. Statistical analysis was performed using the Faultkinwin 5.0 software (Marrett & Allmendinger 1990; Allmendinger *et al.* 2012). This analysis led to the definition of three kinematic populations (Fig. 6a–c) on the basis of the clustering of extension axes (Fig. 6d).

Population A shows kinematic axes (*X:147°/09°, Y:306°/81°, Z:057°/03°*) that reflect NE shortening

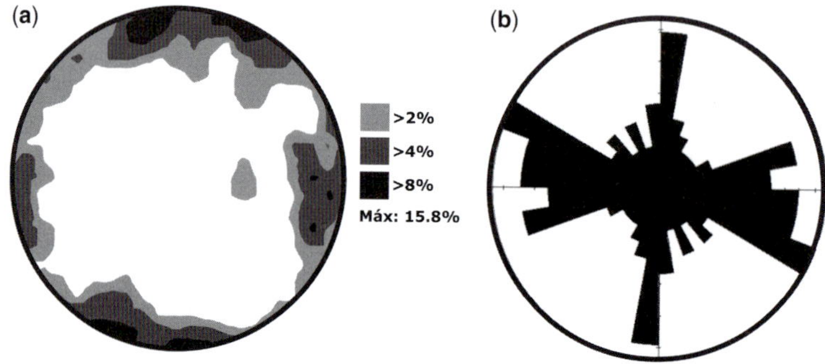

Fig. 4. (**a**) Pole diagram of brittle–ductile shear zones measured in the Hualilán dacites and their wall rock (GEOrient 9.5.0, Holcombe 2011). Equiareal projection, lower hemisphere, $n = 112$. (**b**) Rose diagram showing the WNW mean mode and the two subordinated NNE and ENE modes (frequency intervals of 2%).

related to a dominant strike-slip deformation regime (Fig. 6a). Population B exhibits a similar pattern ($X:206°/03°$, $Y:326°/84°$, $Z:115°/05°$) but reoriented with a WNW shortening direction (Fig. 6b), while in population C ($X:057°/09°$, $Y:290°/74°$, $Z:149°/12°$) X and Z axes are inverted respect to population A (Fig. 6c).

Mutual cross-cutting relationships suggest that population A predates population B. Both these populations represent local kinematics for the study area during the Andean orogeny and reflect a change from NE to WNW shortening by Late Miocene–Pliocene? times. Results obtained for population C are interpreted as being a consequence of orogenic relaxation.

Shear zones striking WNW and ENE are sinistral strike-slip structures in population A. A minor dominantly normal dip-slip is also present. In population B, both WNW and ENE-striking zones show normal dip-slip with sinistral and dextral strike-slip components, respectively. Likewise, applying criteria from Sanderson & Marchini (1984), low

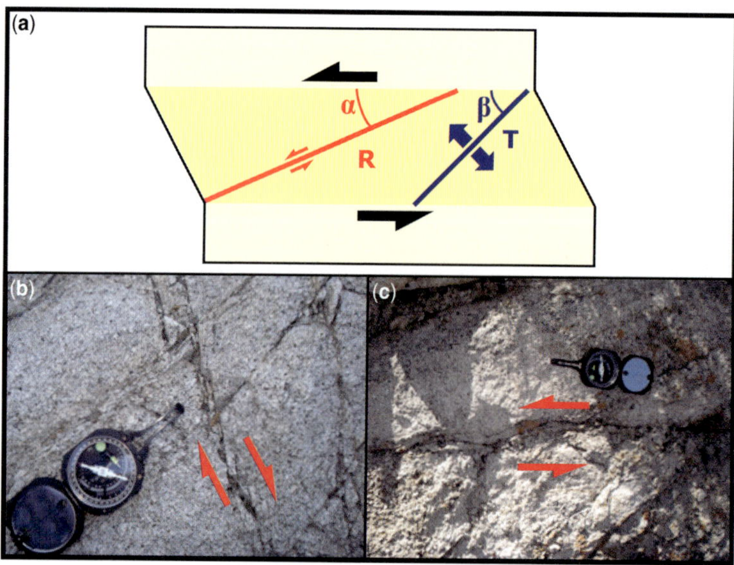

Fig. 5. (**a**) Schematic representation of the brittle–ductile shear zone with Riedel (R) shears and tensional fractures (T). According to Sanderson & Marchini (1984), $\alpha \approx 20°$ and $\beta \approx 45°$ for strike-slip, whereas both values are lower for transtensional deformation. (**b**) Brittle–ductile shear zone with dextral shearing indicated by R shears. (**c**) Transtensional brittle–ductile shear zone (sinistral shearing) filled with carbonates.

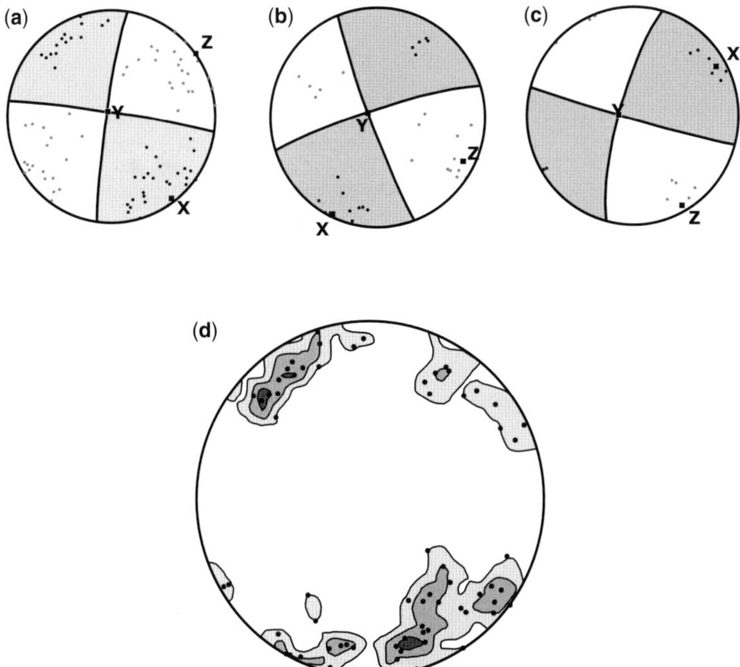

Fig. 6. Kinematic diagrams of extension (black) and shortening (grey) axes obtained using Faultkinwin 5.0 (Marrett & Allmendinger 1990; Allmendinger *et al.* 2012). The main kinematic axes *X*, *Y* and *Z* are represented. (**a**) Population A (*n* = 41). (**b**) Population B (*n* = 16). (**c**) Population C (*n* = 8). (**d**) Extension axes used to define the three different populations (*n* = 65, contour interval 2% per 1% area).

angles between the shear zones and tensional fractures developed within them support transtensional kinematics for cross-strike structures.

Magnetometry

Magnetometric data processing

Aeromagnetometric data were obtained from the Geological Survey (SEGEMAR, Argentina) database. This information was collected at flying heights of 120 m, using north–south lines with spacing of 1000 m and with east–west control lines with spacing of 7500 m. A magnetic anomaly was calculated for the area subtracting the International Geomagnetic Reference Field (IGRF) at the time of the acquisition of the data (Blakely 1995). In order to interpret magnetometric data as the result of geological and structural features, reduction to pole and source parameter imaging (SPI) calculations were made.

Reduction to pole (Baranov 1957) removes the asymmetry caused by the non-vertical magnetization direction so that the anomaly is relocated over the source. The presence of anomalies is interpreted in relation to high-gradient zones over their

source. First vertical and horizontal derivatives were calculated over the magnetic anomaly reduced to pole in order to recognize discontinuities that could be compared with structural data.

The SPI method is based on the complex analytical signal theory (Nabighian 1972, 1974, 1984; Roest *et al.* 1992) applied to determine the attributes of the signal. The local wavenumber *k* is calculated considering the amplitude and phase of the analytical signal, and the algorithm proposed by Blakely & Simpson (1986) led to the calculation of *k* maximum values. These *k* values reflect the contacts between different units and the inverse of *k* is an estimate of the depth of the magnetic source, which is related to high-gradient zones that reflect contacts between different lithologies.

Lithomagnetic domains and lineaments

Lithomagnetic domains were defined on the basis of significant variations in magnetic field parameters, such as intensity and gradient of the field and the geometry of magnetic anomalies. High-gradient zones can be related to the presence of contacts between different units, fractures or faults so that definition of areas of homogeneous magnetic signature led to the recognition of two lithomagnetic units. In this area, the map of the magnetic anomaly reduced to pole (Fig. 7a) reveals a regional variation from NNE to SSW. Magnetization decreases to the SSW with a minimum located at the latitude of the Hualilán swamp. Likewise, many anomalies of high magnetization (>76 nT), short wavelength (1–5 km) and high amplitude (>300 nT) are concentrated in two well-defined areas. The first one is located in the northwestern corner and corresponds to the Cerro Negro de Iglesia area whereas the second cluster is placed in the Hualilán district. Contrasting magnetic properties between Miocene igneous rocks cropping out at both localities and their sedimentary wall rock could satisfactorily explain these anomalies, allowing the recognition of two lithomagnetic domains (Fig. 7b).

Magnetic lineaments were identified on the basis of the horizontal derivative from Figure 7c, though the vertical derivative exhibits a similar pattern (Fig. 7d). Two main groups can be defined according to their orientation (Fig. 7b). The first group consists mainly of NNE lineaments that can be locally oriented in a north to NNW direction. They are continuous along-strike and cross-cut the variation of the regional magnetic field recognized from a NNE to SSW direction (Fig. 7a). In contrast, the second group includes WNW to ENE lineaments, with those of WNW being more frequent. Locally, these oblique lineaments seem to cross-cut the NNE to NNW features.

There is high correlation between the surface structural data described in the section 'Tectonic fabric analysis' and the magnetic lineaments. NNE to NNW lineaments can be interpreted as the result of the main Andean thrusts that made up the Precordillera fold-and-thrust belt while the WNW to ENE lineaments could be related to oblique structures with strike-slip displacement. Those lithomagnetic domains defined as Miocene igneous rocks are located where both sets of oblique lineaments are present. These units are distributed along a major WNW lineament similar to that which cross-cuts the northern Sierra de la Invernada (Fig. 2). A blind thrust, WNW and ENE lineaments and several subsurface igneous bodies can also be recognized below the Holocene deposits in the Pampa de Hualilán (Fig. 2).

Source parameter imaging (SPI)

The map of k maximum values and calculated depths is shown in Figure 8. Wide NNE domains of high density k values can be observed regionally. However, strong oblique segmentation of these domains also occurs and gives rise to along-strike differences between high and low density areas of k.

NNE domains show good correlation with the main NNE to locally NNW lineaments defined by vertical and horizontal derivatives, while the major WNW to ENE lineaments previously defined seem to control the oblique segmentation. It is notable that those lithomagnetic domains representing volcanic and subvolcanic rocks show few k maximum values. These results will be discussed in the following section (see *Discussion*).

Estimated depths show values mostly up to 3 km despite the presence of some areas with depths between 3–5 km. Rarely, data from more than 5 km depth are also present. Along-strike differences in depth can be observed in the NNE domains and they could be related to the oblique lineaments. According to the *c.* 15 km depth estimated by Allmendinger *et al.* (1990) and Cristallini & Ramos (2000) for the detachment level of the Precordillera fold-and-thrust belt, results obtained here with the SPI method suggest that structures recognized in the Hualilán area could be developed within a pre-Miocene basement.

Discussion

Relationship between cross-strike structures and Miocene magmatism

Studies related to cross-strike structural discontinuities have mostly focused on their role as barriers that halt major fault segments and fault propagation during seismic rupture (Pizzi & Galadini 2009; Lin *et al.* 2011). However, their relationship with magmatism has been poorly constrained (Pohn 2000; Tavarnelli *et al.* 2004).

Despite the scarcity of detailed works referring to structural controls on Miocene magmatism in the Andean flat-slab segment, many authors have already mentioned the role of WNW to NW structures during Cenozoic localization of magmatism in other regions of the Central Andes. This relationship has been probed in the Puna by Ré *et al.* (2001), Riller *et al.* (2001), Chernicoff *et al.* (2002) and Petrinovic *et al.* (2005, 2010). Contributions from Urbina *et al.* (1995, 1997), Sruoga *et al.* (1996), Sruoga & Urbina (2008), Urbina & Sruoga (2009) and Japas *et al.* (2010, 2011*a*, *b*) have demonstrated the link between a WNW to NW transtensional belt and the emplacement of Miocene–Pliocene magmatic arc rocks (Tertiary Volcanic Belt of San Luis) during the migration of magmatism associated with the Pampean flat-slab in the Sierras Pampeanas foreland.

Regional tectonic fabric analysis within the Precordillera suggests the presence of WNW to

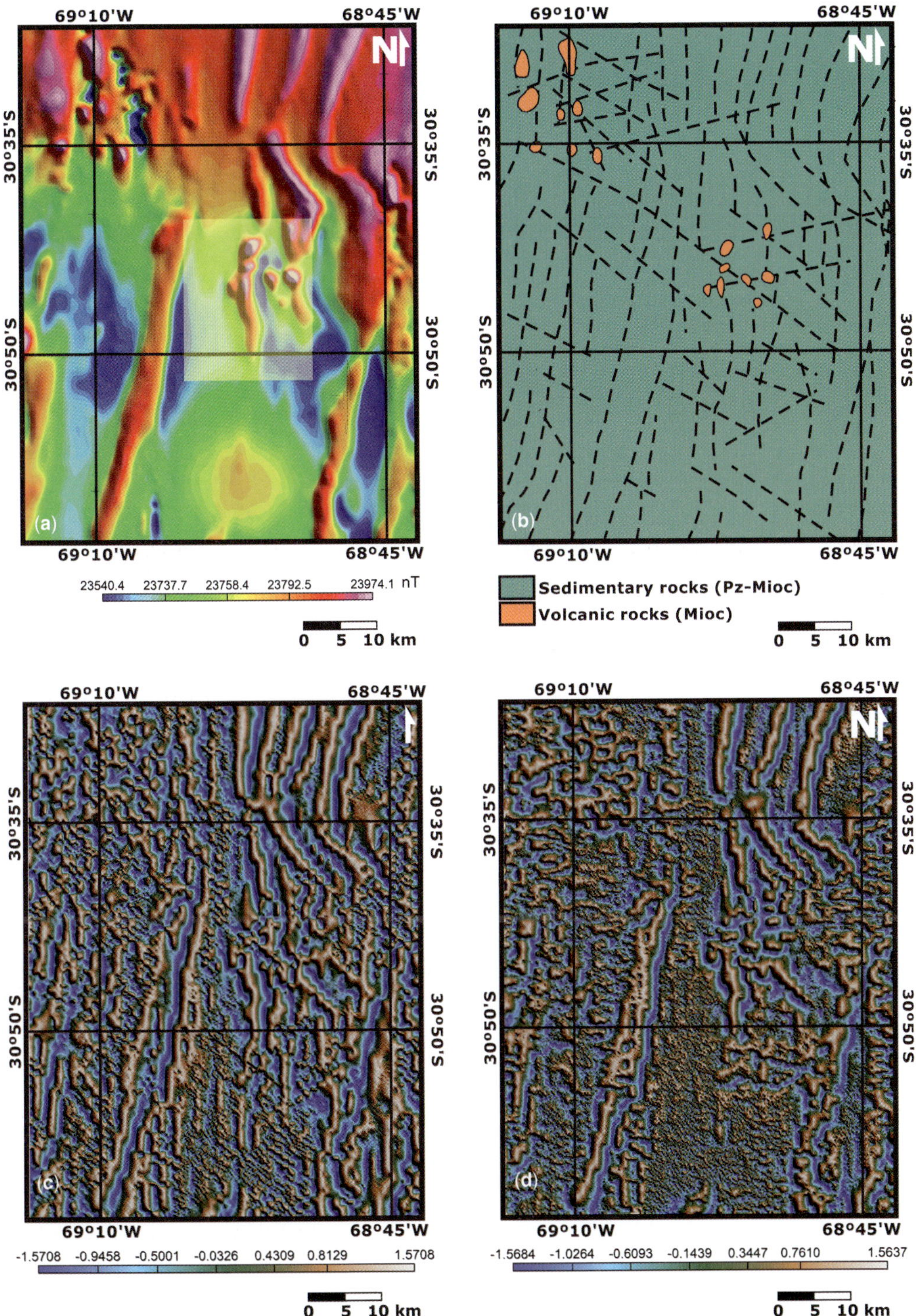

Fig. 7. (**a**) Map of magnetic anomaly reduced to pole. The study area (Fig. 2) is shown with a yellow rectangle. (**b**) Lithomagnetic domains and lineaments interpreted over the magnetic anomaly reduced to pole and first derivatives, respectively. (**c**) First horizontal derivative. (**d**) First vertical derivative.

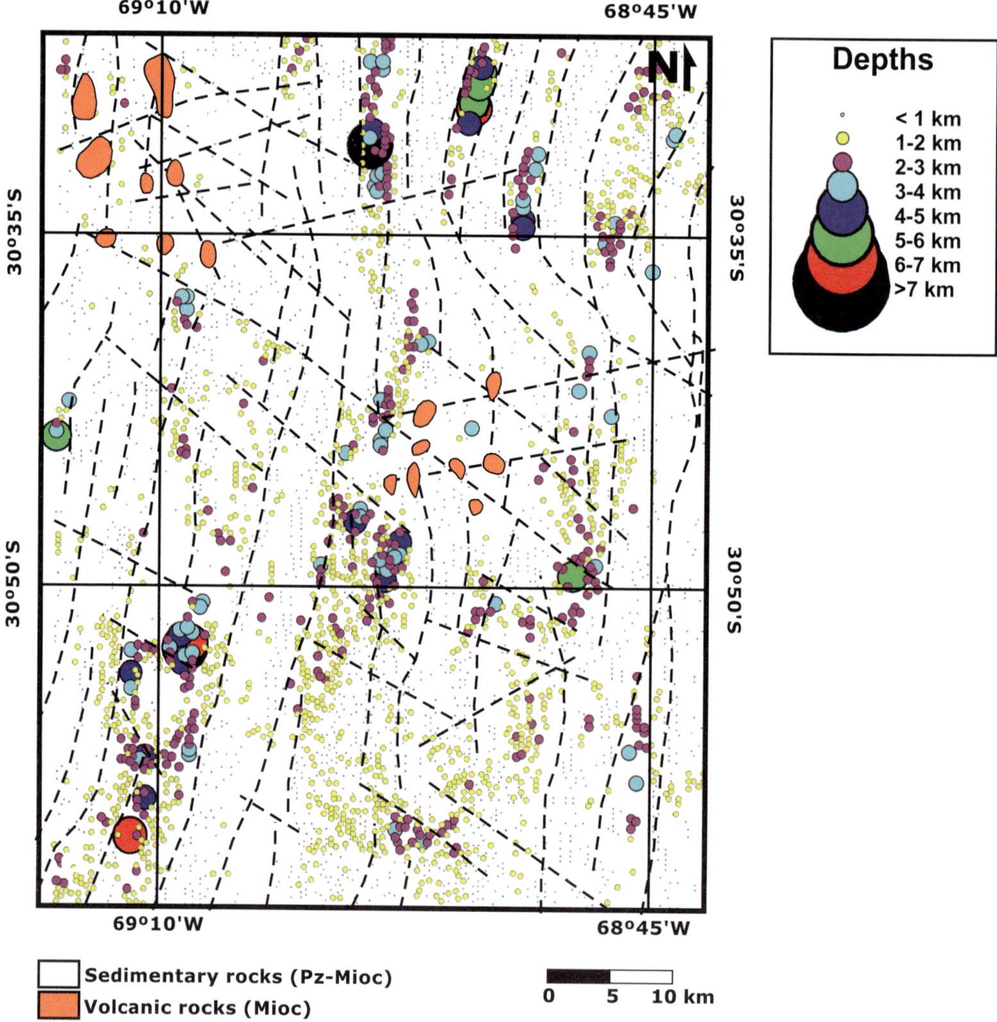

Fig. 8. Lithomagnetic domains and lineaments (modified from Fig. 7b) with estimated depths obtained from source parameter imaging (dots).

NW cross-strike structures in three different areas (Fig. 3b) already recognized by Ré *et al.* (2001). They represent steeply dipping brittle–ductile shear zones that develop at different scales (see 'Structure of the Hualilán area'). Both field relationships and regional observations show that ENE structures are subordinate to those of WNW.

Lithomagnetic domains and lineaments recognized in Figure 7b reveal the presence of two areas where oblique lineaments and Miocene igneous rocks are both concentrated. Both localities are linked by some major WNW lineaments, as shown in Figure 7b, and are aligned in a WNW–NW direction. These results are in agreement with structural data and confirm the existence of

cross-strike structures below the Holocene deposits in the Pampa de Hualilán, supporting proposals from Oriolo (2012) and Oriolo *et al.* (2012) based on geomorphological studies.

SPI solutions are mostly aligned with NNE lineaments (Fig. 8). Therefore, they are interpreted to represent NNE Andean thrusts. Cross-strike structures are only locally defined by alignments of SPI solutions, although WNW and ENE lineaments give rise to strong NNE along-strike segmentation in their grouping and depths. Differences between first derivatives (Fig. 7b) and the SPI method (Fig. 8) regarding the presence of cross-strike structures can be explained by their orientation. They are mostly subvertical structures that cannot be detected easily by the SPI method because of their steep dips (Fig. 4a).

Areas with scarce SPI solutions could be the result of the presence of a sedimentary cover. However, they could also be explained by the presence of several cross-strike structures (Figs 2 & 7b) and the small dimensions of Miocene igneous bodies, which are also well grouped, because these could give rise to interference patterns that would overcome the method resolution.

The geological, structural and geophysical data listed above led to the definition of the presence of WNW and subordinated ENE cross-strike structures in the Central Andes. These may have controlled the emplacement of Miocene magmatism migrating into the foreland due to the flattening of the slab, as field relationships and structural observations have shown. Magmatism could migrate along WNW cross-strike structures whereas it could be emplaced where WNW and ENE features are both present. Despite being dominantly strike-slip structures, minor normal dip-slip kinematic analysis would give rise to the local extension required for volcanic output (Acocella & Funiciello 2010). Moreover, approximately east−west-trending mineralized veins genetically related to Neogene magmatism in the Hualilán area were described by Pelichotti (1976) as supporting local extension. WNW-trending half-graben described by Costa & Cortés (1993) and the WNW transtensional volcanic belt defined by Japas *et al.* (2010) in the Sierras Pampeanas (Fig. 1a) also support the kinematics presented herein.

Cross-strike structures in the study area indicate the presence of the Hualilán Belt (Oriolo *et al.* 2011; Oriolo 2012), which is equivalent to the Talacasto lineament previously defined by Ré *et al.* (2001) and which represents the structural control of the Miocene Central Volcanic Belt of the Precordillera (Oriolo 2012). Based on similar relationships (Fig. 3b), the presence is also proposed, of two other equivalent WNW sinistral transtensional belts controlling Miocene volcanic emplacement in the Precordillera that could be coincident with domain I, sectors A and C (Fig. 3b). This is more evident at 29°45′S where WNW to NW structures and magmatic rocks were documented by Chernicoff & Nash (2002).

The brittle−ductile nature of cross-strike structures could be explained as the result of the Rebinder effect (Karpenko 1974) and hydraulic fracturing that could give rise to strain softening due to the presence of magmatic fluids and thermal activation. However, Hindle & Vietor (2005) suggest that deformation of cross-strike structures is distributed over wide zones made up of minor structures but they do not link these features to the presence of magmatism.

Temporal and kinematic evolution of cross-strike structures

Field data suggest that magmatism and the main tectonic activity of WNW to ENE structures have taken place after the main uplift of the Andean thrusts as Pelichotti (1976) and Oriolo (2012) had previously suggested. Ages between 9 and 6 Ma obtained from other volcanic and subvolcanic rocks from the Precordillera (Kay & Mpodozis 2002) are consistent with this proposal. This supports the relationship between oblique structures and magmatism already mentioned (see 'Discussion').

The regional variation of the magnetic field from a NNE to SSW direction and cross-cutting relationships with those magnetic anomalies interpreted as thrusts (Fig. 7b) suggest a probable 'inherited' nature of the highly oblique lineaments. Analogue modelling and gravimetric data considered by Oriolo (2012) and Oriolo *et al.* (2012) have also supported this hypothesis. Likewise, analogue models from these authors have demonstrated that cross-strike structures do not develop unless they represent pre-existing crustal fabrics but they reactivate after the main shortening in orogenic along-strike structures has occurred. According to results obtained by the SPI method (see 'Source parameter imaging (SPI)'), these structures must be located mostly at depths up to 5 km so they can be developed in a pre-Miocene basement. Moreover, WNW structures in the Hualilán area could be associated with the Pie de Palo Norte lineament, located between the Sierra de Pie de Palo and the Eastern Precordillera (Fig. 1b). Using seismic data, Zapata (1998) has described the lineament as a steeply dipping sinistral strike-slip flower structure in the Bermejo valley that causes left-lateral offsets of Tertiary units in the Eastern Precordillera (Milana 1990).

Siame *et al.* (2005) have provided kinematic data that demonstrate a regional west−WNW shortening for the Precordillera that gives rise to almost north-striking folds and thrusts that have developed since Late Miocene times. On the other hand, kinematic results obtained in this study (see 'Kinematic Analysis') reveal that cross-strike structures show a strong component of strike-slip displacements. Siame *et al.* (2005) have also proposed a clockwise reorientation of kinematic axes in the Precordillera that could have begun between 5 and 2.5 Ma; their proposal is also supported by data obtained in this work (see 'Kinematic Analysis'). This reorientation can be restricted to the Late Miocene−Pliocene? for the study area, as the temporal relationship between magmatism (9 to 6 Ma according to Kay & Mpodozis 2002) and its structural controls suggests (activation of cross-strike structures since *c.* 9 Ma post-dating the main tectonic activity of thrusts; see Oriolo 2012 and references therein). Therefore, kinematics showing minor extension associated with cross-strike structures obtained in Hualilán are local and resulted from the reactivation of ancient structures during an Andean transpressional regime in the Precordillera (Siame *et al.* 2005). These results are significantly different from regional ones but they are consistent with the regional Andean deformation. The similar reorientation pattern between both local and regional kinematic axes (Siame *et al.* 2005) also supports this idea.

Conclusions

The data presented in this paper have led to an increased understanding of the kinematics and development of cross-strike structures in the study area and its regional framework:

(1) The presence of cross-strike structures in the Precordillera is confirmed. They represent WNW brittle–ductile belts with sinistral strike-slip displacements and are probably the result of reactivation of pre-existing crustal fabrics during the Andean orogeny. Particularly, the Hualilán Transtensional Belt is defined for the study area.

(2) Cross-strike structures represent the main structural control in the emplacement of Miocene magmatism migrating into the foreland due to the Andean flat-slab subduction.

(3) Kinematics of cross-strike structures show mostly strike-slip displacements in spite of approximately north-striking thrusting and folding observed at regional scale, though they are all consistent with regional Andean transpressional deformation. Moreover, the regional clockwise reorientation of kinematic axes has also affected the local pattern of deformation since Late Miocene–Pliocene? times.

The authors wish to thank projects PIP CONICET 11420100100334 (M. S. Japas), PICT 2010-1441 and UBACyT 20020100100855 (E. O. Cristallini). J. Cortés, H. Vizán and J. Sellés-Martínez are acknowledged for previous discussions related to this work. We thank the editors and L. Giambiagi and a second anonymous reviewer for helpful comments and corrections that enabled us to improve this manuscript. We also thank R. Allmendinger and R. Holcombe for free access to software.

References

ACOCELLA, V. & FUNICIELLO, F. 2010. Kinematic setting and structural control of arc volcanism. *Earth and Planetary Science Letters*, **289**, 43–53.

ALLMENDINGER, R. W., FIGUEROA, D., SNYDER, D., BEER, J., MPODOZIS, C. & ISACKS, B. L. 1990. Foreland shortening and crustal balancing in the Andes at 30°S latitude. *Tectonics*, **9**, 789–809.

ALLMENDINGER, R. W., CARDOZO, N. & FISHER, D. 2012. *Structural Geology Algorithms: Vectors and Tensors in Structural Geology*. Cambridge University Press, Cambridge.

BALDIS, B. & VACA, A. 1985. Megafracturas relacionadas con el sistema cordillerano. *Primeras Jornadas sobre Geología de Precordillera, Actas*, **1**, 204–208.

BARANOV, V. 1957. A new method for interpretation of aeromagnetic maps: pseudo-gravimetric anomalies. *Geophysics*, **22**, 359–383.

BASTIAS, H. 1985. *Fallamiento cuaternario en la región sismotectonica de Precordillera*. PhD thesis, Universidad Nacional de San Juan.

BASTÍAS, H., WEIDMANN, N. & PÉREZ, M. 1984. Dos zonas de fallamiento plio–cuaternario en la Precordillera de San Juan. *IX Congreso Geológico Argentino, Actas*, **2**, 329–341.

BLAKELY, R. J. 1995. *Potential Theory in Gravity and Magnetic Applications*. Cambridge University Press, Cambridge.

BLAKELY, R. J. & SIMPSON, R. W. 1986. Approximating edges of source bodies from magnetic or gravity anomalies. *Geophysics*, **51**, 1494–1498.

BROWN, M. & SOLAR, G. S. 1999. The mechanism of ascent and emplacement of granite magma during transpression: a syntectonic granite paradigm. *Tectonophysics*, **312**, 1–33.

CAHILL, T. & ISACKS, B. L. 1992. Seismicity and the shape of the subducted Nazca plate. *Journal of Geophysical Research*, **97**, 17503–17529.

CHERNICOFF, C. J. & NASH, C. R. 2002. Geological interpretation of Landsat TM imagery and aeromagnetic survey data, northern Precordillera region, Argentina. *Journal of South American Earth Sciences*, **14**, 813–820.

CHERNICOFF, C. J., RICHARDS, J. P. & ZAPPETTINI, E. O. 2002. Crustal lineament control on magmatism and mineralization in northwestern Argentina: geological, geophysical, and remote sensing evidence. *Ore Geology Reviews*, **21**, 127–155.

COSTA, C. & CORTÉS, J. M. 1993. Tectónica extensional en el extremo sur de la sierra de San Luis. *XII Congreso Geológico Argentino, Actas*, **3**, 113–118.

CRISTALLINI, E. O. & RAMOS, V. A. 2000. Thick-skinned and thin-skinned thrusting in La Ramada fold and thrust belt: crustal evolution of the High Andes of San Juan, Argentina (32° SL). *Tectonophysics*, **317**, 205–235.

DE SAINT BLANQUAT, M. & TIKOFF, B. 1997. Development of magmatic to solid-state fabrics during syntectonic emplacement of the Mono Creek Granite, Sierra Nevada Batholith. *In*: BOUCHEZ, J. L., HUTTON, D. & STEPHENS, D. E. (eds) *Granite: From Segregation of Melts to Emplacement Fabrics*. Kluwer Academic, Dordrecht, 231–252.

DE SAINT BLANQUAT, M., TIKOFF, B., TEYSSIER, C. & VIGNERESSE, J. L. 1998. Transpressional kinematics and magmatic arcs. *In*: HOLDSWORTH, R. E., STRACHAN, R. A. & DEWEY, J. F. (eds) *Continental Transpressional and Transtensional Tectonics*.

Geological Society, London, Special Publications, **135**, 327–340.

GANSSER, A. 1973. Facts and theories on the Andes: Twenty-sixth William Smith Lecture. *Journal of the Geological Society, London*, **129**, 93–131.

GÓMEZ RIVAROLA, L. 2007. *Descripción e interpretación de las rocas volcánicas, volcaniclásticas y facies sedimentarias asociadas en la región de Cerro Negro de Iglesia, Precordillera Occidental, Provincia de San Juan.* Trabajo Final de Licenciatura, Universidad de Buenos Aires.

GUTSCHER, M., MAURY, R., EISSEN, J. P. & BOURDON, E. 2000. Can slab melting be caused by flat subduction? *Geology*, **28**, 535–538.

HEREDIA, N., RODRÍGUEZ FERNÁNDEZ, L. R., GALLASTEGUI, G., BUSQUETS, P. & COLOMBO, P. 2002. Geological setting of the Argentine Frontal Cordillera in the flat slab segment (30°00′–31°30′S latitude). *Journal of South American Earth Sciences*, **15**, 79–99.

HINDLE, D. & VIETOR, T. 2005. 3D strain modeling of tear fault analogues. Abstract #S53A-1091, presented at the *American Geophysical Union, Fall Meeting 2005*.

HOLCOMBE, R. 2011. GEOrient 9.5.0. Stereographic Projections and Rose Diagrams Plots. http://www.holcombecoughlinoliver.com/holcombe/.

HUDNUT, K., SEEBER, L., ROCKWELL, T., GOODMACHER, J., KLINGER, R., LINDVALL, S. & MCELWAIN, R. 1989. Surface ruptures on cross-faults in the 24 November 1987 Superstition Hills, California, earthquake sequence. *Bulletin of the Seismological Society of America*, **79**, 282–296.

JAPAS, M. S. 1998. Aporte del análisis de fábrica deformacional al estudio de la faja orogénica andina. Homenaje al Dr. Arturo J. Amos. *Revista de la Asociación Geológica Argentina*, **53**, 15.

JAPAS, M. S. & RÉ, G. H. 2005. Geodynamic impact of arrival and subduction of oblique aseismic ridges. In: *VI International Symposium on Andean Geodynamics, abstracts*, 408–410.

JAPAS, M. S. & RÉ, G. H. 2012. Neogene tectonic block rotations and margin curvature at the Pampean flat slab segment (28°–33° SL, Argentina). *Geoacta*, **37**, 1–4.

JAPAS, M. S., RÉ, G. H. & BARREDO, S. P. 2002a. Lineamientos andinos oblicuos (entre 22° y 33° S): definidos a partir de fábricas tectónicas. I. Fábrica deformacional y de sismicidad. *XV Congreso Geológico Argentino, Actas*, **I**, 326–331.

JAPAS, M. S., RÉ, G. H. & BARREDO, S. P. 2002b. Lineamientos andinos oblicuos (entre 22° y 33° S) definidos a partir de fábricas tectónicas. III. Modelo cinemático. *XV Congreso Geológico Argentino, Actas*, **I**, 340–343.

JAPAS, M. S., CORTÉS, J. M. & PASINI, M. 2008. Tectónica extensional triásica en el sector norte de la cuenca Cuyana: primeros datos cinemáticos. *Revista de la Asociación Geológica Argentina*, **63**, 213–222.

JAPAS, M. S., URBINA, N. E. & SRUOGA, P. 2010. Control estructural en el emplazamiento del volcanismo y mineralizaciones neógenas, distrito Cañada Honda, San Luis. *Revista de la Asociación Geológica Argentina*, **67**, 494–506.

JAPAS, M. S., SRUOGA, P., KLEIMAN, L. E., GAYONE, M. R., MALOBERTI, A. & COMITO, O. 2011a. Cinemática de la extensión jurásica vinculada a la provincia silícea Chon Aike, Santa Cruz, Argentina. *XVIII Congreso Geológico Argentino, Actas*, 97–98.

JAPAS, M. S., URBINA, N. E., SRUOGA, P. & GALLARD, M. C. 2011b. La Carolina pull-apart in western Tertiary Volcanic Belt, Pampean Flat Slab (33°S), Argentina. *XXII Lateinamerika Kolloquium*.

JAPAS, M. S., SRUOGA, P., KLEIMAN, L. E., GAYONE, M. R., MALOBERTI, A. & COMITO, O. 2013. Cinemática de la extensión jurásica vinculada a la provincia silícea Chon Aike, Santa Cruz, Argentina. *Revista de la Asociación Geológica Argentina*, **70**, 16–30.

JOHNSON, C. A., SATTAR, M. A., ROSELL, R., AL–SHEKAILI, F., AL–ZAABI, N. & GOMBOS, A. 2002. Structure and regional context of onshore Fields in Abu Dhabi, UAE. *Society of Petroleum Engineers, Conference Paper*, http://dx.doi.org/10.2118/78488-MS.

KARPENKO, G. V. 1974. The 45th anniversary of the Rebinder effect. *Materials Science*, **10**, 3–4.

KAY, S. M. & ABBRUZZI, J. M. 1996. Magmatic evidence for Neogene lithospheric evolution of the central Andean 'flat-slab' between 30°S and 32°S. *Tectonophysics*, **259**, 15–28.

KAY, S. M. & MPODOZIS, C. 2002. Magmatism as a probe to the Neogene shallowing of the Nazca plate beneath the modern Chilean flat-slab. *Journal of South American Earth Sciences*, **15**, 39–57.

LARA, L. E., LAVENU, A., CEMBRANO, J. & RODRÍGUEZ, C. 2006. Structural controls of volcanism in transversal chains: resheared faults and neotectonics in the Cordón Caulle–Puyehue area (40.5°S), Southern Andes. *Journal of Volcanology and Geothermal Research*, **158**, 70–86.

LEVERATTO, M. A. 1976. Edad de intrusivos cenozoicos en la Precordillera de San Juan y su implicancia estratigráfica. *Revista de la Asociación Geológica Argentina*, **31**, 53–58.

LIMARINO, C. O., FAUQUÉ, L. A., CARDÓ, R., GAGLIARDO, M. L. & ESCOSTEGUY, L. 2002. La faja volcánica miocena de la Precordillera septentrional. *Revista de la Asociación Geológica Argentina*, **57**, 289–304.

LIN, J., STEIN, R. S., MEGHRAOUI, R., TODA, S., AYADI, A., DORBATH, C. & BELABBES, S. 2011. Stress transfer among en echelon and opposing thrusts and tear faults: triggering caused by the 2003 Mw = 6.9 Zemmouri, Algeria, earthquake. *Journal of Geophysical Research*, **116**, B03305, http://dx.doi.org/10.1029/2010JB007654.

LOGAN, M. A. 2000. Mineralogy and geochemistry of the Argentina. *Ore Geology Reviews*, **17**, 113–138.

MARRETT, R. A. & ALLMENDINGER, R. W. 1990. Kinematic analysis of fault-slip data. *Journal of Structural Geology*, **12**, 973–986.

MILANA, J. P. 1990. *Sedimentología y magnetoestratigrafía de formaciones cenozoicas en el área de Mogna, y su incersión en el marco tectosedimentario de la Precordillera Oriental.* PhD thesis, Universidad Nacional de San Juan.

MUELLER, K. & TALLING, P. 1997. Geomorphic evidence of tear faults accommodating lateral propagation of an active fault-bend fold, Wheeler Ridge, California. *Journal of Structural Geology*, **19**, 397–411.

NABIGHIAN, M. N. 1972. The analytical signal of two-dimensional bodies with polygonal cross-section: its properties and use for automated interpretations. *Geophysics*, **37**, 780–786.

NABIGHIAN, M. N. 1974. Additional comments on the analytical signal of two-dimensional bodies with polygonal cross-section. *Geophysics*, **39**, 85–92.

NABIGHIAN, M. N. 1984. Towards a three-dimensional automatic interpretation of potential field data via generalized Hilbert transforms: fundamental relations. *Geophysics*, **49**, 780–786.

ORIOLO, S. 2012. *Análisis de la deformación en la región de Hualilán, Precordillera de San Juan*. Trabajo Final de Licenciatura, Universidad de Buenos Aires.

ORIOLO, S., JAPAS, M. S. & CRISTALLINI, E. O. 2011. Transtensional megashear zones in the Central Andes: the Hualilán Belt. *Deformation Mechanisms, Rheology and Tectonics Conference Proceedings*, 95.

ORIOLO, S., CRISTALLINI, E. O. & JAPAS, M. S. 2012. Modelos análogos aplicados a sistemas transpresivos y transtensivos. Jornada Abierta del Instituto de Geociencias IGEBA, Tectónica de Desplazamiento de Rumbo, actas, **12**.

PELICHOTTI, R. 1976. *Estudio geológico–económico y proyecto de exploración del Distrito Minero Gualilán, Dpto. Ullúm, Pcia. de San Juan*. Servicio Minero Nacional, Plan San Juan.

PETRINOVIC, I., RILLER, U. & BROD, J. A. 2005. The Negra Muerta Volcanic Complex, southern Central Andes: geochemical characteristics and magmatic evolution of an episodically active volcanic centre. *Journal of Volcanology and Geothermal Research*, **144**, 295–320.

PETRINOVIC, I., MARTÍ, J., AGUIRRE–DÍAZ, G. J., GUZMÁN, S., GEYER, A. & SALADO PAZ, N. 2010. The Cerro Aguas Calientes caldera, NW Argentina: an example of a tectonically controlled, polygenetic, collapse caldera, and its regional significance. *Journal of Volcanology and Geothermal Research*, **194**, 15–26.

PILGER, R. H. 1981. Plate reconstructions, aseismic ridges, and low angle subduction beneath the Andes. *Geological Society of America Bulletin*, **92**, 448–456.

PIZZI, A. & GALADINI, F. 2009. Pre-existing cross-structures and active fault segmentation in the northern-central Apennines (Italy). *Tectonophysics*, **476**, 304–319.

POHN, H. A. 2000. *Lateral ramps in the folded Apalaches and in overthrust belts worldwide – a fundamental element of thrust-belt architecture*. US Geological Survey Bulletin, **2163**.

RAGONA, D., ANSELMI, G., GONZÁLEZ, P. & VUJOVICH, G. 1995. *Mapa Geológico de la Provincia de San Juan. Escala 1:500 000*. SEGEMAR, Buenos Aires.

RAMOS, V. A., CRISTALLINI, E. O. & PÉREZ, D. J. 2002. The Pampean flat-slab of the Central Andes. *Journal of South American Earth Sciences*, **15**, 59–78.

RAMSAY, J. G. & HUBER, M. I. 1987. *The Techniques of Modern Structural Geology (2: Fold and Fractures)*. Academic Press, London.

RÉ, G. H. & JAPAS, M. S. 2004. Andean oblique megashear zones: paleomagnetism contribution. *GSA Abstract with Programs*, **36**, Abstract No. 79033.

RÉ, G. H., JAPAS, M. S. & BARREDO, S. P. 2000. Análisis de fábrica deformacional (AFD): el concepto fractal cualitativo aplicado a la definición de lineamientos cinemáticos neógenos en el noroeste argentino. *X Reunión sobre Microtectónica, Resúmenes*, 12.

RÉ, G. H., JAPAS, M. S. & BARREDO, S. P. 2001. Análisis de fábrica deformacional (AFD): el concepto fractal cualitativo aplicado a la definición de lineamientos cinemáticos neógenos en el noroeste argentino. *Revista de la Asociación Geológica Argentina Serie D*, Publicación Especial, **5**, 75–82.

RILLER, U., PETRINOVIC, I., RAMELOW, J., STRECKER, M. & ONCKEN, O. 2001. Late Cenozoic tectonism, collapse caldera and plateau formation in the central Andes. *Earth and Planetary Science Letters*, **188**, 299–311.

ROEST, W. R., VERHOEF, J. & PILKINGTON, M. 1992. Magnetic interpretation using the 3–D analytical signal. *Geophysics*, **51**, 116–125.

ROMEO, I., CAPOTE, R., TEJERO, R., LUNAR, R. & QUESADA, C. 2006. Magma emplacement in transpression: the Santa Olalla Igneous Complex (Ossa–Morena Zone, SW Iberia). *Journal of Structural Geology*, **28**, 1821–1834.

ROSSELLO, E. A., MOZETIC, M. E., COBBOLD, P. R., DE URREIZTIETA, M. & GAPAIS, D. 1996. El espolón Umango-Maz y la conjugación sintaxial de los Lineamientos Tucumán y Valle Fértil (La Rioja, Argentina). *XIII Congreso Geológico Argentino y III Congreso de Exploración de Hidrocarburos, actas*, **2**, 187–194.

SAHA, T., CHOWDHURY, B. K., CHAKRABORTHY, D., PRUSTY, S. K., KHATRI, B. L. & REDDY, G. V. 2006. Understanding structural configuration on the basis of 3D seismic attributes in Dhansiri Valley, Assam. *Geohorizons*, **2**, 28–31.

SALFITY, J. A. 1985. Lineamientos transversales al rumbo andino en el Noroeste Argentino. *IV Congreso Geológico Chileno, Actas*, **2**, 119–137.

SANDERSON, D. J. & MARCHINI, W. R. D. 1984. Transpression. *Journal of Structural Geology*, **6**, 449–458.

SIAME, L. L., BELLIER, O., SEBRIER, M. & ARAUJO, M. 2005. Deformation partitioning in flat subduction setting: case of the Andean foreland of western Argentina (28°S–33°S). *Tectonics*, **24**, http://dx.doi.org/10.1029/2005TC001787.

SRUOGA, P. & URBINA, N. E. 2008. Volcanismo en ambiente de flat-slab: Cañada Honda, Sierras Pampeanas de San Luis (32° 50′S, 66° 00′O). *XVII Congreso Geológico Argentino, Actas*, **1**, 238–239.

SRUOGA, P., URBINA, N. E. & MALVICINI, L. 1996. El Volcanismo Terciario y los depósitos hidrotermales (Au,Cu) asociados en La Carolina y Diente Verde, San Luis, Argentina. *XIII Congreso Geológico Argentino, Actas*, **3**, 89–100.

TAVARNELLI, E., BUTLER, R. W. H., DECANDIA, F. A., CALAMITA, F., GRASSO, M., ALVAREZ, W. & RENDA, P. 2004. Implications of fault reactivation and structural inheritance in the Cenozoic tectonic evolution of Italy. *In*: CRESCENTI, U., D'OFFIZI, S., MERLINI, S. & SACCHI, R. (eds) *The Geology of Italy*. Societa Geologica Italiana, Rome, Special Volume, 209–222.

TWISS, R. J. & MOORE, E. M. 1992. *Structural Geology*. Freeman & Co, New York.

URBINA, N. E. & SRUOGA, P. 2009. La faja metalogenética de San Luis: mineralización y geocronología en el contexto metalogenético regional. *Revista de la Asociación Geológica Argentina*, **64**, 635–645.

URBINA, N. E., SRUOGA, P. & MALVICINI, L. 1995. El volcanismo Mioceno y la mineralización aurífera asociada en La Carolina y Diente Verde, provincia de San Luis, Argentina. *IX Congreso Latinoamericano de Geología, Actas*, 1–13.

URBINA, N. E., SRUOGA, P. & MALVICINI, L. 1997. Late tertiary gold-bearing volcanic belt in the Sierras Pampeanas of San Luis, Argentina. *International Geology Review*, **39**, 287–306.

URREIZTIETA, M. 1996. *Tectonique Néogène et basins transpressifs en bordure méridionale de l'Altiplano-Puna (27°S), Nord-Ouest argentin*. Memoires, Geosciences Rennes. PhD thesis, Université de Rennes I.

VAN GOOL, J. A. M. & PIAZOLO, S. 2006. Presentation and interpretation of structural data from the Nagssugtokidian orogen using a GIS platform: general trends and features. *Geological Survey of Denmark and Greenland Bulletin*, **11**, 125–144.

WHEELER, R. L. 1980. Cross-strike structural discontinuities: possible exploration tool for natural gas in Appalachian overthrust belt. *American Association of Petroleum Geologists Bulletin*, **64**, 2166–2178.

ŽÁK, J., HOLUB, F. V. & VERNER, K. 2005. Tectonic evolution of a continental magmatic arc from transpression in the upper crust to exhumation of mid-crustal orogenic root recorded by episodically emplaced plutons: the Central Bohemian Plutonic Complex (Bohemian Massif). *International Journal of Earth Sciences*, **95**, 385–400.

ZAPATA, T. R. 1998. Crustal structure of the Andean thrust front at 30°S latitude from shallow and deep seismic reflection profiles, Argentina. *Journal of South American Earth Sciences*, **11**, 131–151.

Post-emplacement thermo-rheological history of a granite intrusion and surrounding rocks: the Monte Capanne pluton, Elba Island, Italy

ALFREDO CAGGIANELLI[1]*, GIORGIO RANALLI[2], ALESSIO LAVECCHIA[1,3], DOMENICO LIOTTA[1] & ANDREA DINI[4]

[1]*Dipartimento di Scienze della Terra e Geoambientali, Bari University, via Orabona 4, 70125 Bari, Italy*

[2]*Department of Earth Sciences, Carleton University, Ottawa K1S 5B6, Canada*

[3]*Department of Computational Geoscience, Simula Research Laboratory, 1364 Fornebu, Norway*

[4]*Institute for Geosciences and Earth Resources, via Moruzzi 1, 56124 Pisa, Italy*

**Corresponding author (e-mail: alfredo.caggianelli@uniba.it)*

Abstract: A thermo-rheological model of the Monte Capanne pluton, Elba Island, Italy is proposed as having general relevance for the thermal and tectonic evolution of upper crustal granites and their surrounding rocks in extensional regions. The thermal evolution of the pluton and country rocks is followed for 1 myr after emplacement, which occurred at *c.* 6.9 Ma. The pluton completely crystallized in *c.* 210 kyr ($\pm$20%). The adjacent rocks reached a thermal peak of 550 °C ($\pm$10%), maintaining a temperature higher than 500 °C for *c.* 100 kyr.

The temperature distribution is used to construct a model for the time-dependent rheology of the pluton and surrounding rocks. A series of 2D cross-sections shows an upward migration of the regional brittle − ductile transition, and the formation of a ductile horizon above the pluton. The former is a combined effect of unroofing and middle crust heating; the latter is the result of temperature increase in rheologically weak country rocks. This ductile horizon has a potential role in the tectonic evolution of the region, since it could favour the formation of upper crustal shear zones and listric faults rooting in the transient brittle − ductile transition and playing a major role in further post-emplacement extension.

The post-collisional history of orogenic belts can be affected by extensional tectonics with formation of metamorphic core complexes and co-eval magmatism (e.g. Lister & Baldwin 1993; Parsons & Thompson 1993; Hill *et al.* 1995). Ascent and emplacement of granitoid magmas favours doming and uplift of rocks (Rosenbaum *et al.* 2008 and references therein), and modifies the pre-existing thermal and rheological structure (Gans *et al.* 1989; Corti *et al.* 2003). Consequently, it plays an important role in the dynamics of extension by weakening parts of the system and producing new rheological heterogeneities (Buck 2006; Müntener & Manatschal 2006).

At the regional scale, this evolution has been modelled in several papers based on numerical approaches (e.g. McKenzie & Bickle 1988; Bown & White 1995; see Gerya 2010 for an exhaustive treatment of geodynamic numerical modelling) predicting that relatively large volumes of melts would be produced by extension-related mantle decompression, thus influencing melt-productivity and rheology (Piccardo *et al.* 2010, and references therein).

However, the rheological evolution induced by magmatic heat advection is transient, thus implying continuous changes in the rheology of the structural aureole rocks, with possible effects on the spatio-temporal localization of the co-eval extensional structures. To contribute to the analysis of this process, we present a time-dependent thermo-rheological model of the post-emplacement history of the Late Miocene (6.9 Ma) Monte Capanne pluton and surrounding wall rocks in Elba Island, Tuscan Magmatic Province (Fig. 1). Emplacement of the pluton took place in an extensional setting during the Messinian, and its exhumation was determined mainly through unroofing normal faults and erosion. Thus, the coupled effects of exhumation and heating from the magma made the rheology significantly time-dependent, within a relatively short period. Model results are compared with the

From: LLANA-FÚNEZ, S., MARCOS, A. & BASTIDA, F. (eds) 2014. *Deformation Structures and Processes within the Continental Crust*. Geological Society, London, Special Publications, **394**, 129–143.
First published online November 20, 2013, updated February 5, 2014, http://dx.doi.org/10.1144/SP394.1

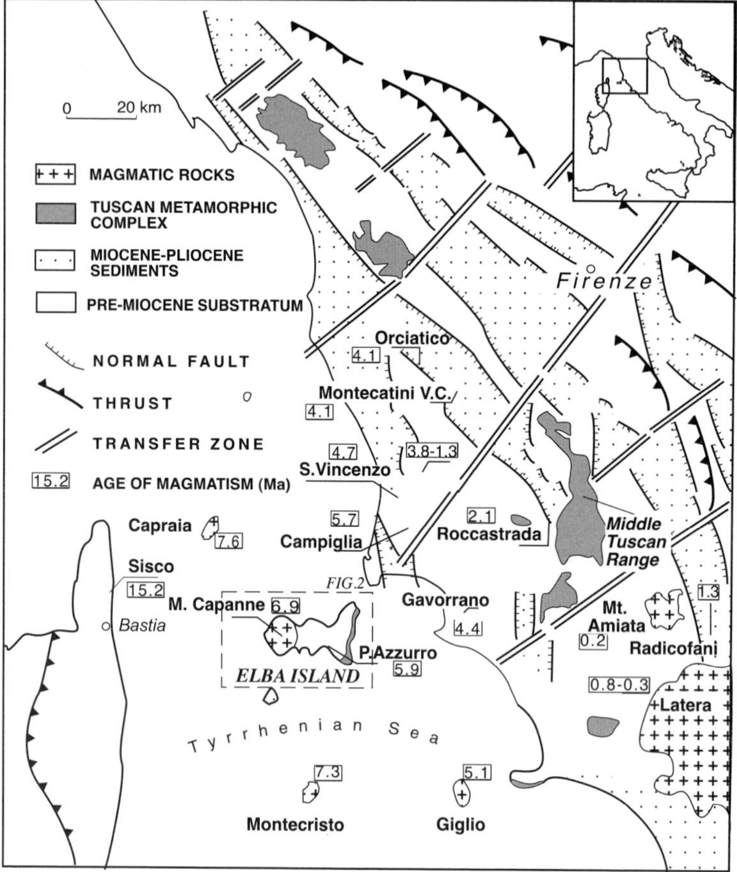

Fig. 1. Elba Island in the context of the inner Northern Apennines. Age of magmatic rocks in Ma is shown.

metamorphic and deformational features of the wall rocks, and discussed in light of the tectonic evolution of the area.

The present study of the Monte Capanne pluton, although constrained and focused by local evidence, has general implications for the analysis of the post-emplacement thermal evolution of plutons and surrounding rocks in post-orogenic extensional settings.

Geological setting

Elba Island (Fig. 2) is part of the inner Northern Apennines (i.e. the present northern Tyrrhenian Sea and southern Tuscany). The structure of the island reflects the evolution of this orogenic belt, with an east-verging nappe pile (Trevisan 1953), built up during the convergence (Late Cretaceous–Eocene) and collision (Eocene–Early Miocene) of the Adria microplate and the Sardinia-Corsica block. This process resulted in the stacking of tectonic units derived from the oceanic and continental palaeogeographical domains belonging to the Northern Apennines (Keller & Pialli 1990; Pertusati *et al.* 1993).

Five tectonostratigraphic complexes have been recognized in Elba, numbered I–V from the bottom to the top of the tectonic pile, according to the classical paper by Trevisan (1953). The lower three complexes (I–III) pertain to the Tuscan continental palaeogeographical domain and consist of the Palaeozoic metamorphic basement and its Late-Triassic–Oligocene sedimentary cover, affected in the inner part by metamorphism in greenschist facies conditions. The two upper levels (IV–V) pertain to the Ligurian oceanic palaeodomain and make up the host rocks of the Monte Capanne pluton. Complex IV consists of Jurassic oceanic lithosphere (peridotite, gabbro, pillow basalt and ophiolite sedimentary breccia) overlain by Late Jurassic–Middle Cretaceous cherts, limestones

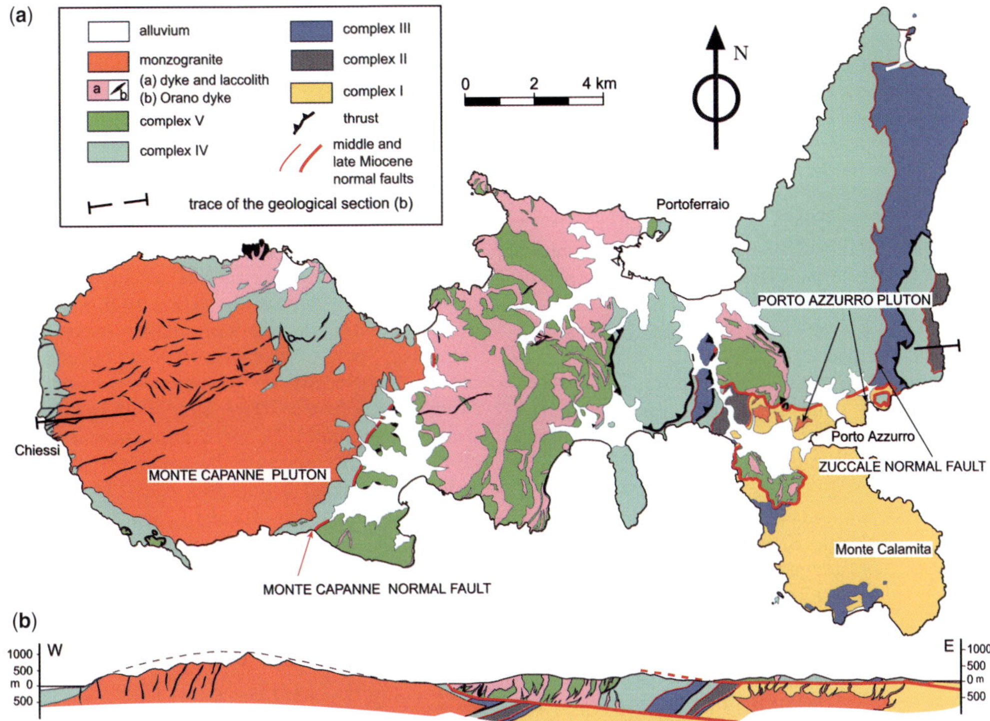

Fig. 2. Elba Island: (**a**) schematic geological map and (**b**) cross-section. The Monte Capanne pluton is exposed for *c.* 67 km^2.

and argillites interbedded with siliceous limestone. Complex V consists of two members: the lower one is a strongly tectonized Paleocene–Eocene succession (argillites, calcarenites and sandy marls) interbedded with ophiolitic breccia; the upper one is represented by a Cretaceous flysch made up of limestone beds, calcareous shales and feldspathic sandstone (Keller & Pialli 1990).

Since Early–Middle Miocene, the whole inner Northern Apennines have been affected by extensional tectonics (Carmignani *et al.* 1994; Brogi & Liotta 2008) and eastward-migrating magmatism (Serri *et al.* 1993; Dini *et al.* 2008). During Middle–Late Miocene, extension produced the lateral segmentation of the previously stacked units through several events of normal faulting. In Elba, this determined a general westward-dipping attitude of the stratigraphic tectonic complexes. Isotope geochronology shows that magmatism in Elba occurred in three main episodes, from 8 to 5.9 Ma (Westerman *et al.* 2004). The emplacement of the Monte Capanne pluton and Orano dykes occurred during the second episode, around 7 Ma.

During Late Miocene, a new faulting event was associated with igneous activity (Bouillin *et al.* 1993; Serri *et al.* 1993; Jolivet *et al.* 1994), producing the Monte Capanne and Zuccale normal faults, associated with the Monte Capanne (6.9 Ma) and Porto Azzurro (5.9 Ma) plutons, respectively (Fig. 2). In comparison to the Porto Azzurro pluton, the Monte Capanne pluton is extensively exposed and its geometry is well constrained (Westerman *et al.* 2004). Several laccoliths and dykes predated and post-dated the emplacement of the Monte Capanne monzogranite. The Orano dykes (Fig. 2), dated at 6.85 Ma (Dini *et al.* 2008), are cross-cut by the Monte Capanne fault.

Shortly after the emplacement of the whole magmatic sequence, the upper part of the igneous-sedimentary complex was tectonically translated eastwards (Monte Capanne fault, Fig. 2) through an extensional top-to-the-east fault surface (Westerman *et al.* 2004; Benvenuti *et al.* 2006). The minimum amount of displacement is estimated to be *c.* 8 km and fault activity occurred in the 6.8–5.9 Ma time interval. According to recent hypotheses, this eastward movement was due, at least partially, to gravitational instability associated with the surface doming produced by the emplacement of laccoliths and the Monte Capanne pluton (Dini *et al.* 2008).

The Monte Capanne pluton and the contact aureole

The Monte Capanne pluton is the largest intrusion exposed in the Tuscan Magmatic Province (Fig. 1). In map view it has a broadly circular outline (Fig. 2) and the 3D shape can be described as a flat truncated cone with a thickness of *c.* 3 km and a diameter ranging from *c.* 8–9 km (top) to *c.* 11 km (bottom). It has a monzogranitic composition (Westerman *et al.* 2003) and is slightly peraluminous (Farina *et al.* 2010). Depth of emplacement, measured to the top of the pluton, is estimated as *c.* 6 km (Westerman *et al.* 2004). The source region of the Monte Capanne monzogranite was mainly the hot lower continental crust, undergoing biotite dehydration melting, with minor contributions from mantle-derived magmas (Westerman *et al.* 2003).

The monzogranite is mainly composed of plagioclase (28–45%), K-feldspars (22–25%), commonly as large megacrysts, quartz (24–27%) and biotite (<10%), with accessory minerals apatite, zircon, tourmaline, monazite, allanite and ilmenite (Gagnevin *et al.* 2005; Farina *et al.* 2010 and references therein). A widespread feature of the pluton is the occurrence of mafic microgranular enclaves, with composition ranging from granodiorite to monzogranite. Statistical analysis on the size and distribution of K-feldspar megacrysts, together with structural, geochemical and isotopic data, allows three different facies to be recognized and suggests that pluton growth took place incrementally (Farina *et al.* 2010) by a process of under-accretion (Annen 2011). However, constraints obtained by numerical thermal models suggest a fast growth process, completed in less than 10 kyr (Farina *et al.* 2010).

Host rocks, belonging to complexes IV and V, exhibit a shear fabric acquired during the Apennines deformational events. As a consequence of the emplacement of the pluton, they were further deformed and underwent thermo-metamorphism (e.g. Daniel & Jolivet 1995). Peak assemblages in calc-silicates include clinopyroxene, wollastonite and grossular, whereas in metapelites they consist of andalusite, cordierite and biotite. Petrological analysis of these assemblages constrains peak metamorphic conditions in a *P–T* window of 150–200 MPa and 575–625 °C (Rossetti *et al.* 2007).

Models and results

In this section the results of the thermal and rheological models are described. The thermal evolution represents the basis for the estimation of the time-dependent rheological behaviour of the pluton and surrounding rocks.

Thermal model

The thermal model assumes conduction in a moving medium and takes into account the latent heat of magma crystallization. Although heat advection by circulating hydrothermal fluids must certainly have played a role (Rossetti *et al.* 2007), an assessment of its importance is precluded by lack of precise knowledge of the geometry of magmatic fluid circulation in the surrounding rocks. Consequently, our temperature values outside the pluton may locally be minimum estimates. The model is based on a 2D geometry with radial symmetry to simulate 3D space dimensionality. The pluton is approximated by the geometry of a truncated cone having a thickness of 3.2 km and top and bottom diameters of 8 and 11 km, respectively. The numerical solution is obtained using the Stella® 9.1 code implementing the fourth-order Runge-Kutta algorithm in order to solve the heat transfer equation:

$$\frac{\partial T}{\partial t} = \frac{1}{\rho\, C_p}\left[\frac{\partial}{\partial r}\left(k(T)\frac{\partial T}{\partial r}\right) + \frac{\partial}{\partial z}\left(k(T)\frac{\partial T}{\partial z}\right)\right] + \frac{A}{\rho\, C_p} - v_z \frac{\partial T}{\partial z} \tag{1}$$

where T is temperature; t is time; r and z are horizontal and vertical (positive downwards) coordinates; ρ and C_p are rock density and specific heat, respectively; $k(T)$ is thermal conductivity; A is radiogenic heat generation rate; and $v_z = dH/dt$ is the (negative) rate of change of the depth H of the pluton due to erosion and tectonic denudation (unroofing rate), which contributes to effective heat advection. Additional information and parameter values are given in Table 1.

The temperature dependence of thermal conductivity (Clauser & Huenges 1995) is assumed to follow the empirical equation of Zoth & Hänel (1988). The use of parameter values independent of lithology reflects the dominant effect of temperature on thermal conductivity for the range of rocks involved in the modelling (Zoth & Hänel 1988; Vosteen & Schellschmidt 2003). Resulting values of conductivity at $T \geq 200–300$ °C are ≤ 2.0 W m^{-1} °C^{-1}, in accordance with experimental determinations (see the compilation by Clauser 2009).

The heat generation rate has been conventionally assumed as decreasing exponentially with depth (see e.g. Ranalli 1995). Since heat generation rates of different rock types generally decrease with decreased silica content (see e.g. Vilà *et al.* 2010),

Table 1. *Parameters adopted in the thermal model*

Parameter	Symbols and equations	Values
Thermal conductivity [W m^{-1} °C^{-1}]	Crust $k(T) = A + B/[350 + T(°C)]$	$k(T = 10 °C) = 2.71$ $A = 0.75$ W m^{-1} °C^{-1} $B = 705$ W m^{-1}
	Mantle $k = $ const	$k = 3.35$
Density [kg m^{-3}]	ρ	2750 (crust) 3300 (mantle)
Specific heat [J kg^{-1} °C^{-1}]	C_P	1000 (crust) 1100 (mantle)
Heat generation rate [μW m^{-3}]	$A = A_0\, e^{(-z/D)}$	$A_0 = 2$ $D = 12000$ m
Unroofing rate [m a^{-1}]	$v_z = \mathrm{d}H/\mathrm{d}t = -c\,H$	$v_{z(t=0)} = 5 \times 10^{-3}$ $H_{(t=0)} = 6000$ m $c = 8.33 \times 10^{-7}$ a^{-1}

exponential decrease with increasing depth is an acceptable approximation for the present model, particularly considering that heat production in rocks of the intermediate and lower crust in the Elba region is unknown. Even so, the length of the simulation (1 myr) makes the effect of radiogenic heat production on temperature variation well within error margins. The vertical velocity, equivalent to the unroofing rate, has been constrained by geochronological data (Bouillin *et al.* 1994). It decreases through time from an initial value of 5 mm a^{-1}.

The release of latent heat of crystallization has been taken into account for the cooling granitic magma by considering an effective specific heat instead of the true specific heat for the interval of crystallization between 850 and 650 °C (see Spear 1993). The adopted value of the effective specific heat (2500 J kg^{-1} °C^{-1}) corresponds to a latent heat of crystallization of 300 kJ kg^{-1}. Apart from the crystallizing magma, the value of the specific heat is fixed to 1000 and 1100 J kg^{-1} °C^{-1} in crust and mantle rocks, respectively.

On the basis of previous studies (Farina *et al.* 2010), we have assumed that pluton growth was completed incrementally by under-accretion of multiple magma pulses in a short time interval (*c.* 10 kyr). Thickness of crust and lithosphere at the time of emplacement (28 and 56 km, respectively) have been estimated taking into account the present thickness values of *c.* 22 and *c.* 50 km (Calcagnile & Panza 1980; Cassinis *et al.* 2003) and the original depth (*c.* 6 km) of emplacement of the pluton.

The initial geotherm has been computed assuming $T = 850$ °C and 1225 °C at the bottom of crust and lithosphere, respectively. The former value is compatible with the dehydration melting temperature of biotite in the lower crust and with the initial temperature of magma. The latter value

represents a boundary condition for the model, whereas T at the base of the crust and the heat flow from the mantle both change during the model evolution. The thermal perturbation produced by the feeding dyke system has not been considered since, given the rapid growth of the pluton, completed in less than 10 kyr (Farina *et al.* 2010), it was of negligible effect.

Total time-run lasts 1 myr after magma emplacement. Results are presented on T–t diagrams and P–T paths for selected locations in and around the pluton (Fig. 3) and as vertical temperature cross-sections at different times (Fig. 4).

Magma cooling history is outlined for two points within the pluton (Fig. 3a). The flexure in the cooling curves at 650 °C reflects the end of crystallization and consequently the end of latent heat release. The presence of residual melt inside the pluton is predicted for the first *c.* 210 kyr after emplacement. Points closer to the pluton bottom undergo faster initial cooling and then, because of their deeper level, lose heat at a slower pace. This is the reason for the cross-cutting of the two cooling curves.

Thermal history of the pluton aureole is shown for four points at increasing distance from the intrusive contact (Fig. 3b). At a distance of 50 m, temperature reaches a maximum of 550 °C after 25 kyr, and is maintained above 500 °C for 100 kyr. Temperature peaks progressively decrease to 490 °C ($t = 35$ kyr), 440 °C ($t = 50$ kyr) and 400 °C ($t = 75$ kyr) at distances from the intrusive contact of 250, 500 and 750 m, respectively.

The P–T path for a point at 50 m from the intrusive contact (Fig. 3c) shows the fast heating of the wall rock followed by a progressively slower cooling which reflects also the concurrent unroofing effect. The P–T path crosses the reaction responsible for the formation of the cordierite-

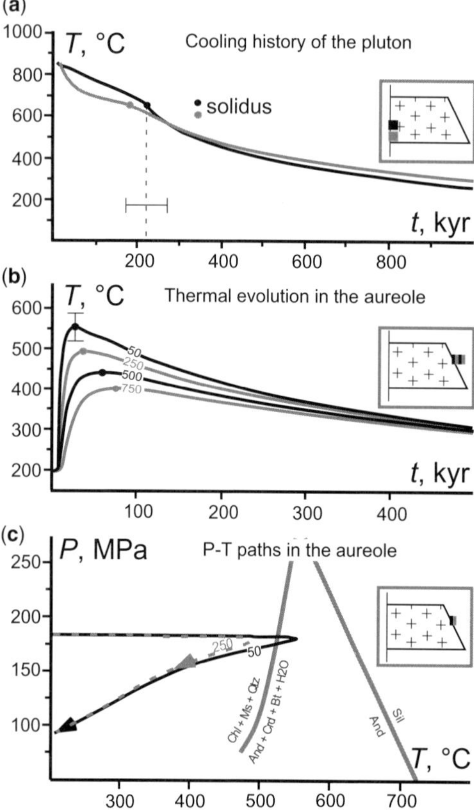

Fig. 3. (**a**) Thermal evolution of two points along the axis of the pluton at distances of 400 m (grey line) and 1200 m (black line) from its bottom. The vertical dashed line and the horizontal bar mark the end of crystallization and relative uncertainties. (**b**) Thermal evolution for rocks in the aureole close to the flank of the pluton at distances of 50, 250, 500 m and 750 m from the intrusive contact. The vertical bar shows the uncertainty in peak temperature for the point closest to the contact. (**c**) Diagram showing *P–T* paths of the two points at a distance of 50 and 250 m from the intrusive contact for 1 myr after magma emplacement. The And–Sil equilibrium (Holdaway 1971) and the And–Crd forming univariant reaction in the K_2O-FeO-MgO-Al_2O_3-SiO_2-H_2O system (see Pattison *et al.* 2002) are shown for reference. Mineral symbols: Chl, chlorite; Ms, muscovite; Qtz, quartz; And, andalusite; Crd, cordierite; Bt, biotite; Sil, sillimanite.

andalusite-biotite assemblage observed in hornfels. Nonetheless, the estimated peak *T* is slightly lower than that inferred for metamorphism in the aureole (Rossetti *et al.* 2007), probably because our model does not consider the contribution of circulating hydrothermal fluids.

Vertical cross-sections through the axis of the pluton are shown in Figure 4 for selected time frames up to 1 myr. The first frame is extended down to a depth of 15 km. Then the bottom becomes progressively shallower, owing to the effect of unroofing.

Magma emplacement produces a perturbation in the thermal structure of the upper and middle crust. After 100 kyr the 400 °C isotherm, originally at *c.* 15 km, has reached a depth of *c.* 4 km. Thermal relaxation then produces a progressive sinking of the isotherm to depths greater than 8 km after 1 myr.

In shallower crustal levels (i.e. above the pluton) heat loss is enhanced by unroofing and by the temperature dependence of thermal conductivity. Unroofing increases thermal gradients, while thermal conductivity is higher at low temperatures. The resulting effect is evident up to 500 kyr, as highlighted by the strongly inhomogeneous isotherm spacing. By 1 myr the isotherms have recovered a more regular pattern, with weak curvature and spacing slightly increasing with depth.

In order to test the sensitivity of the results to model parameters, surface radiogenic heat production, latent heat of crystallization, thermal conductivity and initial magma temperature T_0 have been independently varied. Variations of A_0 of the order of $\pm 10\%$ produce negligible effects on temperature distribution. Variations of $\pm 10\%$ in the value of latent heat and ± 50 °C in T_0 increase or decrease the time required for a complete crystallization of the magma by 20 and 50 kyr, respectively (Fig. 3a). Assuming a constant value of thermal conductivity ($k = 2.0$ W m^{-1} °C^{-1}), the time for total crystallization of the pluton is reduced by about 20 kyr, corresponding to a variation of *c.* 10%. However, as mentioned previously, the introduction of temperature dependence of thermal conductivity is more in line with experimental determinations in both rocks and partially crystallized magmas (Clauser 2009; Bea 2010).

Similar considerations can be applied to the contact aureole: *a* $\pm$ 10% variation of latent heat results in a thermal peak change of *c.* ± 15 °C, while a variation of T_o (± 50 °C) produces a fluctuation in the thermal peak of *c.* ± 35 °C (Fig. 3b). Assuming a constant value of 2.0 W m^{-1} °C^{-1} for the thermal conductivity, the peak temperature increases by *c.* 20 °C. Although the model parameters are subject to unquantifiable uncertainties, the results of the thermal model are within tolerable error margins.

Rheological model

The simplified lithological section adopted for the model is given in Figure 5. It takes into account, although only schematically, the tectonic

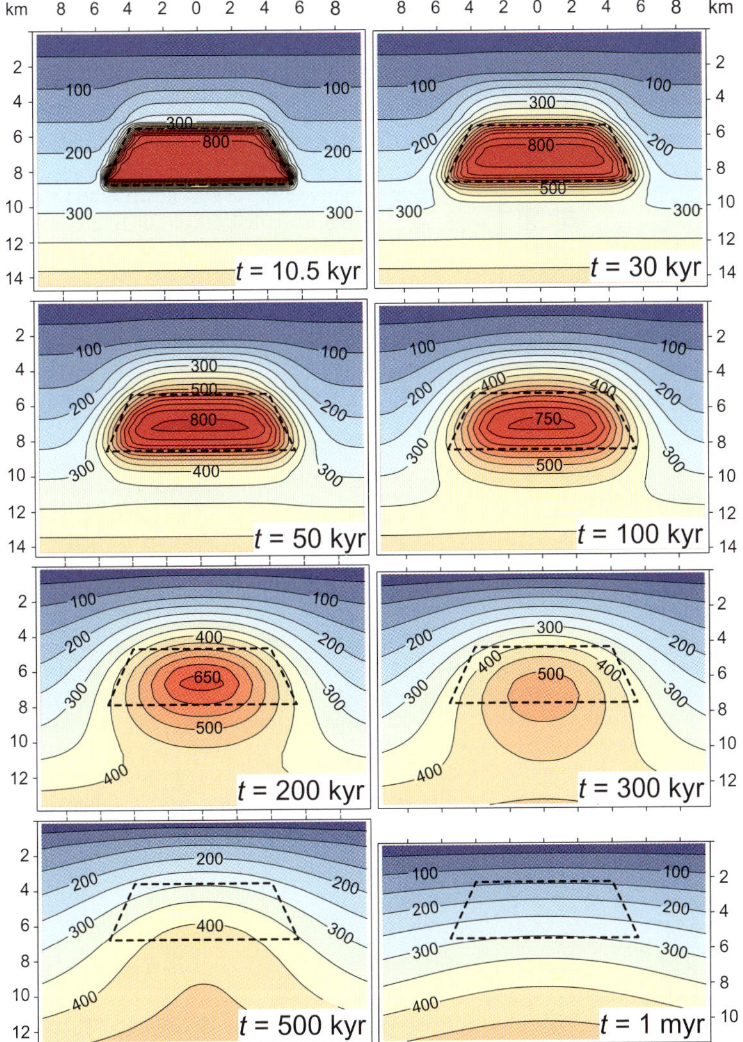

Fig. 4. Temperature (°C) cross-sections at various times during the first 1 myr after magma emplacement. The pluton margins are shown by the black dashed line.

architecture outlined by Trevisan (1953) and confirmed by several authors dealing with the Elba structure.

Two-dimensional strength profiles have been obtained using the frictional criterion for the brittle field and the power-law creep equation for the ductile field. For the cohensionless frictional criterion we use the equation (Sibson 1974):

$$\sigma = \sigma_1 - \sigma_3 = \beta(1 - \lambda)\rho g z \qquad (2)$$

where $\sigma = \sigma_1 - \sigma_3$ is the critical stress difference (compression is positive), ρ is the mean rock density above depth z, g is gravity, β is a dimensionless parameter depending on the frictional coefficient and deformational regime, and λ is the pore fluid factor (ratio of pore fluid pressure to lithostatic pressure).

Assuming a uniform friction coefficient $\mu = 0.75$ for all rocks (Ranalli 1995), with the exception of serpentinite ($\mu = 0.3$, Escartin *et al.* 1997) and an extensional tectonic regime results in $\beta = 0.75$ and 0.45 for other rocks and serpentinite, respectively (see Ranalli 1995). A common density value of 2750 kg m^{-3} was assigned to complexes V, IV and I. Different values ranging between 2650 and 2800 kg m^{-3} according to the

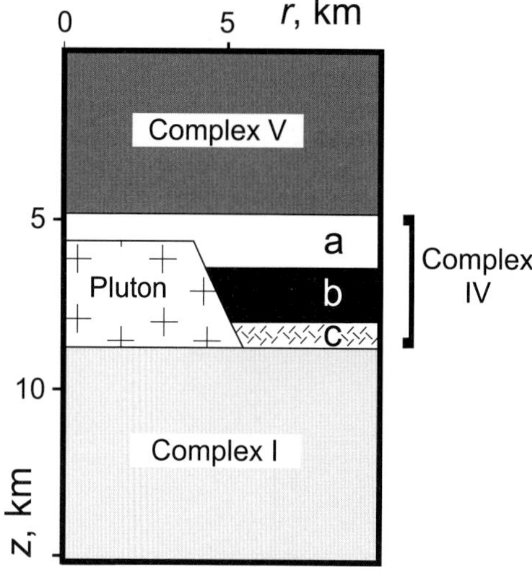

Fig. 5. Simplified geological structure adopted for the symmetrical rheological model. See text for details and compare with Table 2.

margin. As for the pluton itself, its behaviour when melt is present ($T > 650\ °C$) has been considered ductile and equation (2) has not been applied, since the viscosities of melts, even if partially crystallized, are orders of magnitude lower than the corresponding solid rock (see e.g. Dingwell 1995).

The power-law dislocation creep equation is (see e.g. Ranalli 1995):

$$\dot{\varepsilon} = A_c \sigma^n \exp\left(-\frac{E}{RT}\right) \tag{3}$$

where $\dot{\varepsilon}$ is strain rate, σ is stress difference or shear stress, A_c, n and E are creep parameters, R is the gas constant and T is the temperature.

Equation (3) can be expressed in terms of stress as:

$$\sigma = \left(\frac{\dot{\varepsilon}}{A_c}\right)^{\frac{1}{n}} \exp\left(\frac{E}{nRT}\right)$$
$$= \left[A_c^{\frac{-1}{n}} \exp\left(\frac{E}{nRT}\right) \dot{\varepsilon}^{\frac{1-n}{n}}\right] \dot{\varepsilon}. \tag{4}$$

The term in square brackets is the temperature- and strain rate-dependent viscosity of the material. Viscosities of some relevant minerals as a function of temperature, for a constant strain rate of $10^{-14}\ \text{s}^{-1}$, are compared to that of the pluton (when in the solid state) in Figure 6. The creep parameters for the latter were estimated for iso-strain rate conditions from the volume fractions of quartz, K-feldspar and plagioclase, following the procedure

different lithologies of the complexes resulted in negligible variations of critical stress. A depth-dependent relation was adopted for λ (Sibson 1990), with λ linearly increasing from 0.4 at the surface to 0.8 at the base of the crust. Exceptions to this choice are the cells surrounding the pluton, where a value of 0.6 has been used, to account for a possible increase of pore fluid pressure near the

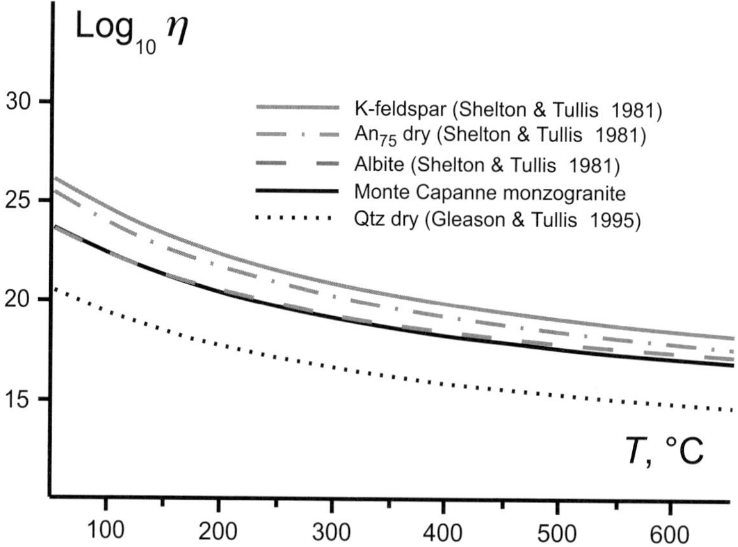

Fig. 6. Temperature dependence of viscosity η (Pa s, strain rate $10^{-14}\ \text{s}^{-1}$) for some relevant minerals and the Monte Capanne monzogranite. Data from Shelton & Tullis (1981) and Gleason & Tullis (1995).

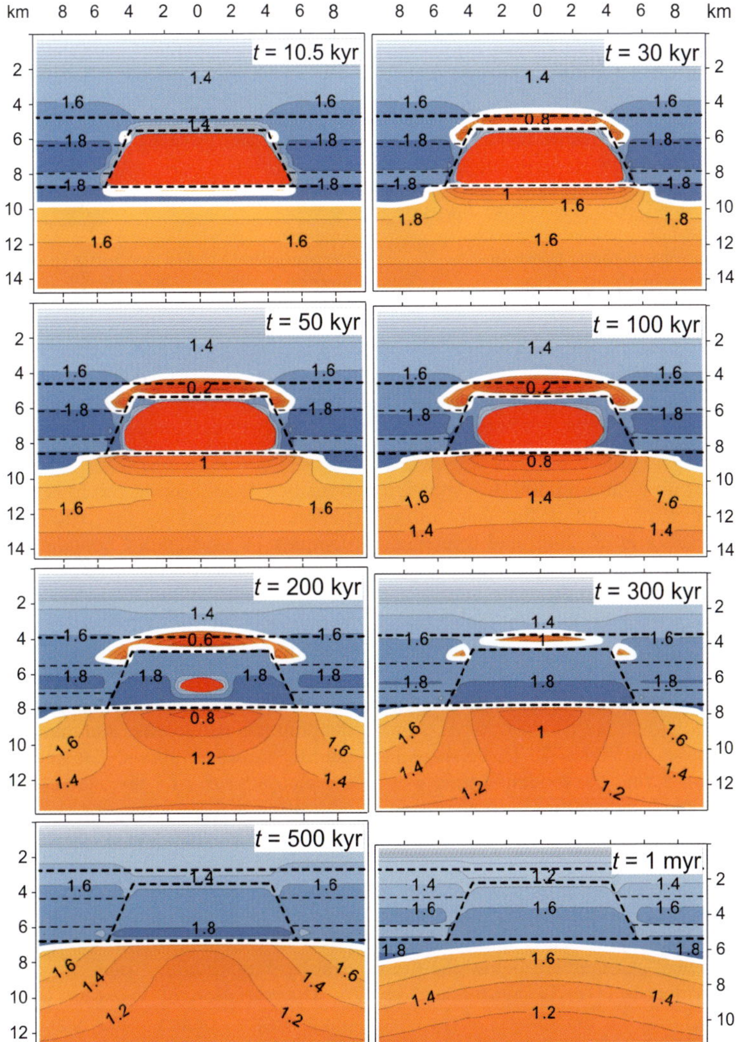

Fig. 7. Rheological cross-sections at selected times after intrusion. Values on the contour lines denote Log$_{10}$ (strength, MPa). Orange shades: ductile domain; blue shades: brittle domain; red: part of the pluton in hypersolidus conditions ($T > 650$ °C). The thick white lines show the locations of brittle − ductile transitions. Black dashed lines indicate margins of the pluton and complexes.

for multiphase rocks suggested by Ji *et al.* (2003). As expected, monzogranite viscosity lies between feldspar and quartz viscosities.

To produce strength-vertical sections (Fig. 7), brittle and ductile strengths (equations 2 & 4) have been compared, using the temperature distribution obtained from the thermal model and a strain rate of 10^{-14} s^{-1}, *c.* 0.3 per million years, corresponding to an extension velocity of *c.* 3 cm a^{-1} distributed over a width of *c.* 100 km. The properties of the main rocks making up complexes V, IV and I have been modelled using the creep parameters listed

in Table 2. The choice is severely limited by the narrow range of minerals and rocks whose creep properties are known. The parameters that have been used reflect as much as possible the rheology of the main constituents of the rocks, and alternate choices (within the limits imposed by the lithology) do not alter the results significantly. The serpentinite making up the basal layer of complex IV has been considered to be brittle, on the basis of experimental results (Raleigh & Paterson 1965; Escartin *et al.* 1997) showing that the main serpentine minerals exhibit brittle behaviour in low-pressure

Table 2. *Rheological parameters related to the simplified lithological scheme of Figure 5*

	Lithology	Rheological analog	A_c [MPa^{-n} s^{-1}]	n	E [J mol^{-1}]
Complex V	Clay-rich sediments, felsic magmatic rocks	Wet quartzite	3.2×10^{-4}	2.3	1.54×10^5
Complex IV	(a) Radiolarites, clay-rich sediements, minor carbonates, felsic intrusions	Wet quartzite	3.2×10^{-4}	2.3	1.54×10^5
	(b) Mafic metamorphic rocks	Diabase	2×10^{-4}	3.4	2.6×10^5
	(c) Serpentinized peridotite	Serpentinite	Always brittle in relevant conditions		
Complex I	Micaschist	Granite	1.8×10^{-9}	3.2	1.23×10^5
Pluton	Monzogranite	Calculated from Qtz, Kfs and Pl contents	1.19×10^{-11}	3.6	2.33×10^5

conditions, at least up to around 500 °C. Although ductile behaviour is possible if the dehydrating rock is drained (Rutter *et al.* 2009 and references therein), in our case a brittle behaviour is still reasonable at higher temperature (550 ± 10% °C) since a continuous fluid flow from the cooling magma contributes to maintaining a high value of pore fluid factor (λ), thus inhibiting ductile deformation of the host serpentinite rocks in the heating stage (Fig. 4).

Results are shown in Figure 7. The change in shape of the pluton during the modelled time window is negligible and has therefore been neglected. A marked evolution of the rheology with time is observed. At 10.5 kyr, the pluton is still almost completely molten, except for a thin shell close to the roof and flanks, where the rheological behaviour is essentially brittle. At this early stage, the only significant rheological changes are below the pluton, where a thin ductile layer is present, and in small areas located near the top corners of the pluton. The regional brittle–ductile transition is still undisturbed, at a depth of *c.* 10 km. The decrease in brittle strength, particularly evident above the pluton, but discernible also in the flanks, is a consequence of pore fluid pressure, assumed higher in proximity of the intrusive contact.

Between 30 and 100 kyr, the pluton continues to crystallize inwards, and a ductile domain develops above it and around the top corners, within the sedimentary cover of complex IV and in the lowermost part of complex V. The progressive crystallization of the pluton is accompanied by an upward deflection of the regional brittle–ductile transition, which touches the base of the pluton.

In the 100–300 kyr interval, the ductile zone above the pluton tends to shrink, while the brittle strength progressively decreases as a consequence of unroofing. The pluton is almost totally crystallized by 200 kyr. The regional brittle-ductile

transition is still along the lower boundary of the pluton, controlled by the pluton–complex I and complex IV–complex I boundaries. This is partly a consequence of the assumed brittle behaviour of serpentinite in the lower layer of complex IV. However, a similar result is obtained if this layer is modelled with the rheology of wet peridotite.

At 500 kyr, the ductile domain capping the pluton disappears. The thermal perturbation below the pluton is still intense and maintains the brittle–ductile transition almost flat and coinciding with the pluton base. At 1 myr, the cooling of the crustal section is more advanced and this produces the deepening of the brittle–ductile transition that acquires a convex-up profile. At the end of the simulation the brittle–ductile transition is shallower than in the initial condition, owing to the effect of unroofing that has reduced crustal thickness by 3400 m after 1 myr.

The variation of viscosity with time at three points in the country rocks near the intrusive contact is shown in Figure 8. After the emplacement of magma, an initial viscosity drop occurs within

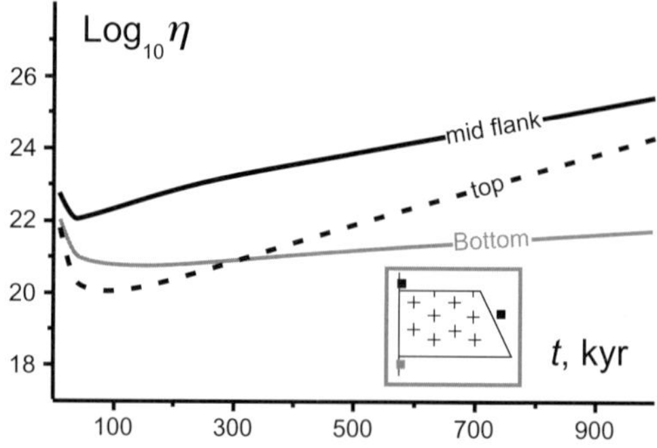

Fig. 8. Time variation of viscosity η (Pa s) at selected points around the pluton.

c. 40 kyr near the flank of the pluton, *c.* 100 kyr at the top and *c.* 200 kyr at the bottom. After reaching the minimum value, viscosity increases with the highest rate at the top and the slowest rate at the bottom. This behaviour is affected both by lithology and by the different thermal evolution in each location, which depends on conduction and unroofing. The viscosity drop is more pronounced where metasediments are present (top and flank), while the increase takes place more slowly where the thermal perturbation is more persistent (bottom).

Tectonic implications

The results of the thermal model are validated, within margins of error, by the metamorphic assemblages occurring in the thermal aureole (Rossetti *et al.* 2007). The results of the rheological model have implications for the regional post-emplacement tectonic evolution, and more in general for the relation between low-angle normal faults and magmatism in regions undergoing extension.

The well-known Andersonian theory of faulting (see e.g. Ranalli 1995) provides a first-order framework for normal faulting, but is unable to account for variations in the fault geometry with depth as commonly observed in low-angle listric faults. Proposed explanations include the change in rheological behaviour between the upper (brittle), and the lower (viscous) crust, resulting in the localization of shearing along the rheological transition (e.g. Lister & Davis 1989), and/or a different orientation of the stress field in the upper crust (Westaway 1999).

Low-angle normal faulting is commonly associated with magmatism at upper crustal levels, particularly in regions affected by a severe post-collisional extension, where metamorphic core complexes developed (Gans *et al.* 1989; Lister & Davis 1989; Hill *et al.* 1995), as is the case of the inner Northern Apennines (Carmignani & Kligfiled 1990; Carmignani *et al.* 1994). Parsons & Thompson (1993) have suggested that the close association in space and time between low-angle normal faults and magmatism is a consequence of inflation of magmas at a rate faster than the regional extension, thus increasing the horizontal stress in the crust and, consequently, altering and rotating the regional stress field. A consequence of this process is the space accommodation of the magma intrusion and its buoyancy effect, both features favouring the curvature of the low-angle shear zones and the pluton unroofing (Brun & Van Den Driessche 1994; Corti *et al.* 2003 and references therein). A similar evolution has been envisaged for the emplacement of the Porto Azzurro pluton and the doming of the Calamita promontory (Fig. 2), in southern

Fig. 9. Examples of rheological transitions in the contact aureole of the Monte Capanne pluton. (**a**) Shallow level felsic dyke predating the Monte Capanne emplacement, with evidence of a subsolidus ductile deformation, shown by a shape-preferred orientation of minerals and by the presence of σ- and δ-type feldspar porphyroclasts. (**b**) Hornfels transected by a network of hydrothermal veins, after transition to brittle behaviour. Similar evolution has been documented by Rossetti *et al.* (2007) for calc-silicates outcropping near the eastern border of the pluton. (**c**) Gabbro in the contact aureole of the Monte Capanne pluton: the rock, affected exclusively by brittle deformation, is cross-cut by two generations of hydrothermal veins.

Elba (Pertusati *et al.* 1993; Garfagnoli *et al.* 2005; Smith *et al.* 2010).

A further link between pluton emplacement and low-angle normal faults and shear zones is suggested by the results of the present thermo-rheological model for the Monte Capanne pluton, where ductile behaviour is predicted in the time interval 30–300 kyr in wall rocks immediately above the intrusion, mechanically controlled by weak minerals such as quartz (Figs 7 & 9a). However, this time interval may be shorter if the contribution to heat loss through hydrothermal-fluid circulation is significant (Maineri *et al.* 2003).

The occurrence of a soft ductile horizon can lead to the localization of shear, evolving into a low-angle fault at shallower depth. Therefore, the unroofing process and the formation of listric low-angle faults can be favoured by transient rheology, and the faults formed this way would root at the local brittle – ductile transition above the pluton, and extend to the regional brittle – ductile transition on the sides. This process is sketched in

Figure 10. A slightly asymmetrical shape of the pluton (Farina *et al.* 2010) may be responsible for the development of the eastern-dipping unroofing normal fault.

In this view, the Monte Capanne fault (Fig. 2) is not considered critical in controlling the emplacement level of the pluton, as hypothesized by Jolivet *et al.* (1998), but alternatively to be a consequence of the mechanical weakening of the heated wall rocks. Thus the Monte Capanne fault was mostly active during the post-emplacement time period, as suggested by Westerman *et al.* (2004). This interpretation is also supported by the lack of any magnetic fabrics indicating a tectonically controlled pluton emplacement (Cifelli *et al.* 2012).

The timing of this suggested sequence fits the regional evidence in Elba. Westerman *et al.* (2004) have shown that the Monte Capanne fault (Fig. 2) post-dated the Orano dykes (6.85 Ma; Dini *et al.* 2008) and preceded the development of the Zuccale normal fault, dated at 5.9 Ma, synchronous

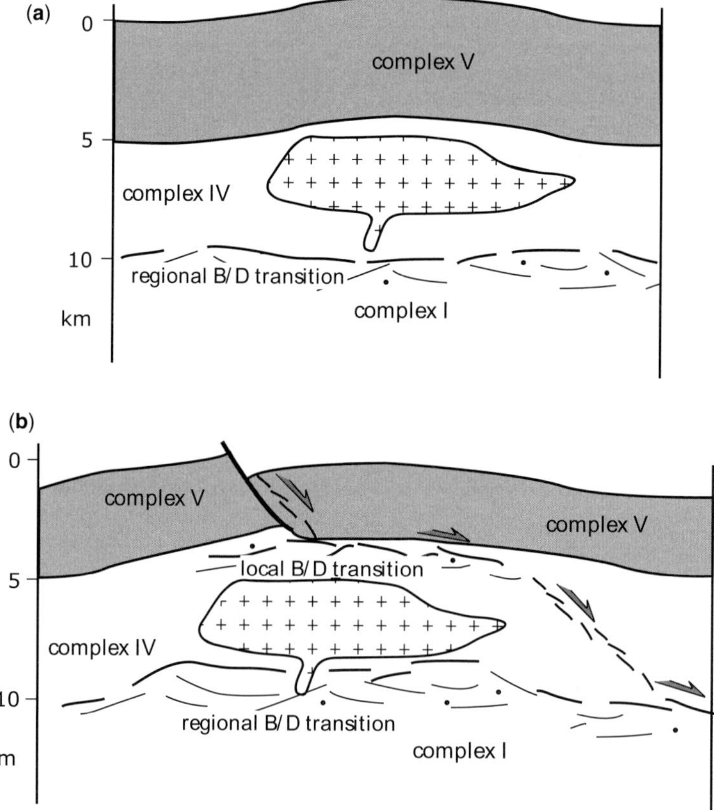

Fig. 10. Rheologically induced extensional unroofing: (**a**) situation at the time of pluton emplacement; (**b**) after *c.* 100 kyr, when the ductile horizon above the pluton favours extensional shearing.

with the Porto Azzurro pluton emplacement. Therefore, the unroofing of the Monte Capanne pluton occurred within 1 myr of pluton emplacement. During further cooling, rocks above the pluton returned into the brittle domain (Fig. 9b) and underwent fracturing and hydrothermal-fluid circulation (Rossetti *et al.* 2007). On the other hand, gabbroic rocks remained in the brittle domain during cooling (Fig. 9c).

Conclusions

The main results emerging from the present model can be summarized as follows:

(1) The Monte Capanne pluton became totally crystallized by *c.* 210 kyr after emplacement (probable error margins ± 10–20%). The thermal perturbation in the surrounding rocks generated by the intrusion became negligible after 1 myr.

(2) At a distance of 50 m from the pluton, a thermal peak of *c.* 550 °C (± 5–10%) was reached in the aureole after *c.* 25 kyr, in almost isobaric conditions (pressure *c.* 180 MPa). Here temperature remained higher than 500 °C for *c.* 100 kyr and is compatible with the observed cordierite–andalusite–biotite assemblage in spotted schist and hornfels.

(3) As the Monte Capanne monzogranite cooled and crystallized, a relatively fast transition (≤ 210 kyr) occurred from melt-present conditions to brittle behaviour. This situation prevented the formation of significant ductile deformation in the pluton, in accordance with field studies (Westerman *et al.* 2004; Farina *et al.* 2010).

(4) A rather long-lived (*c.* 300 kyr) soft ductile cap is predicted above the pluton, where weaker sedimentary rocks were originally present. This rheological feature was probably critical, both in space and time, in the unroofing of the pluton and the formation of large listric normal faults.

(5) The brittle – ductile transition evolves in time, both locally and regionally, since it is controlled by both lithology and temperature. While strongly affected by the former, it does not always coincide with lithological boundaries.

The final message of this paper is to underline that heat transfer to the upper crust through magmatic advection and subsequent conduction produces a time-dependent rheology which implies migration of the brittle – ductile transition and transient formation of soft ductile layers. The latter can favour the formation of shear zones and listric normal faults rooted in the brittle – ductile transition. In turn, the presence of these faults and shear zones will render the extensional tectonics more effective, producing rapid exhumation of the pluton and surrounding rocks. As a consequence of this heat transfer towards the surface, significant hydrothermal circulation and mineral deposition in fractured host rocks can be favoured. While we have used the Monte Capanne pluton as a natural laboratory to constrain our model, we think that this type of rheological and tectonic history could be applicable to other regions showing a similar tectonomagmatic evolution.

We are grateful to S. Llana Funez, J. Escuder-Viruete, M Fernández and an anonymous reviewer for the constructive comments that helped to improve the manuscript. G. Ranalli's research is supported by a grant from NSERC (Natural Sciences and Engineering Research Council of Canada). Financial support for the research of A. Caggianelli and D. Liotta comes from Bari University funds (Fondi di Ateneo).

References

ANNEN, C. 2011. Implications of incremental emplacement of magma bodies for magma differentiation, thermal aureole dimensions and plutonism-volcanism relationships. *Tectonophysics*, **500**, 3–10.

BEA, F. 2010. Crystallization dynamics of granite magma chambers in the absence of regional stress: multiphysics modeling with natural examples. *Journal of Petrology*, **51**, 1541–1569.

BENVENUTI, M., COSTAGLIOLA, P., DINI, A., LATTANZI, P., RUGGIERI, G., VASELLI, O. & TANELLI, G. 2006. Evolution of the hydrothermal system at La Crocetta 'feldspar mine': fluid inclusion and stable isotope constraints on the environment of late stage veins. *Periodico di Mineralogia*, **75**, 39–50.

BOUILLIN, J. P., BOUCHEZ, J. L., LESPINASSE, P. & PECHER, A. 1993. Granite emplacement in an extensional setting; an AMS study of the magmatic structures of Monte Capanne (Elba, Italy). *Earth and Planetary Science Letters*, **118**, 263–279.

BOUILLIN, J. P., POUPEAU, G. & SABIL, N. 1994. Etude thermo-chronologique de la dénudation du plutone du Monte Capanne (île d'Elbe, Italie) par les traces de fission. *Bulletin de la Société Géologique de France*, **165**, 19–25.

BOWN, J. & WHITE, R. 1995. Effect of finite extension rate on melt generation at rifted continental margins. *Journal of Geophysical Research*, **100**, 18 011–18 029.

BROGI, A. & LIOTTA, D. 2008. Highly extended terrains, lateral segmentation of the substratum, and basin development: the middle-late Miocene Radicondoli Basin (inner northern Apennines, Italy). *Tectonics*, **27**, TC5002, 20.

BRUN, J.-P. & VAN DEN DRIESSCHE, J. 1994. Extensional gneiss domes and detachment fault systems: structure and kinematics. *Bulletin de la Societé Geologique de France*, **165**, 519–530.

BUCK, W. R. 2006. The role of magma in the development of the Afro-Arabian Rift System. *In*: YIRGU, G.,

EBINGER, C. J. & MAGUIRE, P. K. H. (eds) *The Afar Volcanic Province within the East African Rift System*. Geological Society, London, Special Publications, **259**, 43–54.

CALCAGNILE, G. & PANZA, G. F. 1980. The main characteristics of the lithosphere-asthenosphere system in Italy and surrounding regions. *Pure and Applied Geophysics*, **119**, 865–879.

CARMIGNANI, L. & KLIGFIELD, R. 1990. Crustal extension in the northern Apennines: the transition from compression to extension in the Alpi Apuane core complex. *Tectonics*, **9**, 1275–1303.

CARMIGNANI, L., DECANDIA, F. A., FANTOZZI, P. L., LAZZAROTTO, A., LIOTTA, D. & MECCHERI, M. 1994. Tertiary extensional tectonics in Tuscany (Northern Apennines, Italy). *Tectonophysics*, **238**, 295–315.

CASSINIS, R., SCARASCIA, S. & LOZEJ, A. 2003. The deep crustal structure of Italy and surrounding areas from seismic refraction data. A new synthesis. *Bollettino Società Geologica Italiana*, **122**, 365–376.

CIFELLI, F., MINELLI, L., ROSSETTI, F., URRU, G. & MATTEI, M. 2012. The emplacement of the Late Miocene Monte Capanne intrusion (Elba Island, Central Italy): constraints from magnetic fabric analyses. *International Journal of Earth Sciences*, **101**, 787–802.

CLAUSER, C. 2009. Heat transport processes in the Earth's crust. *Surveys in Geophysics*, **30**, 163–191.

CLAUSER, C. & HUENGES, E. 1995. Thermal conductivity of rocks and minerals. *In*: AHRENS, T. J. (ed.) *Rock Physics and Phase Relations. A Handbook of Physical Constants*. AGU Reference Shelf 3, American Geophysical Union, Washington DC, 105–126.

CORTI, G., BONINI, M., CONTICELLI, S., INNOCENTI, F., MANETTI, P. & SOKOUTIS, D. 2003. Analogue modelling of continental extension: a review focused on the relations between the patterns of deformation and the presence of magma. *Earth Science Reviews*, **63**, 169–247.

DANIEL, J. M. & JOLIVET, L. 1995. Detachment faults and pluton emplacement: Elba Island (Tyrrhenian Sea). *Bulletin de la Société Géologique de France*, **166**, 341–354.

DINGWELL, D. B. 1995. Viscosity and anelasticity of melts. *In*: AHRENS, T. J. (ed.) *Mineral Physics & Crystallography A Handbook of Physical Constants*. AGU Reference Shelf 2, American Geophysical Union, Washington DC, 209–217.

DINI, A., WESTERMAN, D. S., INNOCENTI, F. & ROCCHI, S. 2008. Magma emplacement in a transfer zone: the Miocene mafic Orano dyke swarm of Elba Island, Tuscany, Italy. *In*: THOMSON, K. & PETFORD, N. (eds) *Structure and Emplacement of High-Level Magmatic Systems*. Geological Society, London, Special Publications, **302**, 131–148.

ESCARTIN, J., HIRTH, G. & EVANS, B. 1997. Effects of serpentinization on the lithospheric strength and the style of normal faulting at slow-spreading ridges. *Earth and Planetary Science Letters*, **151**, 181–189.

FARINA, F., DINI, A., INNOCENTI, F., ROCCHI, S. & WESTERMAN, D. S. 2010. Rapid incremental assembly of the Monte Capanne Pluton (Elba Island, Tuscany) by downward stacking of magma sheets. *Geological Society of America Bulletin*, **122**, 1463–1479.

GAGNEVIN, D., DALY, J. S., POLI, G. & MORGAN, D. 2005. Microchemical and Sr isotopic investigation of zoned K-feldspar megacrysts: insights into the petrogenesis of a granitic system and disequilibrium crystal growth. *Journal of Petrology*, **46**, 1689–1724.

GANS, P. B., MAHOOD, G. A. & SCHERMER, E. 1989. Synextensional magmatism in the Basin and Range Province: a case study from the eastern Great Basin. *Geological Society of America Special Paper*, **233**, 53.

GARFAGNOLI, F., MENNA, F., PANDELI, E. & PRINCIPI, G. 2005. The Porto Azzurro Unit (Mt. Calamita promontory, south-eastern Elba Island, Tuscany): stratigraphic, tectonic and metamorphic evolution. *Bollettino della Società Geologica Italiana*, **3**, 119–138.

GERYA, T. V. 2010. *Introduction to Numerical Geodynamic Modelling*. Cambridge University Press, New York.

GLEASON, G. C. & TULLIS, J. 1995. A flow law for dislocation creep of quartz aggregates determined with the molten salt cell. *Tectonophysics*, **247**, 1–23.

HILL, E. J., BALDWIN, S. L. & LISTER, G. S. 1995. Magmatism as an essential driving force for formation of active metamorphic core complexes in eastern Papua New Guinea. *Journal of Geophysical Research*, **100**, 10 441–10 451.

HOLDAWAY, M. J. 1971. Stability of andalusite and the aluminosilicate phase diagram. *American Journal of Science*, **271**, 97–131.

JI, S., ZHAO, P. & XIA, B. 2003. Flow laws of multiphase materials and rocks from end-member flow laws. *Tectonophysics*, **370**, 129–145.

JOLIVET, L., DANIEL, J. M., TRUFFET, C. & GOFFE', B. 1994. Exhumation of deep crustal metamorphic rocks and crustal extension in arc and back-arc regions. *Lithos*, **33**, 3–30.

JOLIVET, L., FACCENNA, C. *ET AL.* 1998. Midcrustal shear zone in postorogenic extension: example from the northern Tyrrhenian Sea. *Journal of Geophysical Research*, **103**, 12 123–12 160.

KELLER, J. V. A. & PIALLI, G. 1990. Tectonics of the Island of Elba: a reappraisal. *Bollettino Società Geologica Italiana*, **109**, 413–425.

LISTER, G. S. & BALDWIN, S. L. 1993. Plutonism and the origin of metamorphic core complexes. *Geology*, **21**, 607–610.

LISTER, G. S. & DAVIS, G. A. 1989. The origin of metamorphic core complexes and detachment faults formed during Tertiary continental extension in the northern Colorado River region, USA. *Journal of Structural Geology*, **11**, 65–93.

MAINERI, C., BENVENUTI, M., COSTAGLIOLA, P., DINI, A., LATTANZI, P., RUGGIERI, G. & VILLA, I. M. 2003. Sericitic alteration at the La Crocetta deposit (Elba Island, Italy): interplay between magmatism, tectonics and hydrothermal activity. *Mineralium Deposita*, **1**, 67–86.

MCKENZIE, D. & BICKLE, M. J. 1988. The volume and composition of melt generated by extension of the lithosphere. *Journal of Petrology*, **29**, 342–532.

MÜNTENER, O. & MANATSCHAL, G. 2006. High degrees of melt extraction recorded by spinel harzburgite of the Newfoundland margin: the role of inheritance and consequences for the evolution of the southern North

Atlantic. *Earth and Planetary Science Letters*, **252**, 437–452.

PARSONS, T. & THOMPSON, G. A. 1993. Does magmatism influence low-angle normal faulting? *Geology*, **21**, 247–250.

PATTISON, D. R. M., SPEAR, F. S., DEBUHR, C. R., CHENEY, J. T. & GUIDOTTI, C. V. 2002. Thermodynamic modeling of the reaction muscovite + cordierite → Al2SiO5 + biotite + quartz + H2O: constraints from natural assemblages and implications for the metapelitic petrogenetic grid. *Journal of Metamorphic Geology*, **20**, 99–118.

PERTUSATI, P. C., RAGGI, G., RICCI, C. A., DURANTI, S. & PALMIERI, R. 1993. Evoluzione post-collisionale dell'Elba centro-orientale. *Mineralogical Magazine*, **72**, 925–940.

PICCARDO, G. B., RANALLI, G. & GUARNIERI, L. 2010. Seismogenic shear zones in the lithospheric mantle: ultramafic pseudotachylytes in the Lanzo Peridotite (Western Alps, NW Italy). *Journal of Petrology*, **51**, 81–100.

RALEIGH, C. B. & PATERSON, M. B. 1965. Experimental deformation of serpentinite and its tectonic implications. *Journal of Geophysical Research*, **70**, 3965–3985.

RANALLI, G. 1995. *Rheology of the Earth*. Chapman & Hall, London.

ROSENBAUM, G., WEINBERG, R. F. & REGENAUER-LIEB, K. 2008. The geodynamics of lithospheric extension. *Tectonophysics*, **458**, 1–8.

ROSSETTI, F., TECCE, F., BILLI, A. & BRILLI, M. 2007. Patterns of fluid flow in the contact aureole of the Late Miocene Monte Capanne pluton (Elba Island, Italy): the role of structure and rheology. *Contributions to Mineralogy and Petrology*, **153**, 743–760.

RUTTER, E. H., LLANA FÙNEZ, S. & BRODIE, K. H. 2009. Dehydration and deformation of intact cylinder of serpentinite. *Journal of Structural Geology*, **31**, 29–43.

SERRI, G., INNOCENTI, F. & MANETTI, P. 1993. Geochemical and petrological evidence of the subduction of delaminated Adriatic continental lithosphere in the genesis of the Neogene–Quaternary magmatism of central Italy. *Tectonophysics*, **223**, 117–147.

SHELTON, G. & TULLIS, J. 1981. Experimental flow laws for crustal rocks. *Transactions of the American Geophysical Union*, **62**, 396.

SIBSON, R. H. 1974. Frictional constraints on thrust, wrench and normal faults. *Nature*, **249**, 542–544.

SIBSON, R. H. 1990. Conditions for fault-valve behavior. *In*: KNIPE, R. J. & RUTTER, E. H. (eds) *Deformation Mechanisms, Rheology and Tectonics*. Geological Society, London, Special Publications, **54**, 15–28.

SMITH, S. A. F., HOLDSWORTH, R. E. & COLLETTINI, C. 2010. Interactions between low-angle normal faults and plutonism in the upper crust: insights from the Island of Elba, Italy. *Geological Society of America Bulletin*, **123**, 329, http://dx.doi.org/10.1130/B30200.1

SPEAR, F. S. 1993. *Metamorphic Phase Equilibria and Pressure-Temperature-Time Paths*. Mineralogical Society of America, Washington DC.

TREVISAN, L. 1953. La 55a riunione estiva della Società Geologica Italiana. Isola d'Elba.18–23 Settembre 1951. *Bollettino della Società Geologica Italiana*, **70**, 434–472.

VILÀ, M., FERNÀNDEZ, M. & JIMÈNEZ-MUNT, I. 2010. Radiogenic heat production variability of some common lithological groups and its significance to lithospheric thermal modelling. *Tectonophysics*, **490**, 152–164.

VOSTEEN, H. D. & SCHELLSCHMIDT, R. 2003. Influence of temperature on thermal conductivity, thermal capacity and thermal diffusivity for different types of rock. *Physics and Chemistry of the Earth*, **28**, 499–509.

WESTAWAY, R. 1999. The mechanical feasibility of low-angle normal faulting. *Tectonophysics*, **308**, 407–443.

WESTERMAN, D. S., DINI, A., INNOCENTI, F. & ROCCHI, S. 2003. When and where did hybridization occur? The case of the Monte Capanne pluton, Italy. *Atlantic Geology*, **39**, 147–162.

WESTERMAN, D. S., DINI, A., INNOCENTI, F. & ROCCHI, S. 2004. Rise and fall of a nested Christmas-tree laccolith complex, Elba Island, Italy. *In*: PETFORD, N. & BREITKREUTZ, C. (eds) *Physical Geology of High-Level Magmatic systems*. Geological Society, London, Special Publications, **234**, 195–213.

ZOTH, G. & HÄNEL, R. 1988. *Appendix. In*: HÄNEL, R., RYBACH, L. & STEGENA, L. (eds) *Handbook of Terrestrial Heat Flow Density Determination*. Kluwer, Dordrecht, 449–466.

Using titanium-in-quartz geothermometry and geospeedometry to recover temperatures in the aureole of the Ballachulish Igneous Complex, NW Scotland

D. J. MORGAN[1], M. C. JOLLANDS[1,2], G. E. LLOYD[1]* & D. A. BANKS[1]

[1]*School of Earth and Environment, University of Leeds, Leeds LS2 9JT, UK*

[2]*Research School of Earth Sciences, The Australian National University, Canberra ACT 0200, Australia*

**Corresponding author (e-mail: G.E.Lloyd@leeds.ac.uk)*

Abstract: 'Titanium-in-Quartz' geothermometry suggests quartzites could yield reliable temperature estimates. We here apply four calibrations of the titanium-in-quartz geothermometer to contact-metamorphosed quartzites surrounding the Ballachulish Igneous Complex, Scotland. Two agree broadly with thermal modelling and pre-existing geothermometry; two give temperatures consistently too low. As reported in earlier studies, the technique suffers from difficulties in analysing low titanium (Ti) levels with high spatial and analytical precision. However, this study finds that the critical problem is one of Ti heterogeneity, which poses difficulties in constraining the chemical activity of Ti during quartz growth under metamorphic conditions. Scanning electron microscope-cathodoluminescence (SEM-CL) textures support an interpretation of extensive Ti disequilibrium despite the presence of rutile, indicating dynamic interplay between grain boundary diffusion, fluid/melt percolation and grain growth. The strong zonation suggests a possible geothermometer based on apparent volume diffusion of Ti-in-quartz to derive grain growth histories. Analysis of rutile–quartz interaction implies peak contact temperatures of 645 ± 12 °C, precise but reliant on external estimates of cooling rate from thermal models. Our conclusions support caution in applying Ti-in-quartz geothermometry in aureole settings. However, rutile–quartz juxtaposition prior to heating to >600 °C defines a Ti diffusion couple, employable as a thermometer if cooling rates are constrained by other means.

Quartz, whilst common in crustal rocks, is generally of limited use in temperature determinations due to its restricted chemistry. However, the ability to derive accurate temperature estimates in essentially pure quartzites would assist many geological investigations (e.g. Law *et al.* 2004; Thigpen *et al.* 2010*a*, *b*, 2013). Attempts to recover temperature information have used diverse methods, such as: aluminium in quartz (Dennen *et al.* 1970; Götze *et al.* 2001, 2004), 'fractal' geometry (Kruhl & Nega 1996; Wu *et al.* 2006; Mamtani & Greiling 2010), thermoluminescence (Sawakuchi *et al.* 2011), *c*-axis opening angles (e.g. Kruhl 1998; Law *et al.* 2004, 2010; Thigpen *et al.* 2010*a*, *b*, 2013), recrystallization mechanism (Stipp *et al.* 2002*a*, *b*) and, importantly for this study, titanium (Ti) abundance in quartz (Wark & Watson 2006; Grujic *et al.* 2011). However, these studies have had mixed and variable success.

Titanium-in-quartz geothermometry

Wark & Watson (2006) suggested the original Ti-in-quartz ('TitaniQ') geothermometer based on empirical measurements of Ti incorporated into synthetic quartz formed between 600 and 1000 °C at 1 GPa in the presence of rutile. The presence of rutile ensures that the system is saturated with respect to TiO_2; that is, the chemical activity of Ti can be assumed to be unity (i.e. a_{TiO_2}). Under such conditions, Wark & Watson (2006) proposed that the following relationship holds,

$$T = \frac{-3765}{\log_{10}(X_{Ti}^{Qtz}) - 5.69} - 273, \qquad (1)$$

where T is temperature in °C and X_{Ti}^{Qtz} is the concentration of Ti in the quartz lattice in parts per million by mass (ppm).

This calibration was devised and used originally for temperature determinations in hydrothermal quartz (Wark & Watson 2006; Rusk *et al.* 2008; Müller *et al.* 2010). However, it has been applied also to ascertain temperatures in granulites (Sato & Santosh 2007), rhyolites (Vazquez *et al.* 2009; Wark *et al.* 2007), migmatites (Spear & Wark 2009; Storm & Spear 2009), mylonites (Kohn & Northrup 2009; Grujic *et al.* 2011; Korchinski *et al.*

From: LLANA-FÚNEZ, S., MARCOS, A. & BASTIDA, F. (eds) 2014. *Deformation Structures and Processes within the Continental Crust*. Geological Society, London, Special Publications, **394**, 145–165.
First published online November 25, 2013, http://dx.doi.org/10.1144/SP394.8

2012), porphyry quartz veins (Müller *et al.* 2010) and granites (Wiebe *et al.* 2007).

The original geothermometer of Wark & Watson (2006) (equation 1) was later refined to include a pressure coefficient (*P*) by Thomas *et al.* (2010),

$$T = \frac{(60952 \pm 31222) + (1741 \pm 63)P}{((1.52 \pm 0.04) - R \ln X_{TiO_2}^{Qtz} + R \ln a_{TiO_2})} - 273, \tag{2}$$

where *P* is measured in kilobars, a_{TiO_2} is the activity of rutile (assumed to be unity if rutile is present), $X_{TiO_2}^{Qtz}$ is the mole fraction of the TiO_2 component in the crystal and *R* is the gas constant. This refinement reflects the results of further experiments by Thomas *et al.* (2010), which indicated a significant pressure dependence of Ti incorporation into quartz. In principle, the refinement allows Ti-in-quartz thermometers to yield reliable temperature determinations for samples from any known pressures.

In a parallel work, Kawasaki & Osanai (2008) proposed another calibration for use in granulitic samples based upon natural samples from Antarctica. In their calibration,

$$T = \left(\frac{-5895}{\ln X_{TiO_2}^{Qtz} + 1.729}\right) - 273, \tag{3}$$

as in Thomas *et al.* (2010), $X_{TiO_2}^{Qtz}$ denotes mole fraction TiO_2.

In the course of evaluating the validity of the Thomas *et al.* (2010) calibration (equation 2), Wilson *et al.* (2012) considered the difficulties of determining and reconciling the main parameters (i.e. *T*, a_{TiO_2} and *P*) within the model. In addition, Huang & Audétat (2012) report that Ti incorporation into quartz in hydrothermal settings has strong growth-rate dependence, such that in their view, the calibration of Thomas *et al.* (2010) strongly overestimates pressure. They proposed therefore a recalibration,

$$T = \frac{-0.27943 \times 10^4 - 660.53 P^{0.35}}{\log_{10} Ti_{ppm} - 5.6459} - 273. \tag{4}$$

In addition to direct geothermometry methods, it may be possible to also employ the diffusion of Ti within quartz to investigate temperature–time processes operating in the aureole. The diffusion coefficient for this was determined by Cherniak *et al.* (2007) and, like all diffusion coefficients, is strongly temperature-dependent. Previous attempts to use diffusion–partitioning relationships to establish temperature–time relationships have met with some success (e.g. Lasaga & Jiang 1995), as have methods based on differential diffusion (Morgan & Blake 2006; Reiners 2009).

The present study compares the outputs of the four calibrations (equations 1–4), by application to a suite of thermally well-constrained contact metamorphosed rocks from the Ballachulish Igneous Complex (BIC), NW Scotland. We also evaluate the possible use of kinetic methods in temperature constraint.

Geological setting

Regional geology

The BIC is located near Fort William, NW Scotland (Fig. 1). It lies within the Grampian Highlands geological province, adjacent to the Great Glen Fault, which forms its northernmost limit (Voll 1991). The BIC consists of granite, monzodiorite and quartz diorite, intruded as an approximately concentric cylindrical body, with granite in the centre and quartz diorite at the periphery (Buntebarth 1991). The BIC is related to the waning stage of the Ordovician–Devonian Caledonian Orogeny (Voll 1991), with igneous activity post-dating regional metamorphism (Bailey & Maufe 1960). Early attempts at dating the BIC gave 412 ± 28 Ma (Weiss & Troll 1991). This date was later confirmed and further constrained by $^{207}Pb/^{206}Pb$ and $^{206}Pb/^{238}U$ dating, which yielded 427 ± 1 Ma and 423 ± 0.3 Ma respectively (Fraser *et al.* 2004; via aliquots and single grain analyses of zircon and monazite).

The BIC (Fig. 1) intruded at a depth of around 10 km into a range of Dalradian supergroup metasediments (Voll 1991), including most importantly for the purposes of this study the Appin Quartzite, part of the Ballachulish Subgroup of the Lower Dalradian Appin Group (Pattison & Voll 1991). The Appin Quartzite comprises *c.* 85% quartz, with the remaining *c.* 15% consisting of predominantly alkaline feldspars and accessory apatite, tourmaline, zircon, titanomagnetite, monazite and, significantly, rutile (Buntebarth & Voll 1991). The quartzite has a bimodal grain-size distribution (Hoernes & Voll 1991), with cross-bedding and graded bedding common (Pattison & Voll 1991). The depositional palaeoenvironment is suggested to have been a NE-prograding tidal delta formed in a tectonically stable environment (Voll 1991; Lind 1996).

Aureole thermal conditions

Thermal modelling of the aureole to the BIC (Buntebarth 1991), together with geothermometry data

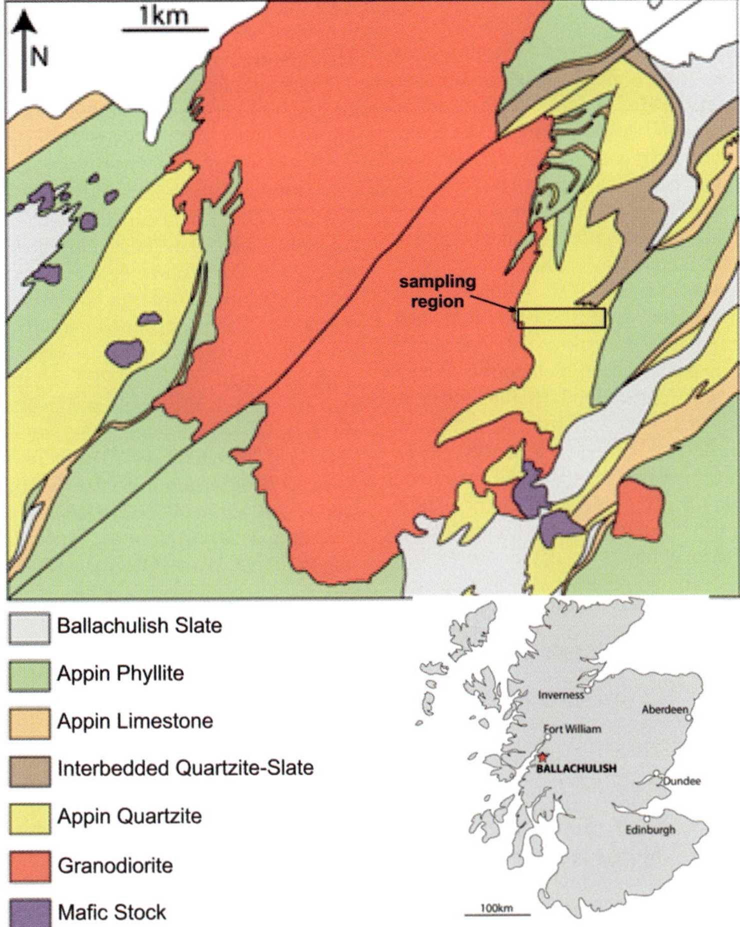

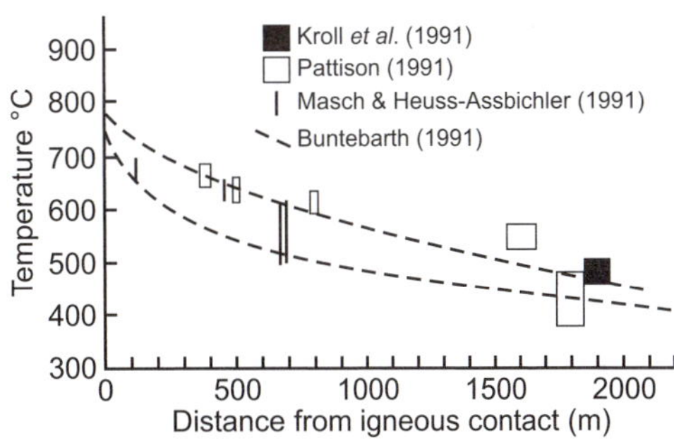

Fig. 1. Simplified geological map of the Ballachulish Igneous Complex (BIC), NW Scotland, showing the region of Appin Quartzite sampled (inset, location map).

(Kroll *et al.* 1991; Masch & Heuss-Assbichler 1991; Pattison 1991), show an expected exponential decrease in peak metamorphic temperature progressing away from the igneous contact (Fig. 2). A depth of intrusion of *c.* 10 km (3.0 $\pm$ 0.5 kilobars, Pattison & Harte 1988; Pattison 1989; from metapelitic phase assemblages) suggests the country rocks were at 250–300 °C pre-emplacement, assuming a normal geothermal gradient. The maximum temperatures attained within the igneous body were *c.* 1100 °C (Weiss & Troll 1989), leading to contact metamorphism of the country rock at temperatures of up to 850 °C (Lind 1996). However, metamorphic temperature estimates vary depending on lithology, location on the contact and the geometry of predictive thermal models (see Pattison & Harte 1997; Bergman & Piazolo 2012). Contact metamorphism is thought to have lasted for around 0.5 myr, including *c.* 0.27 myr

during which conditions were hot enough to cause partial melting in the innermost aureole (Pattison & Harte 1997).

Fig. 2. Thermal profile in the contact aureole of the BIC constrained by modelling and geothermometry.

Sampling

The samples of Appin Quartzite used in this study were collected by Lind (1996). Their locations on the eastern side of the BIC, relative to the vertical igneous contact, are shown in Figure 1. The temperatures of the aureole (Fig. 2) have been modelled previously as a function of distance from the igneous contact to an uncertainty within a factor of four, dependent on model geometry and boundary conditions (e.g. Buntebarth 1991; Kroll *et al.* 1991; Masch & Heuss-Assbichler 1991; Pattison 1991; summarized in Pattison & Harte 1997; and later revisited by Bergman & Piazolo 2012).The temperature profile in the adjacent metapelitic assemblages is very well constrained (many zones to ± 10 °C and at worst ± 25 °C) based on extensive study (Pattison 1991; Pattison & Harte 1997). Thus, the sample suite provides a well-constrained target for a comparative geothermometric study. By measuring Ti concentrations in quartz within each sample, we can compare results for the various Ti-in-quartz calibrations (equations 1–4) to assess their validity against external results.

Methodology

Polished and carbon-coated (e.g. Lloyd 1987) blocks of Appin Quartzite were imaged using backscattered electron (BSE) atomic number (Z) contrast imaging on a FEI Quanta 650 scanning electron microscope (SEM). Areas consisting of quartz with rutile were identified in BSE images via their large Z-contrast difference (e.g. Fig. 3). Major and minor phases present were identified using an attached Oxford X-max 80 energy-dispersive X-ray spectrometer (EDS). Samples were also imaged simultaneously via cathodoluminescence (CL) using a KE Centaurus panchromatic CL detector (e.g. Fig. 4).

The rims of quartz grains from four samples showing high CL (56183, 56167, 56158 and 56148; Fig. 4) were analysed subsequently via laser ablation-inductively coupled plasma-mass spectrometry (LA-ICP-MS) using a Geolas Q optical delivery system with Lambda Physik 193 nm ArF Excimer laser coupled to an Agilent 7500c ICP-MS. A spot size of 100 μm and beam energy of 15 J cm^{-2} at the sample surface were necessary to derive measurable quantities of Ti whilst maintaining acceptable spatial resolution. A laser pulse repetition rate of 5 Hz for 200 pulses was used following 20–30 s of background measurement. Data were processed using the SILLS software package (Guillong *et al.* 2007) and the NIST SRM610 glass standard (Pearce *et al.* 1997) for calibration. As well as Ti and Si, Na, K, Ca and Al concentrations

were monitored to ensure quartz ablation rather than that of another mineral (e.g. feldspar). The detection limit of Ti in the quartz was 0.15 ppm.

Four samples (56169, 56183, 56148 and 56158; three of which were analysed via ICP-MS) were analysed using a JEOL JXA-8230 electron microprobe. Analytical parameters were set according to the 'optimum' methodology reported by Storm & Spear (2009): 5μm beam diameter, 15 kV accelerating voltage and 50 nA beam current. Counting times were set as 200 s on-peak and 100 s at each background position above and below the peak. Ti concentration was determined simultaneously using three wavelength dispersive spectrometers (two employed high intensity PET crystals and one a high intensity LiF crystal) and averaged using the JEOL ZAF software package, reporting concentrations in increments of 10 ppm TiO$_2$ (10^{-3} wt%). A fixed concentration of 100% silica was assumed. The 6-sigma detection limit of the instrument is 22 ppm Ti, equivalent to 35 ppm TiO$_2$, comparable to that of Wark & Watson (2006).

Results

Scanning electron microscopy

Six samples were imaged using SEM BSE Z-contrast (Fig. 3). These were selected due to their distance from the igneous contact (Table 1) and relative purity of quartzite. For the purpose of this study, the positive identification of rutile (TiO$_2$) among and within the quartz is important, suggesting that the chemical activity of Ti (a_{TiO_2}) should be unity if the grains remained in communication. This simplifies the use of the Ti-in-quartz geothermometer as corrections for activity need not be made (Wark & Watson 2006). We will return to the question of the validity of this assumption later in this contribution.

Cathodoluminescence

Following exploratory SEM BSE Z-contrast analysis (Fig. 3), quartz-rich areas within the six samples were imaged using SEM-CL. This approach revealed a complex variety of intra- and intergranular textures in the quartz (e.g. Fig. 4). Similar textures were reported from Appin Quartzites within the Ballachulish aureole by Lind (1996) and subsequently a refined classification based on interpreted origins was suggested by Holness & Watt (2001; see also Piazolo *et al.* 2005; Bergman & Piazolo 2012). This classification is described in Table 2 (see Fig. 5a–e for examples). In addition to the five types reported by Holness & Watt (2001), two further types of quartz CL texture

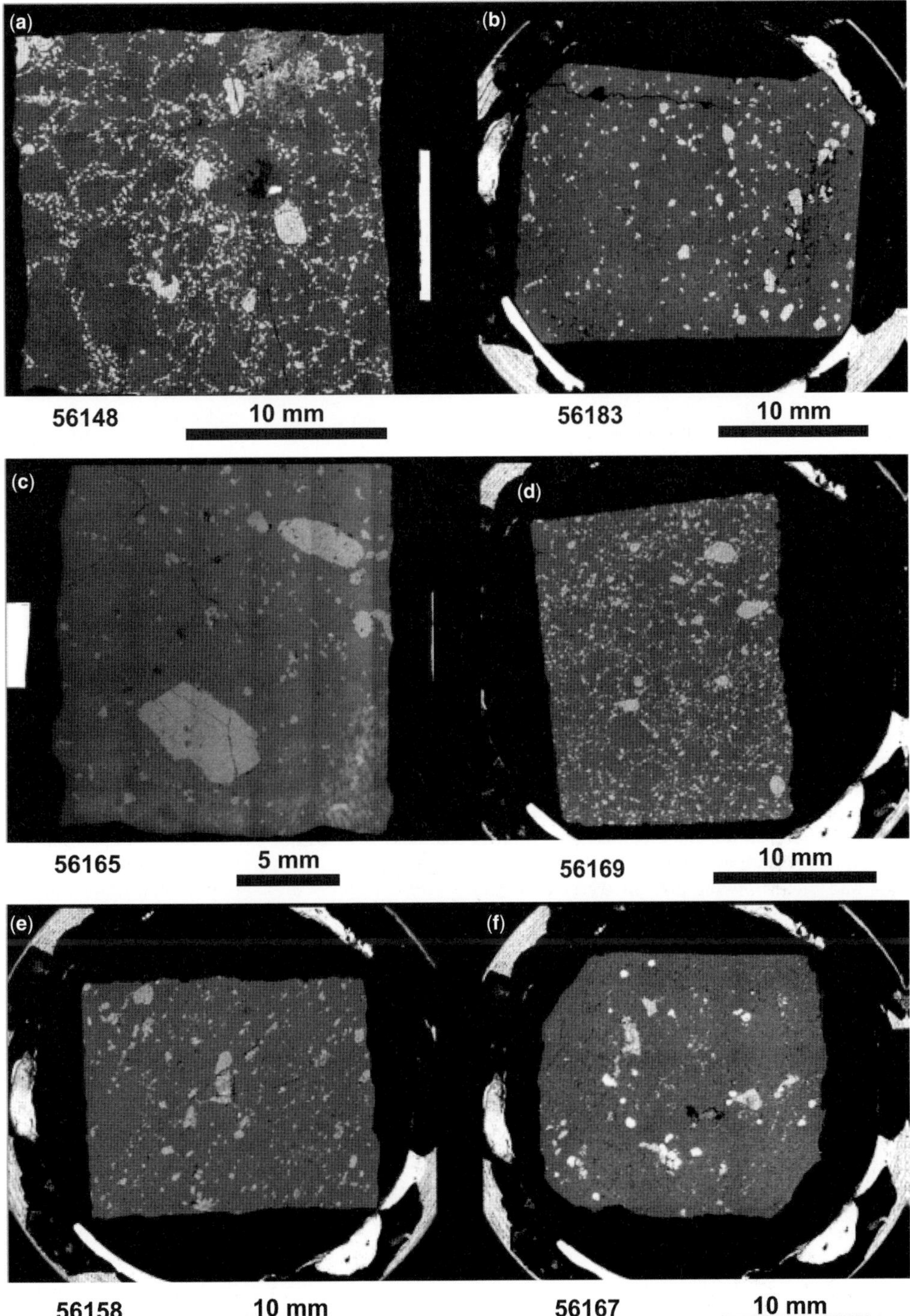

Fig. 3. BSE Z-contrast montages of all six quartzite samples (**a**) to (**f**) selected for analysis. The predominant mid-grey phase is quartz, with the secondary paler grey phase being feldspar (generally alkali; the minor plagioclase has a similar contrast to quartz). Bright white phases are a variety of accessory phases (e.g. apatite, titanite, muscovite, rutile, zircon, monazite, titanomagnetite and Fe-oxides). Black areas generally represent surface imperfections.

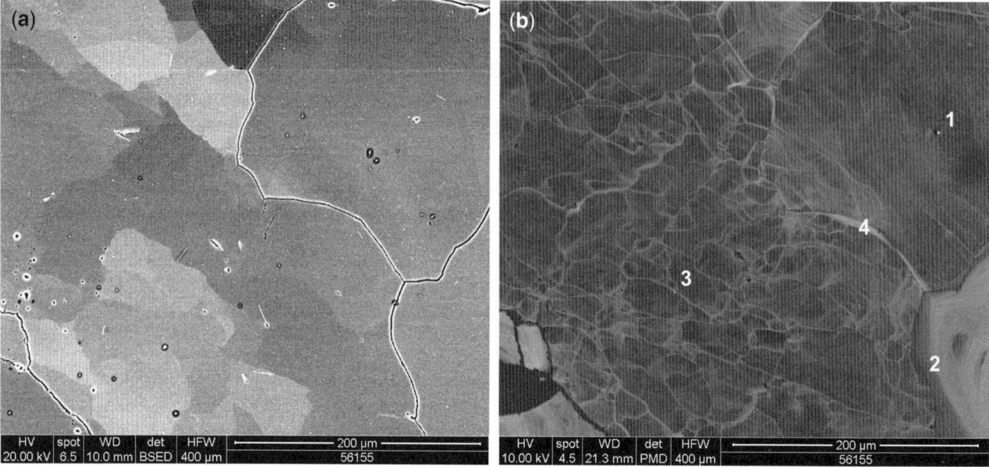

Fig. 4. Comparison between (**a**) SEM BSE orientation contrast (SEM-OC) and (**b**) SEM cathodoluminescence (SEM-CL) images of exactly the same field of view. The subgrain microstructure clearly visible in (a) is absent from (b). In contrast, the microstructures sensitive to differences in quartz impurity and/or point defect content are clearly visible (b) but absent in (a). The numbers in (b) refer to different types of quartz SEM-CL textures (see Fig. 5).

have been recognized in this study (see Table 2 and Fig. 5f).

Type 1 quartz (Fig. 5a) exhibits weak or 'mottled' CL, with occasional bright spots and, in general, is restricted to the inner parts of grains. It appears to be overprinted by all other types (2–7) of quartz CL microstructure (Figs 4 & 5). According to Seyedolali *et al.* (1997), metamorphic quartz displays either an indistinct, mottled texture, or nearly uniform (non-differential) CL. We interpret Type 1 quartz therefore as representing the original quartz CL microstructure. This interpretation is in agreement with both Holness & Watt (2001) and Bergman & Piazolo (2012). The latter authors considered Type 1 quartz (their 'signature 1') to result from a combination of a presumed igneous origin of the detrital grains and a subsequent regional metamorphic event (presumably the Caledonian Orogeny) predating intrusion of the BIC.

Type 2 quartz (Figs 4b & 5b) has been interpreted as representing migration recrystallization of quartz grain boundaries due to heating caused by contact metamorphism from the BIC (e.g. Lind 1996; Holness & Watt 2001; Bergman & Piazolo 2012). Thus, if the concentration of Ti-in-quartz is indicative of temperature, Type 2 quartz should exhibit high levels of Ti. We focus our attention therefore on Type 2 quartz.

Data processing: LA-ICP-MS

Laser-ablation analyses were performed on four samples (56183, 56167, 56158 and 56148). Results where ablations inadvertently struck feldspars (elevated Na, Ca or K) and/or rutile needles (bursts of high Ti counts) were discarded. However, it is possible that small rutile inclusions may have escaped notice. Samples showing Ti concentrations within three standard deviations of the detection limits (i.e. 0.15 ppm) were also removed. Data were then processed using the thermometers with $P = 0.3$ GPa where appropriate (equivalent to 10 km depth) and activity of rutile equal to 1 due to the presence of rutile in the samples. The calculated temperatures for the four geothermometers are given in Table 3 and Figure 6a. The results presented suggest that temperatures calculated by the geothermometers of Thomas *et al.* (2010) and Kawasaki & Osanai (2008) are anomalously low relative to those predicted by thermal modelling (e.g. Buntebarth 1991) and by observation of similarly positioned metapelitic aureole samples (Pattison 1991; Pattison & Harte 1997). In contrast, the

Table 1. *Samples imaged using SEM, with distance from surface exposure of the igneous contact*

Sample Number	Distance from BIC contact (m)
56148	2
56165	215
56158	260
56183	325
56169	650
56167	900

Table 2. *Classification of SEM cathodoluminescence quartz grain textures (Types 1 and 3–5, after Holness & Watt 2001; Type 2, after Lind 1996, Holness & Watt 2001; Types 6 and 7, this study)*

Type	Description	Occurrence	Origin
1	'Dark and mottled' quartz, with irregular bright patches (Fig. 5a)	Cores of grains	Relict of regional metamorphism
2	Alternating zones of bright and dark contrast quartz (Fig. 5b)	Generally parallel or sub-parallel to grain margins	Grain boundary migration during growth and/or coarsening at onset of contact metamorphism
3	Bright linear features within quartz grains (Fig. 5c)	No obvious relationship with grain boundary orientation	Cracking and sealing close to the intrusion
4	Brightly luminescing quartz (Fig. 5d)	On grain boundaries but without fine-scale banding	Melt/fluid migration along grain boundaries and cracks
5	Non-luminescent linear features (Fig. 5e)	Cross-cutting all other features	Healed fractures due to late stage and low temperature fluid migration
6	Diffuse 'wavy' banding (Fig. 5f)	Surrounding Type 7	?
7	Straight bands with clear truncation relationships (Fig. 5f)	Surrounded by Type 6	Reminiscent of faceted grains growing into free voids

temperatures from the Wark & Watson (2006) 'TitaniQ' geothermometer, despite calibration for 1 GPa pressure, straddle the modelled values. However, no clear distance − temperature correlation is observed due to divergent results within the same sample (Fig. 6a). We shall discuss possible reasons for this behaviour below (see Fig. 7).

Data processing: EPMA analysis

Electron microprobe analysis was performed on four samples (56169, 56183, 56148 and 56158). Sixty measurements of Ti content (ppm) were taken from each sample in a combination of isolated spot analyses and line analyses with fixed spacing. Spot analyses were located in the brightly luminescing grain boundary regions of 'Type 2 quartz'. The line analyses were generally perpendicular to banding in 'Type 2' or 'Type 6' quartz and usually included non-luminescent grain centres (i.e. 'Type 1' quartz). Data were again processed with $P = 0.3\,\text{GPa}$ where appropriate (equivalent to 10 km depth) and activity of rutile equal to 1.

Average Ti concentrations (ppm) along with calculated crystallization temperatures for each sample are given in Table 4 and shown in Figure 6b. Whilst the determined values bracket the modelled temperatures, there is again no clear distance − temperature relationship within the uncertainty of the analysis, other than the sample closest to the pluton recording a temperature higher than those found more distally.

Concentrations of Ti determined from line analyses, together with the relevant greyscale level along the lines from associated CL images, are shown in Figures 8 and 9. These were measured from a variety of quartz types in different settings and will be discussed in more detail below.

Discussion of methodologies

LA-ICP-MS. Locating the high-CL emission areas for analysis via LA-ICP-MS was challenging in reflected light, such that all LA-ICP-MS analyses were confined to identifiable grain boundary locations that could be recognized optically. Accidental feldspar ablation occurred during some analytical runs, as quartz and feldspar were optically almost indistinguishable within the LA-ICP-MS system. A large spot size yields little control on the selection and/or identification of the 'type' of quartz (following Holness & Watt 2001) being ablated, which changes 'type' at short wavelengths (see Fig. 5). This means that the ablation frequently represents a mixing over multiple types, as the sampling resolution of 100 μm is in general much greater than the scale of zoning visible in quartz (e.g. Müller *et al.* 2003; Fig 5, this study). In samples with short-wavelength variation in CL zoning, LA-ICP-MS averages over the ablated volume, creating a convolving effect over several zones, as shown schematically in Figure 7. Better spatial resolution could be achieved at the expense of precision and an increase in the detection limit, but the major problem of location within the sample remains to be resolved.

The existence of marked heterogeneity of Ti distribution within the quartz implies that there must be disequilibrium relative to the rutile crystals that are present. This also implies that the assumption that $a_{TiO_2} = 1$ is not true in most cases. It is

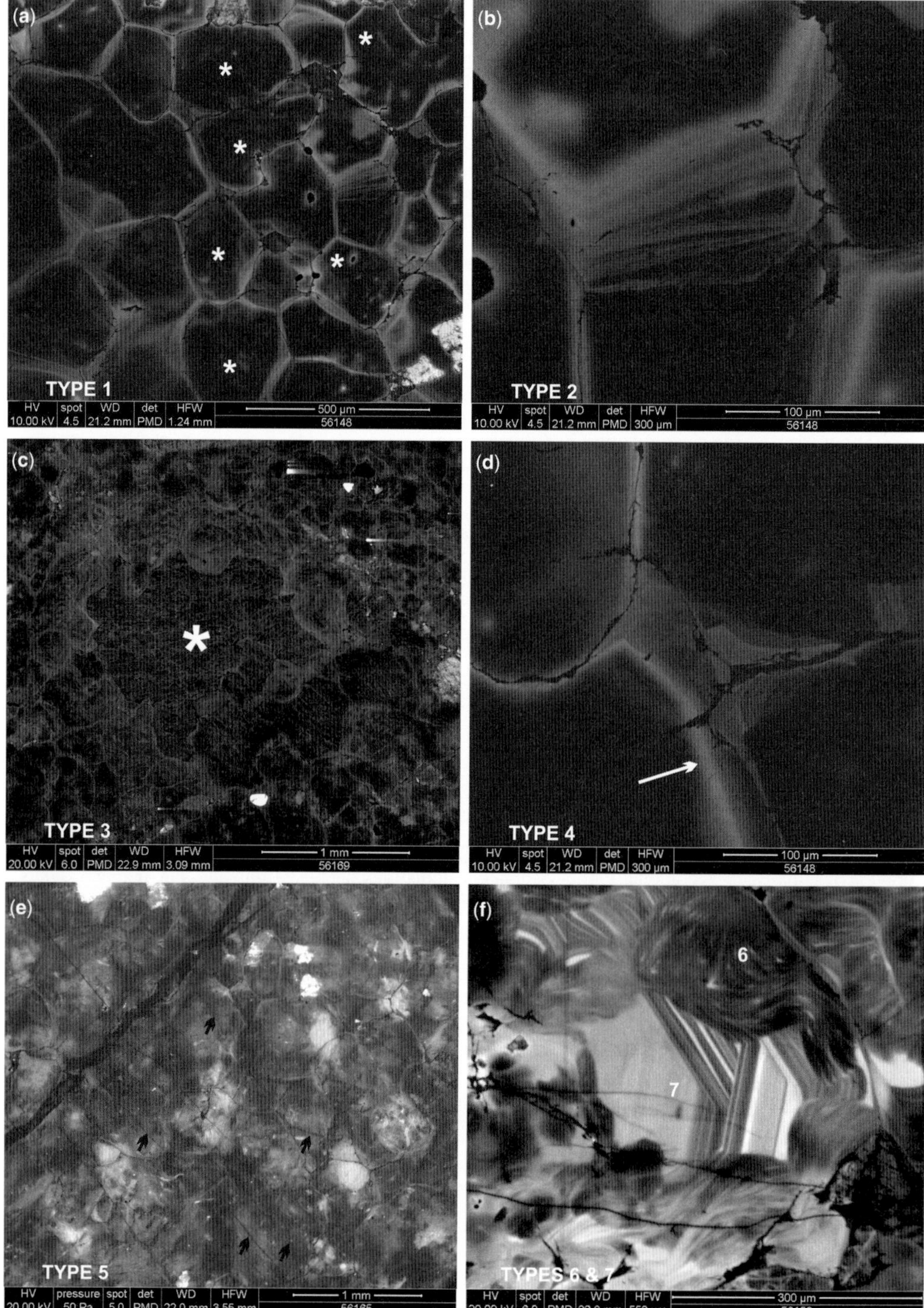

Fig. 5. CL images of various quartzite samples (**a**) to (**f**) illustrating the different 'types' of quartz texture present. Compare with Table 2 for descriptions.

Table 3. *Ti-in-quartz as determined by LA-ICPMS, along with temperatures calculated from the four geothermometers. Significantly more analyses than this were taken originally but subsequently were discarded due to contamination by feldspar or rutile, or where Ti detection in the mass spectrometer was within error of the background measurement. Note that internal variability within a single sample is very high and accounts for much of the variation that can be reported. LA-ICPMS necessarily samples multiple crystal zones at the scale of an ablation pit, which means that anomalously high-Ti and low-Ti zones, as well as micro-inclusions of rutile, can affect the final result. The existence of heterogeneity in Ti distribution at length-scales finer than 100 micrometres also shows that the samples are recording a transient disequilibrium and that diffusion had not homogenised crystal interiors*

Sample	Ti (ppm)	Geothermometer temperatures (°C)			
		Wark & Watson (2006)	Thomas *et al.* (2010)	Kawasaki & Osanai (2008)	Huang & Audétat (2012)
56148	166	812	646	601	826
56148	114	764	607	555	776
56148	133	783	623	573	796
56148	20	586	466	392	593
56148	27	612	487	415	620
56158	25	604	481	409	613
56158	24	602	479	406	610
56158	38	644	512	444	653
56158	18	576	458	384	584
56158	31	623	496	427	633
56183	40	649	517	449	658
56183	35	634	504	437	645
56183	46	662	527	461	672
56183	61	691	550	488	702
56183	36	638	508	439	658
56167	37	642	511	442	650

likely this would cause marked underestimates of temperature but cannot be assessed directly. The method of utilizing a sample with external temperature constraint to establish a reference a_{TiO_2}, as employed by Grujic *et al.* (2011), would here seem to be impossible due to the large and real heterogeneities in Ti content within the samples and the lack of any Ti phases other than rutile against which to correlate within the samples. As we discuss later, the extreme variations in Ti content seen would indicate that a_{TiO_2} was highly variable during contact metamorphism, and so accurately deploying a Ti-in-quartz thermometer in isolation is exceedingly difficult.

EPMA. EPMA poses different analytical challenges to LA-ICP-MS. Correlating CL images between SEM and the electron microprobe is relatively straightforward. The electron microprobe employed a focused spot of a nominal 5 μm width, ideal given the CL zonation, although electron scattering within the sample will make analyses less spatially precise. EPMA has lower analytical precision (to the nearest *c.* 10 ppm Ti) and a

considerably higher detection limit; the 6-sigma value is reliably above background at 22 ppm Ti, meaning this abundance is required to avoid considerable uncertainties from counting statistics. Secondary X-ray emission is also problematic. Figure 6b recreates the same effect seen by Wark & Watson (2006). EPMA points measured within 100 μm of a rutile grain show that a marked increase in apparent Ti abundance is caused by secondary excitation of the neighbouring rutile. The TiO_2 contents reported by the instrument cannot be attained in quartz at equilibrium unless temperatures exceed 2000 K, which is clearly unfeasible.

Calibration of a microprobe is typically performed on high-Ti phases such as rutile and the calibration is then extrapolated to very low Ti concentrations. This creates an inherent uncertainty based on the gradient of the regression line, whether matrix effects should be considered, and in particular, the exact method of determining and subtracting the background counts.

The absolute accuracy of the EPMA reported here cannot be substantiated other than by cross-comparison against the LA-ICP-MS data, which

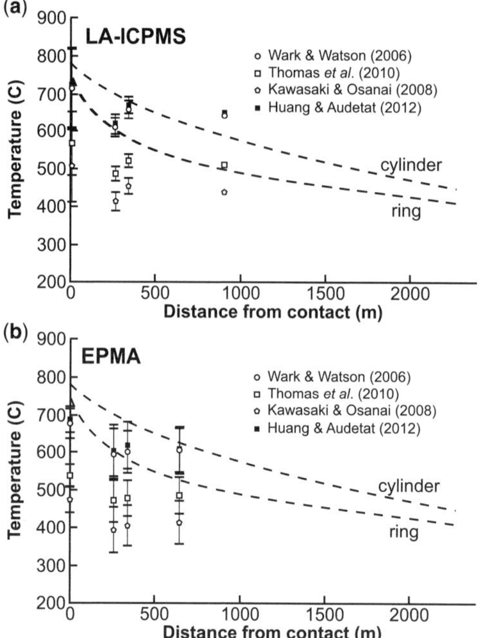

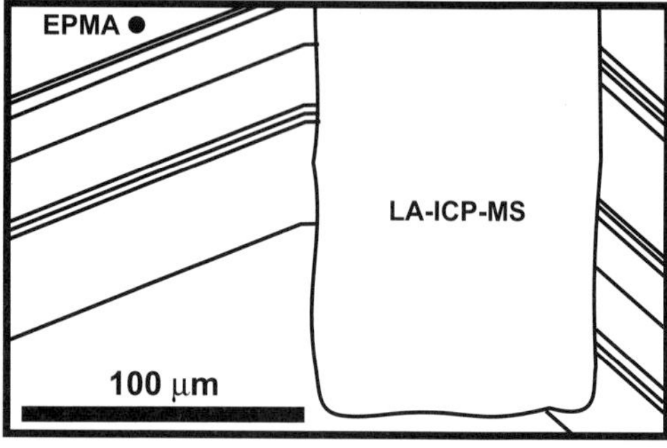

Fig. 7. The difference in sample size taken by LA-ICP-MS and EPMA, relative to representative zoning in a quartz crystal. After Müller *et al.* (2003).

Fig. 6. Temperature determinations using four different geothermometers; error bars are two standard deviations above and below the mean, representing 95% confidence limits. (**a**) LA-ICP-MS analysis. Note extremely large spread closest to the contact representing highly variable Ti contents. The lack of error bars at the furthest point is due to only one Ti measurement being considered valid, hence standard deviation cannot be calculated. (**b**) EPMA analysis. Error in all samples is relatively constant with around ± 100 °C spread around the mean falling within 95% confidence limits.

yield higher values although averaged over larger areas. This unreliability may account for the differences between the EPMA and LA-ICP-MS values and suggest that in routine analysis, an appropriate Ti-rich quartz standard or glass would be superior for calibration purposes. EPMA would be viable to measure Ti contents in quartz that equilibrated at higher temperatures of over *c.* 600 °C, but not reliably at lower temperatures.

Use of CL prior to EMPA/LA-ICP-MS. Cathodoluminescence was used to determine regions of quartz grains for further analysis (e.g. Fig. 3). This approach was based on the assumption (e.g. Rusk *et al.* 2006; Spear & Wark 2009) that the brightest CL regions would be those hosting the greatest Ti concentration, thereby providing the peak temperature of quartz recrystallization. The validity of this assumption has been assessed using data from EPMA line scans across individual 'zoned' quartz grains (Figs 8 & 9).

CL correlation with Ti content. Figures 8 and 9 compare the continuous measurement of CL grey level contrast using a conventional 0–255 scale against point measurements of Ti concentration by EPMA. If CL is Ti-dependent, then there should be a positive, linear correlation between greyscale and Ti concentration. In practice, the degree of correlation with EPMA is highly variable, ranging from strongly positive (Figs 8a, b & 9a, b), through weakly positive (Figs 8c, f & 9c, f) to no obvious correlation whatsoever (Figs 8d, e and 9d, e). In the case of the crystal shown in Figure 8b (see also Fig. 9b), we have significant secondary X-ray emission from nearby rutile. Thus, the correlation and the high values in this image are spurious. Whilst the correlation in Figures 8a and 9a seems quite robust, there are many data above the 50 ppm level, clearly above the EPMA background. In most other cases, measured concentrations of Ti are lower and there is no obvious link with CL emission within the uncertainty of the analysis. There are multiple possible explanations, including: (1) larger uncertainties in measurements; (2) involvement of other CL activators, such as Al and/or Li (Perny *et al.* 1992) or K, Na and Fe (Landtwing 2005); and CL emission may be correlated to lattice defects related to lattice substitutions (Watt 1997).

It has been suggested that Ti only becomes the primary CL activator in high-temperature quartz above *c.* 600 °C (Rusk *et al.* 2008), and at lower temperatures (including those areas away from the contact at Ballachulish) a different CL activator may be responsible. Thus, the assumption that the rims of quartz grains represent the maximum recrystallization temperatures with respect to Ti concentrations in quartz may not be valid in all cases, except perhaps close to the contact where we see definite Ti enrichment correlated to CL emission. We revisit these issues later.

Table 4. *Average Ti concentration in each sample (n = 60 for each sample, values of zero discarded) as determined by EPMA, along with calculated temperatures from each geothermometer. Values in brackets represent one standard deviation based upon variability in measured Ti contents*

Sample	Ti (ppm)	Geothermometer temperatures			
		Wark & Watson (2006)	Thomas *et al.* (2010)	Kawasaki & Osanai (2008)	Huang & Audétat (2012)
56148	53	676 (39)	538 (31)	475 (35)	687 (40)
56158	22	593 (69)	472 (56)	395 (61)	602 (69)
56183	24	601 (57)	478 (47)	406 (52)	618 (64)
56169	26	607 (62)	485 (50)	414 (56)	609 (59)

Geospeedometry

Use of kinetic arrest to determine peak metamorphic temperatures

There are numerous examples of rutile − quartz interaction preserved in the samples from the Ballachulish aureole, but they are best developed in immediate proximity to the pluton contact. Sample 56148 originates just 2 m from the contact zone and consistently yields the highest metamorphic temperatures in this study.

In this sample, we have a set of low-Ti crystals which in their marginal zones display strong CL, indicative of higher Ti contents by correlated EPMA analysis (Figs 8a & 9a). High-Ti zones record ghost growth zones and show the migration of the grain boundaries during annealing (see also Bergman & Piazolo 2012). Occasional rutile grains sit within the grain boundaries, such that the grain boundary network should in theory be buffered to $a_{TiO_2} = 1$. Further, there are included grains of rutile within low-Ti quartz which are interacting directly with their hosts via volume diffusion. The fine-scale oscillations in the CL image show that LA-ICP-MS lacks the spatial precision relative to sample heterogeneity, while the abundant rutile inclusions preclude accurate EPMA work due to the phenomenon of secondary X-ray fluorescence as mentioned earlier and demonstrated in Figure 7 (compare with Fig. 8). It is difficult therefore to accurately employ a Ti-in-quartz thermometer for this sample. We will instead use the diffusion coefficient to ascertain the temperature of peak diffusion and compare against other results.

Characterization of diffusion

Diffusion processes can be characterized in their extent by their diffusivity − time (Dt) product. This has units of metres-squared and is often normalized by the length-scale of the process, squared to yield a dimensionless number that describes the extent of diffusion.

The initial scenario here is the inverse of that discussed in the original paper presenting the diffusion coefficient by Cherniak *et al.* (2007). They discussed the possibility of rutile needles acting as localized Ti sinks upon cooling. The increase in temperature due to a contact metamorphic event will instead drive additional Ti to partition from rutile into the host quartz, with each rutile grain acting as an effectively infinite source of Ti into the surrounding crystal. We can be fairly sure that this happened at close to peak metamorphic conditions, as the partition coefficient (upon which the Ti-in-quartz thermometers are based) and diffusion speed will be most favourable at this point to the establishment of a source. As the temperature drops, the diffusion slows such that the rutile grain cannot recover its Ti. This is because upon cooling, the diffusion speed will be sufficiently sluggish to take considerably longer (14 orders of magnitude longer − unfeasible by any means) for a quartz crystal to lose by diffusion at 250 °C Ti acquired by diffusion at 800 °C. Therefore, this process is effectively one-way and records the peak metamorphic event and the highest temperature.

The diffusion haloes that are biggest within the sample are those associated with fully enclosed rutile inclusions. This geometry is effectively an infinite source into a semi-infinite medium and has a relatively simple solution (Crank 1975), which we chose to model in one dimension.

Analysing the diffusion halo, it represents a total amount of diffusion equivalent to a Dt product of 3.34×10^{-11} m^2. If we consider a single temperature for diffusion, and for the moment ignore uncertainties on the initial diffusion coefficient (D_0) and the activation energy of diffusion (ΔH), this would be equivalent to *c.* 300 years at 800 °C (upper bound temperature for the aureole) or *c.* 42 600 years at 650 °C (what we would consider close to a lower bound based upon the modelling

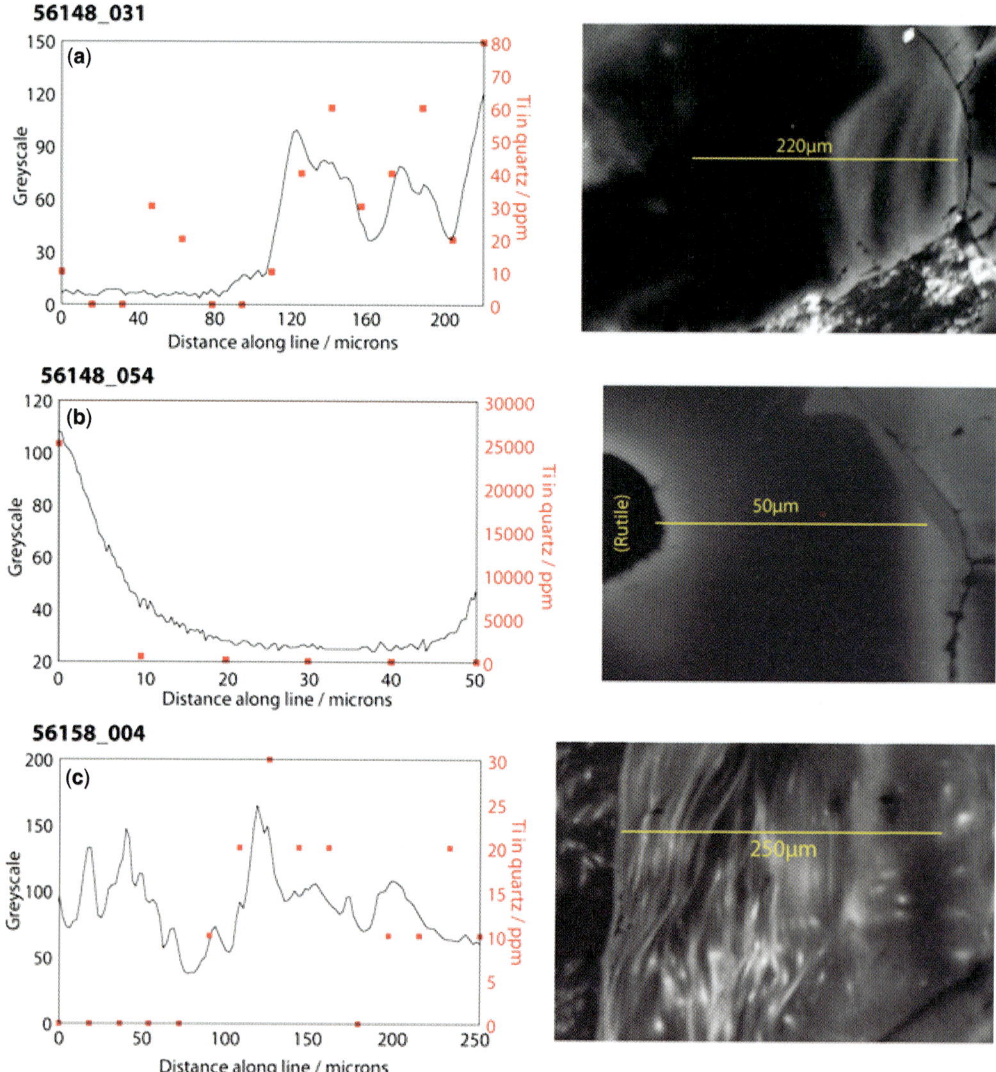

Fig. 8. Right, locations (yellow lines) of EPMA line scans on SEM-CL images of zoned quartz grains for samples (**a**) to (**f**). Left, plots of greyscale profile (solid black line) and point analyses of Ti-in-quartz (red squares). Note the highly elevated Ti concentration in (b) due to sampling directly from the edge of a rutile grain.

of Pattison & Harte 1997; in this area). This, however, neglects that the sample would be cooling continuously from peak metamorphic conditions, and as such diffusion becomes progressively more sluggish as the cooling progresses.

Methodology for modelling diffusion with cooling

As the cooling rate is effectively linear, this can be undertaken relatively easily as a simple iterative step procedure (Fig. 10). In terms of uncertainty propagation, the main variables are the uncertainty on ΔH and D_0. These are related through the slope and intercept of a linear relation (on a graph of log diffusivity v. $1/T$(K), the gradient is $-\Delta H/R$ and the intercept on the D axis is D_0). Because of this, these uncertainties cannot be treated independently without strongly overestimating the uncertainty on the diffusion speed; high ΔH should be combined with the higher log D_0 and vice versa.

Implementing the above procedure (Fig. 10), the total Dt product converges to a limit as the

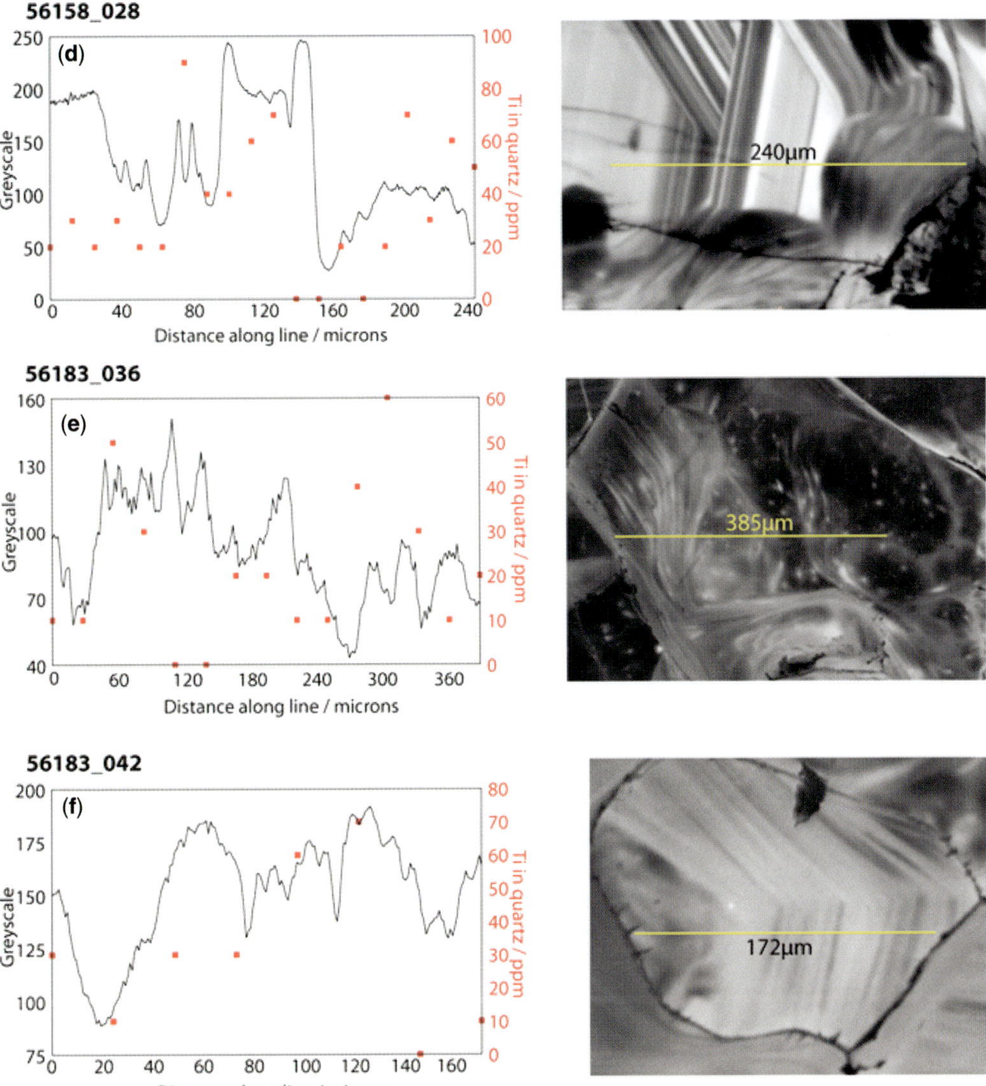

Fig. 8. *Continued.*

temperature decreases and diffusion slows to effectively zero. If we start at 800 °C, 98% of the total diffusion that can occur will have occurred by 680 °C; if we go with the lower bound of 650 °C as a start condition, our 98% diffusion mark would have been achieved by 575 °C. Therefore, the system closes to diffusion quite rapidly.

Thermal modelling of the Ballachulish aureole (Buntebarth 1991; Masch & Heuss-Assbichler 1991; as summarized in Pattison & Harte 1997) gives an approximately linear cooling rate for samples at or near the contact with the pluton, given a cylinder geometry, 1050 °C intrusion temperature and 250 °C ambient temperature in the country rock. The apparent cooling rate for these samples is roughly 200 °C in 435 000 years, or 0.00046 °C per year. By modelling this linear cooling using the method above, we can attempt to use cooling trends based upon observations and thermal models to recreate the observed diffusion distances, in order to constrain the initial temperatures. With a fixed cooling rate, and known diffusion behaviour with temperature, for any given starting temperature, a finite amount of diffusion can occur before quenching happens. In this type of model, our main sources of uncertainty

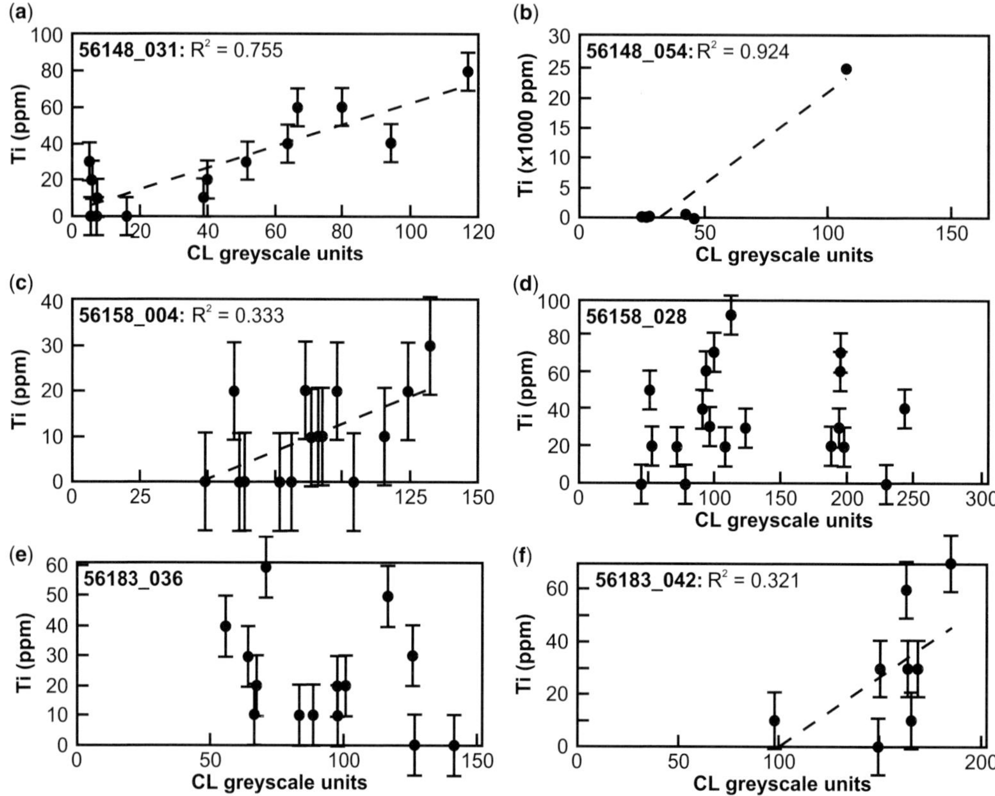

Fig. 9. Measured concentration of Ti (determined by EPMA) against greyscale from CL images averaged over 5 μm for the line analyses shown in Figure 8. Trend lines are determined by the least squares method with R^2 values. Figures (**d**) and (**e**) show no correlation, and (**c**) and (**f**) show very weak correlation. Figures (**a**) and (**b**) both have high R^2 values, with the correlation in (a) considered to be the strongest as that in (b) relies on a single point.

are the activation energy of the diffusion process $(273 \pm 12\ \text{kJ mol}^{-1})$ and the value of D_0 ($\log_{10} D_0 = -7.154 \pm 0.525$), which between them control the variability in D with T (Cherniak *et al.* 2007).

If we take the cooling rate close to the contact as reported by Pattison & Harte (1997), then within the constraints provided by the diffusivity, we require the initial temperature to be constrained between 633 °C and 654 °C, with the likelihood of a peak around 645 °C. This is consistent with values reported in the northeastern and southeastern profiles (Pattison & Harte 1997) and shows that combined with a reasonable estimate of cooling rate, we can here use a diffusion coefficient to constrain temperature relatively precisely, with precision comparable to that of the most precise geothermometers, due to the high sensitivity of diffusion to temperature.

Accounting for possible variations in geometry, the cooling rate might be up to four times faster (ring geometry, Pattison & Harte 1997), but this

has a relatively low impact. Quadrupling the cooling rate to 0.00184 °C per year, would require a compensatory increase in the peak starting temperature. However, the shift is small, as the dependency of diffusion coefficient on temperature is large. In this case, we would need to increase the starting temperature by just 35 °C, to 685 °C. Even incorporating this worst-case scenario as an actual uncertainty, our temperature estimate would be quite robust, comparable with most typical geothermometers and previous results.

We can take this further, to investigate the rate at which the crystals appear to be growing from the diffused width of the grain boundary ghosts. The diffusion geometry here is similar, and can be approximated as a one-dimensional diffusion process across an initial step profile. Again, each boundary will have a characteristic Dt product and we can use this to ascertain where on the cooling curve the appropriate integrated Dt product is encountered. This is illustrated by considering the Type 2 quartz CL texture of alternating zones of

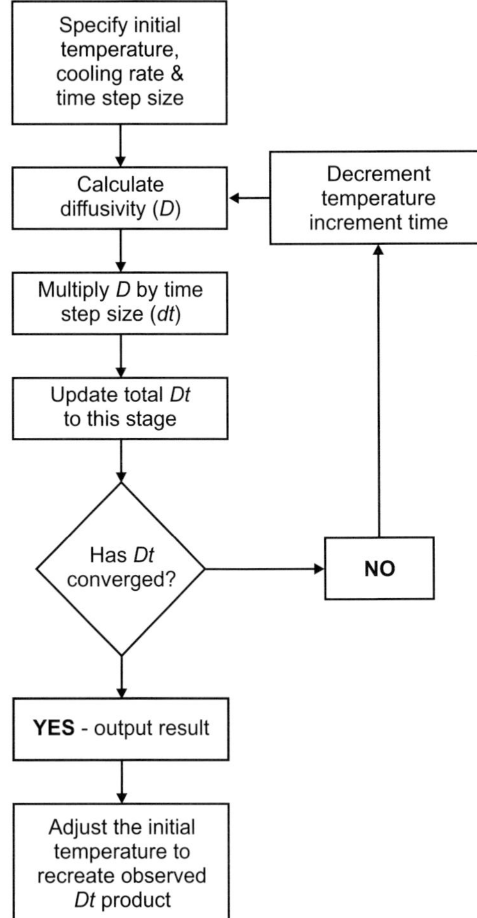

Fig. 10. Simplified flow chart for diffusion (geospeedometry) analysis.

bright and dark contrast quartz that develop generally parallel or sub-parallel to grain margins (Fig. 5b) and are interpreted as resulting from grain

boundary migration during growth and/or coarsening at the onset of contact metamorphism (Lind 1996; Holness & Watt 2001). The results (Table 5 and Fig. 11) show the following: (1) consistent monotonic decrease in extent of diffusion from core to rim (Fig. 11a); (2) consistent decrease in the temperature at which each successive zone must have formed (Fig. 11b); (3) grain boundary migration was slowing to a halt as temperatures went below c. 500 °C (Fig. 11c); and (4) grain boundary migration was limited for the first c. 70 000 years following peak metamorphism (Fig. 11d).

A consideration of the temperature at which a zone formed according to its diffused width, compared to the CL greyscale colour of that zone is shown in Figure 12. This suggests a strong positive correlation between calculated temperature and CL emission, in a manner that could not be unambiguously established via EPMA analyses. This is what we would expect if the main CL activator was Ti, and is consistent with the diffusion halo (O in Fig. 11a) around the rutile inclusion.

The data here would infer that extensive recrystallization did not occur on this grain boundary until the temperature had dropped below c. 600 °C. The fact that the bands showing the greatest diffusion extent are in grain cores and those with the least at the edge show that the grain boundary migration is post-peak, which is contrary to the observations of Holness & Watt (2001), who inferred that the textures developed as the sample heated.

We would conjecture that there was a low abundance of free fluids on the up-temperature path, possibly as they were consumed by melting reactions within the aureole and possibly as the cold Appin Quartzite is plausibly impermeable (Holness 1998). Once the temperature subsided below the peak conditions, the fluids could have been liberated into warmer, more permeable (Holness 1998) country rock, allowing recrystallization to occur. The fluids released could well have been saturated in Ti, allowing a high a_{TiO_2} value, but one that

Table 5. *Summary of diffusion modelling results (see Figs 8 & 9)*

Region	Dt (m^2)	Zone formation temperature (°C)	Start time after peak condition (years)	Growth rate (μm a^{-1})	Thickness or distance grown (μm)
A	5.79×10^{-12}	604	88 700		
B	1.68×10^{-12}	578	147 000	3.47×10^{-4}	20.24
C	3.91×10^{-13}	547	211 000	9.20×10^{-4}	26.13
D	3.23×10^{-13}	544	219 000	1.62×10^{-4}	39.11
E		Zone is convolved with two superimposed profiles			
F	1.96×10^{-13}	535	239 000	9.17×10^{-4}	57.45
G	1.20×10^{-13}	525	260 000	6.93×10^{-4}	72.01
H	5.71×10^{-14}	512	288 000	2.44×10^{-4}	78.83
I	3.64×10^{-14}	504	306 000	1.46×10^{-4}	81.46

(a)

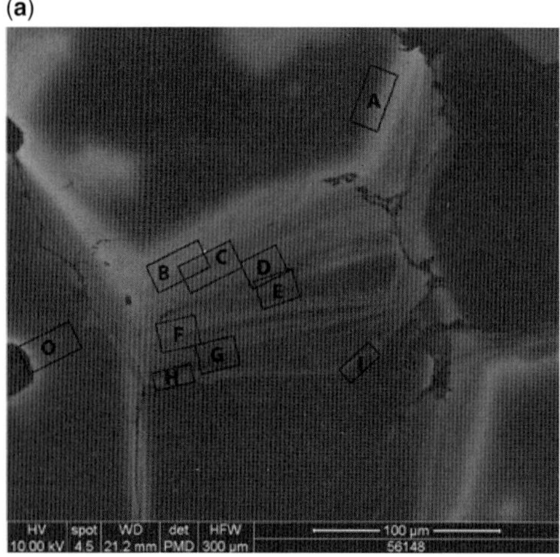

(b)

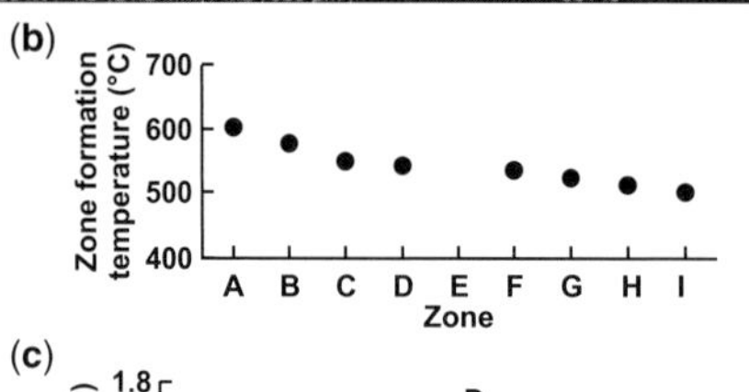

(c) **(d)**

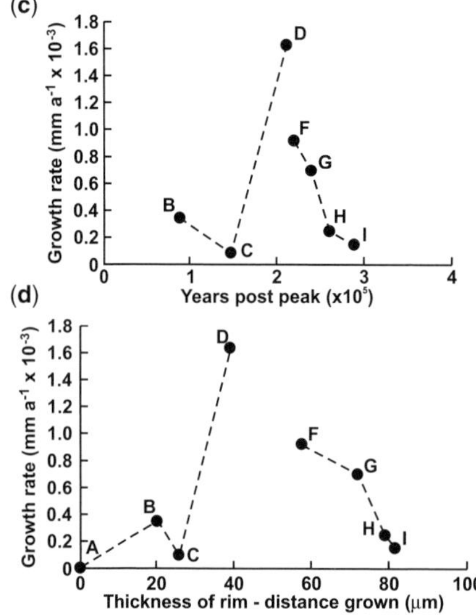

Fig. 11. Diffusion analysis of Type 2 quartz CL-texture zones (see also Table 2). (**a**) Location of measurement zones. (**b**) Calculated temperatures per zone. (**c**) Grain rim growth rates v. time post-peak temperature. (**d**) Grain rim growth rates v. distance grown.

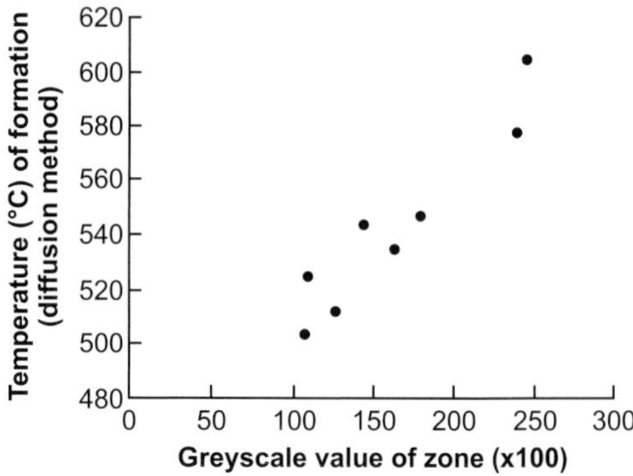

Fig. 12. Temperature of zone formation according to its diffused width plotted against its CL greyscale index.

appears to be transient, as there are clearly low-Ti zones in the crystal rims from LA-ICP-MS and EPMA analysis, despite the presence of nearby rutile. Equivalently, rapid bursts of grain boundary migration during recrystallization may have caused departures from equilibrium with grain boundary rutile allowing zones to grow under conditions of low a_{TiO_2}.

Discussion

This study has employed Ti-in-quartz geothermometry using commonly available analytical tools and has tested the thermometers on thermally well-constrained rocks from the aureole of the Ballachulish Igneous Complex (BIC), NW Scotland. This section considers the implications of our results.

Presence or absence of rutile

In its simplest form (Wark & Watson 2006), the TitaniQ geothermometer assumes that when rutile is present the activity of TiO_2 is unity (i.e. $a_{TiO_2} = 1$) although it can be used also when rutile is absent (i.e. $a_{TiO_2} \neq 1$). However, in the case of our samples, the assumption that $a_{TiO_2} = 1$ may not be valid, despite the presence of rutile. This is evidenced by the rapid CL brightness oscillations in crystal rim zones, the observed Ti heterogeneity by both LA-ICP-MS and EPMA, and the existence of correlation between CL brightness and Ti content for high-Ti crystals. Therefore, temperatures were rapidly oscillating during recrystallization by hundreds of degrees Celsius (effectively impossible), or the effective a_{TiO_2} value was not constant, or crystals are recording disequilibrium Ti contents.

Even in a magmatic system, experimental evidence (Hayden & Watson 2007) suggests that true saturation of rutile within melts is rarely achieved and indeed is unlikely in most conditions where Ti-in-quartz geothermometry is currently employed. In a metamorphic system, variations in a_{TiO_2} can be due to kinetic limitations as Ti from rutile may not be able to diffuse by grain boundary diffusion fast enough to buffer crystal growth before the system is closed during cooling (Spear & Wark 2009). In a metamorphic setting, it would seem therefore to be very hard to ascertain what the a_{TiO_2} value would be during grain growth, making Ti-in-quartz thermometers hard to employ with confidence. Buffering the a_{TiO_2} value in the grain boundaries may be possible due to dynamic processes such as fluid flow, or grain boundary mobilization (Grujic *et al.* 2011), but at lower temperatures, volume and possibly grain boundary diffusion seem too slow.

Pressure effects

Temperatures determined using the pressure-sensitive TitaniQ geothermobarometer (Thomas *et al.* 2010) deviate further from modelled temperatures for the Ballachulish samples than those based on the original 1 GPa calibration (see Fig. 6). This may be coincidental and caused by ablation of low-Ti quartz during analysis, but equally, it may not, as the revised calibration from Huang & Audétat (2012) for 0.3 GPa recovers very similar temperature estimates to the original 1 GPa calibration thermometer of Wark & Watson (2006), and close to the external constraints for the aureole.

Huang & Audétat (2012) propose that the growth rate and environment of the quartz has a significant impact on Ti partitioning, and therefore any recovered temperatures. We can ask also whether the metamorphic environment in the aureole is equivalent to the experimental conditions under which the thermometers have been calibrated.

Which calibration is 'correct' is beyond the scope of this paper. However, it is clear that even ignoring analytical difficulties, the heterogeneity of Ti distribution in quartz creates problems in interpretation and in setting boundary conditions for the use of the thermometers.

Ti within the fluid phase

It has been suggested by previous workers (e.g. Harte *et al.* 1991; Pattison & Harte 1997) that there has been significant infiltration of fluids into the Ballachulish aureole. It is evident also from CL imaging that there has been recrystallization to yield several distinct textures, possibly in the presence of variable and distinct fluids (e.g. Holness & Watt 2001; Bergman & Piazolo 2012). Fluid flow is thought to have been channelled through fracture networks, with only minor grain boundary flow due to permeability contrasts (Holness 1998; Holness & Clemens 1999). Fluid, if present, is likely to be the dominant medium of Ti transportation and mediation in the grain boundary network. Thus, the a_{TiO_2} parameter could be considered to reflect the buffering capacity of the grain boundary network. If dry, it is likely that Ti will be very sluggish, whereas if wet, there is a fast communication medium to allow Ti transport.

Complications from the geological setting

The samples used in the present study were located in terms of the shortest horizontal distance between the outcrop and the igneous contact, based on surface exposures (Lind 1996). Unfortunately, such a location does not account for variations in the three-dimensional geometry of the igneous wall-rock contact. For example, if the contact is outwardly dipping, samples are closer to the contact in three dimensions than they are in two dimensions. However, if the BIC is a cylinder with near-vertical walls (e.g. Buntebarth 1991), the location method used is not problematic. Furthermore, it is known that many other variables may impact on the temperature distribution within an aureole, including the rate and nature of magma convection, the amount of fluid released, and the wall-rock permeability and homogeneity (e.g. Buntebarth 1991).

Ti heterogeneity and crystal growth effects

Whilst there is clearly significant Ti heterogeneity in the recrystallized rim zones, it is not clear what the origins of the heterogeneity are, or what the magnitude is of the impact on Ti-in-quartz thermometry. Observation of the CL images suggests that as grain boundaries migrate, larger steps are associated with lower (darker) CL emission and so fast growth may well lead to differing incorporation of trace elements. This was something considered and shown to operate by Huang & Audétat 2012, via their experiments. It is quite possible that under metamorphic conditions, recrystallizing quartz may record different Ti contents according to exceedingly localized growth rates. This is consistent with behaviours noted in recrystallizing calcite by McCaig *et al.* (2007), where supply of elements to a recrystallizing interface can be limited by the rates of diffusion along grain boundaries.

Determining the main CL activator in quartz

It has been shown here via comparison between CL greyscale contrast and Ti concentration (i.e. Figs 8, 9, 11 & 12) that whilst in some cases Ti may be primarily responsible for CL in quartz (e.g. Figs 8a, 9a & 11), in other cases (e.g. Figs 8d, e & 9d, e) it is not the cause of CL emission. Thus, before using quartz CL micrographs as an indication of Ti content, it is important to ascertain the main potential activators of the CL signal. However, a more rigorous approach is to use a CL monochromator to filter specific emission wavelengths. For example, it is known that CL due to Ti occupies the blue part of the spectrum, in the region of 415 nm wavelength (Spear & Wark 2009).

Interpretation of results

Whilst there are clearly problems with the methodology used to derive Ti concentrations from quartz that have been noted by many authors (e.g. Wark & Watson 2006; Grujic *et al.* 2011), Ti-in-quartz methods can recover temperatures close to those known to have existed in the Ballachulish aureole. However, uncertainties are large due to difficulties in analysis and constraining a_{TiO_2}.

One of the main problems lies not just in measuring the Ti contents of the quartz grains accurately, but also in understanding that the contents are highly heterogeneous. In such cases, which Ti contents to use and which to reject is not entirely clear in an objective sense. For the purposes of our microprobe determinations, we have averaged out our results as they are close to detection limits in an effort to remove noise. Whether this is valid in the light of the CL observations that emission intensity (and by inference, Ti distribution) is highly variable at short length scales is an important consideration. During a contact metamorphic event such as this, the temperature does not become high enough, for long enough, to homogenize the quartz grains to an equilibrium value via volume diffusion. Even from the point of view of cooling plutons (Wiebe *et al.* 2007), Ti heterogeneity is preserved in quartz grains.

We can, however, turn this problem around and employ the fact that the equilibration has been kinetically arrested to investigate cooling histories and indeed peak temperatures with confidence (e.g. Figs 8, 9, 11 & 12).

Conclusions

As pointed out by Wark & Watson (2006) and as demonstrated here, the algorithms for the various thermometers have no internal checks for attainment of equilibrium and so must be applied carefully, and the most appropriate calibration selected with due care. Ti is a slow-diffusing species in quartz and would respond very sluggishly at temperatures below 500 °C. The potential for disequilibrium is very high, and therefore the potential for erroneous geothermometry is significant. We can, however, exploit the preservation of disequilibrium textures and the presence of rutile as these set up natural diffusion couples during contact metamorphism and quartz recrystallization.

If aureole cooling rates are known or are reasonably determined via modelling, the kinetic arrest of diffusion processes can offer a novel method to track the timing and temperature of recrystallization within quartz. Our demonstration here has shown a peak temperature in the Appin Quartzite sample studied of between 645 °C and 685 °C, and that fluid flow seems to have occurred only after peak conditions when temperatures dropped below *c.* 600 °C. However, this method needs an external control on cooling rate to within a factor of four to be a reliable thermometer.

The data collection via SEM, EPMA and LA-ICP-MS, and the initial interpretation within this work were completed by M. C. Jollands as part of an M.Geol., fourth-year-student independent research project at the University of Leeds, supervised by G. E. Lloyd, D. A. Banks and D. J. Morgan. Consideration of diffusion was undertaken by D. J. Morgan during the process of writing this article. Thanks are due to E. Condliffe for assistance with EMPA, SEM and CL analyses. M. C. Jollands also thanks D. Cornford for assistance with statistical analysis. The authors would like to thank D. Grujic and A. Kronenberg for comments that greatly improved the focus and quality of the manuscript. We are grateful also for the editorial assistance provided by S. Llana-Fúnez.

References

BAILEY, E. B. & MAUFE, H. B. 1960. *The Geology of Ben Nevis and Glencoe and the Surrounding Country: Explanation of Sheet 53*. Memoirs of the Geological Survey of Scotland, Edinburgh.

BERGMAN, H. & PIAZOLO, S. 2012. The recognition of multiple magmatic events and pre-existing deformation zones in metamorphic rocks as illustrated by CL signatures and numerical modelling: examples from the Ballachulish contact aureole, Scotland. *International Journal of Earth Sciences*, **101**, 1127–1148.

BUNTEBARTH, G. 1991. Thermal models of cooling. *In*: VOLL, G., TÖPEL, J., PATTISON, D. R. M. & SEIFERT, F. (eds) *Equilibrium and Kinetics in Contact Metamorphism: The Ballachulish Igneous Complex and Its Aureole*. Springer Verlag, Heidelberg, 379–402.

BUNTEBARTH, G. & VOLL, G. 1991. Quartz grain coarsening by collective crystallization in contact quartzites. *In*: VOLL, G., TÖPEL, J., PATTISON, D. R. M. & SEIFERT, F. (eds) *Equilibrium and Kinetics in Contact Metamorphism: The Ballachulish Igneous Complex and Its Aureole*. Springer Verlag, Heidelberg, 251–265.

CHERNIAK, D. J., WATSON, B. E. & WARK, D. 2007. Ti diffusion in quartz. *Chemical Geology*, **236**, 65–74.

CRANK, J. 1975. *The Mathematics of Diffusion*. Clarendon Press, Oxford.

DENNEN, W. H., BLACKBURN, W. H. & QUESADA, A. 1970. Aluminium in quartz as a geothermometer. *Contributions to Mineralogy and Petrology*, **27**, 332–342.

FRASER, G. L., PATTISON, D. R. M. & HEAMAN, L. M. 2004. Age of the Ballachulish and Glencoe Igneous

Complexes (Scottish Highlands), and paragenesis of zircon, monazite and baddeleyite in the Ballachulish Aureole. *Journal of the Geological Society, London*, **161**, 47–462.

GÖTZE, J., PLÖTZE, M., TICHOMIROWA, M., FUCHS, H. & PILOT, J. 2001. Aluminium in quartz as an indicator of the *temperature* of formation of agate. *Mineralogical Magazine*, **65**, 407–413.

GÖTZE, J., PLÖTZE, M., GRAUPNER, T., HALLBAUER, D. K. & BRAY, C. J. 2004. Trace element incorporation into quartz: a combined study by ICP-MS, electron spin resonance, cathodoluminescence, capillary ion analysis, and gas chromatography. *Geochimica et Cosmochimica Acta*, **68**, 3741–3759.

GUILLONG, M., MEIER, D. L., ALLAN, M. M., HEINRICH, C. A. & YARDLEY, B. W. D. 2007. SILLS: a Matlab-based program for the reduction of laser ablation ICP-MS data of homogeneous materials and inclusions. *In*: SYLVESTER, P. (ed.) *Laser Ablation ICP-MS in the Earth Sciences: Current Practices and Outstanding Issues*. Mineralogical Association of Canada Short Course Series, 328–333.

GRUJIC, D., STIPP, M. & WOODEN, J. L. 2011. Thermometry of quartz mylonites: importance of dynamic recrystallisation on Ti-in-quartz re-equilibration. *Geochemistry, Geophysics, Geosystems*, **12**, http://dx.doi.org/10.1029/2010GC003368

HARTE, B., PATTISON, D. R. M., HEUSS-ASSBICHLER, S., HORNES, S., MASCH, L. & WEISS, S. 1991. Evidence of fluid phase behaviour and controls in the intrusive complex and its aureole. *In*: VOLL, G., TÖPEL, J., PATTISON, D. R. M. & SEIFERT, F. (eds) *Equilibrium and Kinetics in Contact Metamorphism: The Ballachulish Igneous Complex and Its Aureole*. Springer Verlag, Heidelberg, 19–38.

HAYDEN, L. & WATSON, B. E. 2007. Rutile saturation in hydrous siliceous melts and its bearing on Ti-thermometry of quartz and zircon. *Earth and Planetary Science Letters*, **258**, 561–568.

HOERNES, S. & VOLL, G. 1991. Detrital quartz and K-feldspar in quartzites as indicators of oxygen isotope exchange kinetics. *In*: VOLL, G., TÖPEL, J., PATTISON, D. R. M. & SEIFERT, F. (eds) *Equilibrium and Kinetics in Contact Metamorphism: The Ballachulish Igneous Complex and Its Aureole*. Springer Verlag, Heidelberg, 315–326.

HOLNESS, M. B. 1998. Contrasting rock permeability in the aureole of the Ballachulish igneous complex, Scottish Highlands: the influence of surface energy? *Contributions to Mineralogy and Petrology*, **131**, 86–94. http://dx.doi.org/10.1007/s004100050380.

HOLNESS, M. B. & CLEMENS, J. D. 1999. Partial melting of the Appin Quartzite driven by fracture-controlled H_2O infiltration in the aureole of the Ballachulish Igneous Complex, Scottish Highlands. *Contributions to Mineralogy and Petrology*, **136**, 154–168.

HOLNESS, M. & WATT, G. 2001. Quartz recrystallisation and fluid flow during contact metamorphism: a cathodoluminescence study. *Geofluids*, **1**, 215–228.

HUANG, R. & AUDETAT, A. 2012. The titanium-in-quartz (TitaniQ) thermobarometer: A critical examination and re-calibration. *Geochimica et Cosmochemica Acta*, **84**, 75–89. http://dx.doi.org/10.1016/j.gca.2012.01.009.

KAWASAKI, T. & OSANAI, Y. 2008. Empirical thermometer of TiO_2 in quartz for ultrahigh-temperature granulites of East Antarctica. *In*: SATISH-KUMAR, M., MOTOYOSHI, Y., OSANAI, Y., HIROI, Y. & SHIRAISHI, K. (eds) *Geodynamic Evolution of East Antarctica: A Key to the East–West Gondwana Connection*. Geological Society, London, Special Publications, **308**, 419–430.

KOHN, M. J. & NORTHRUP, C. J. 2009. Taking mylonites' temperatures. *Geology*, **37**, 47–50.

KORCHINSKI, M., LITTLE, T. A., SMITH, E. & MILLET, M.-A. 2012. Variation of Ti-in-quartz in gneiss domes exposing the world's youngest ultrahigh-pressure rocks, D'Entrecasteaux Islands, Papua New Guinea. *Geochemistry, Geophysics, Geosystems*, **13**, http://dx.doi.org/10.1029/2012GC004230

KROLL, H., KRAUSE, C. & VOLL, G. 1991. Disordering, re-ordering and unmixing in alkali feldspars from contact-metamorphosed quartzites. *In*: VOLL, G., TÖPEL, J., PATTISON, D. R. M. & SEIFERT, F. (eds) *Equilibrium and Kinetics in Contact Metamorphism: The Ballachulish Igneous Complex and Its Aureole*. Springer Verlag, Heidelberg, 267–296.

KRUHL, J. H. 1998. Reply: prism- and basal-plane parallel subgrain boundaries in quartz: a microstructural geothermobarometer. *Journal of Metamorphic Petrology*, **16**, 142–146.

KRUHL, J. H. & NEGA, M. 1996. The fractal shape of sutured quartz grain boundaries: application as a geothermometer. *Geologische Rundschau*, **85**, 38.

LANDTWING, M. R. 2005. Relationships between SEM-cathodoluminescence response and trace-element composition of hydrothermal vein quartz. *American Mineralogist*, **90**, 122–131.

LASAGA, A. C. & JIANG, J. 1995. Thermal history of rocks: P–T–t paths from geospeedometry, petrologic data and inverse theory techniques. *American Journal of Science*, **295**, 697–741.

LAW, R. D., SEARLE, M. P. & SIMPSON, R. L. 2004. Strain, deformation temperatures and vorticity of flow at the top of the Greater Himalayan Slab, Everest Massif, Tibet. *Journal of the Geological Society, London*, **161**, 305–320.

LAW, R. D., MAINPRICE, D., CASEY, M., LLOYD, G. E., KNIPE, R. J., COOK, B. & THIGPEN, J. R. 2010. Moine thrust zone mylonites at the Stack of Glencoul: I-microstructures, strain and the influence of recrystallization on quartz crystal fabric development. *In*: LAW, R. D., BUTLER, R. W. H., HOLDSWORTH, R. E., KRABBENDAM, M. & STRACHAN, R. A. (eds) *Continental Tectonics and Mountain Building: The Legacy of Peach and Horne*. Geological Society, London, Special Publications, **335**, 543–577.

LIND, A. 1996. *Microstructural Stability and the Kinetics of Textural Evolution*. PhD thesis, University of Leeds.

LLOYD, G. E. 1987. Atomic number and crystallographic contrast images with the SEM: a review of backscattered electron techniques. *Mineralogical Magazine*, **51**, 3–19.

MAMTANI, M. A. & GREILING, R. O. 2010. Serrated quartz grain boundaries, temperature and strain rate: testing fractal techniques in a syntectonic granite. *In*: SPALLA, M. I., MAROTTA, A. M. & GOSSO, G. (eds) *Advances in Interpretation of Geological Processes:*

Refinement of Multi-Scale Data and Integration in Numerical Modelling. Geological Society, London, Special Publications, **332**, 35–48.

MASCH, L. & HEUSS-ASSBICHLER, S. 1991. Decarbonation reactions in siliceous dolomites and impure limestones. *In*: VOLL, G., TÖPEL, J., PATTISON, D. R. M. & SEIFERT, F. (eds) *Equilibrium and Kinetics in Contact Metamorphism: The Ballachulish Igneous Complex and Its Aureole.* Springer Verlag, Heidelberg, 211–227.

MCCAIG, A., COVEY-CRUMP, S. J., BEN ISMAIL, W. & LLOYD, G. E. 2007, Fast diffusion along mobile grain boundaries in calcite. *Contributions to Mineralogy and Petrology*, **153**, 159–175.

MORGAN, D. J. & BLAKE, S. 2006, Magmatic residence times of zoned phenocrysts – introduction and application of the Binary Element Diffusion Modelling (BEDM) technique. *Contributions to Mineralogy and Petrology*, **151**, 58–70.

MÜLLER, A., WIEDENBECK, M., VAN DEN KERKHOF, A. M., KRONZ, A. & SIMON, K. 2003. Trace elements in quartz – a combined electron microprobe, secondary ion mass spectrometry, laser-ablation ICP-MS, and cathodoluminescence study. *European Journal of Mineralogy*, **15**, 747–763.

MÜLLER, A., HERRINGTON, R., ARMSTRONG, R., SELTMANN, R., KIRWIN, D. J., STENINA, N. G. & KRONZ, A. 2010. Trace elements and cathodoluminescence of quartz in stockwork veins of Mongolian porphyry-style deposits. *Mineralium Deposita*, **45**, 707–727.

PATTISON, D. R. M. 1989. P–T conditions and the influence of graphite on pelitic phase relations in the Ballachulish Aureole, Scotland. *Journal of Petrology*, **30**, 1219–1244.

PATTISON, D. 1991. P-T-a(H$_2$O) conditions in the thermal aureole. *In*: VOLL, G., TÖPEL, J., PATTISON, D. R. M. & SEIFERT, F. (eds) *Equilibrium and Kinetics in Contact Metamorphism: The Ballachulish Igneous Complex and Its Aureole.* Springer Verlag, Heidelberg, 327–350.

PATTISON, D. R. M. & HARTE, B. 1988, Evolution of structurally contrasting anatectic migmatites in the 3-kbar Ballachulish aureole, Scotland. *Journal of Metamorphic Geology*, **6**, 475–494.

PATTISON, D. R. M. & HARTE, B. 1997. The geology and evolution of the Ballachulish Igneous Complex and Aureole. *Scottish Journal of Geology*, **33**, 1–29.

PATTISON, D. R. M. & VOLL, G. 1991. Regional geology of the Ballachulish area. *In*: VOLL, G., TÖPEL, J., PATTISON, D. R. M. & SEIFERT, F. (eds) *Equilibrium and Kinetics in Contact Metamorphism: The Ballachulish Igneous Complex and Its Aureole.* Springer Verlag, Heidelberg, 19–38.

PEARCE, N. J. G., PERKINS, W. T., WESTGATE, J. A., GORTON, M. P., JACKSON, S. E., NEAL, C. R. & CHENERY, S. P. 1997. A compilation of new and published major and trace element data for NIST SRM 610 and NIST SRM 612 glass reference materials. *Geostandards and Geoanalytical Research*, **21**, 115–144.

PERNY, B., EBERHARDT, P., RAMSEYER, K., MULLIS, J. & PANKRATH, R. 1992. Microdistribution of Al, Li, and Na in α quartz: possible causes and correlation with short-lived cathodoluminescence. *American Mineralogist*, **77**, 534–544.

PIAZOLO, S., PRIOR, D. J. & HOLNESS, M. D. 2005. The use of combined cathodoluminescence and EBSD analysis: a case study investigating grain boundary migration mechanisms in quartz. *Journal of Microscopy*, **217**, 152–161.

REINERS, P. W. 2009. Nonmonotonic thermal histories and contrasting kinetics of multiple thermochronometers. *Geochimica et Cosmochimica Acta*, **73**, 3612–3629.

RUSK, B. G., REED, M. H., DILLES, J. H. & KENT, A. J. R. 2006. Intensity of quartz cathodoluminescence and trace-element content in quartz from the porphyry copper deposit at Butte, Montana. *American Mineralogist*, **91**, 1300–1312.

RUSK, B. G., LOWERS, H. A. & REED, M. H. 2008. Trace elements in hydrothermal quartz: relationships to cathodoluminescent textures and insights into vein formation. *Geology*, **36**, 547–550.

SATO, K. & SANTOSH, M. 2007. Titanium in quartz as a record of ultrahigh-temperature metamorphism: the granulites of Karur, southern India. *Mineralogical Magazine*, **71**, 143–154.

SAWAKUCHI, A. O., DEWITT, R. & FALEIROS, F. M. 2011. Correlation between thermoluminescence sensitivity and crystallization temperatures of quartz: potential application in geothermometry. *Radiation Measurements*, **46**, 51–58.

SEYEDOLALI, A., KRINSLEY, D. H., BOGGS, S., JR, O'HARA, P. F., DYPVIK, H. & GOLES, G. G. 1997. Provenance interpretation of quartz by scanning electron microscope-cathodoluminescence fabric analysis. *Geology*, **25**, 787–790.

SPEAR, F. S. & WARK, D. 2009. Cathodoluminescence imaging and titanium thermometry in metamorphic quartz. *Journal of Metamorphic Geology*, **27**, 187–205.

STIPP, M., STUNITZ, H., HEILBRONNER, R. & SCHMID, S. M. 2002*a*. The eastern Tonale fault zone; a 'natural laboratory' for crystal plastic deformation of quartz over a temperature range from 250 to 700C. *Journal of Structural Geology*, **24**, 1861–1884.

STIPP, M., STUNITZ, H., HEILBRONNER, R. & SCHMID, S. M. 2002*b*. Dynamic recrystallization of quartz: correlation between natural and experimental conditions. *In*: DE MEER, S., DRURY, M. R., DE BRESSER, J. H. P. & PENNOCK, G. M. (eds) *Deformation Mechanisms, Rheology and Tectonics: Current Status and Future Perspectives.* Geological Society, London, Special Publications, **200**, 171–190.

STORM, L. C. & SPEAR, F. S. 2009. Application of the titanium-in-quartz thermometer to pelitic migmatites from the Adirondack Highlands, New York. *Journal of Metamorphic Geology*, **27**, 479–494.

THIGPEN, J. R., LAW, R. D., LLOYD, G. E., BROWN, S. J. & & COOK, B. 2010*a*. Deformation temperatures, vorticity of flow and strain symmetry in the Loch Eriboll mylonites, NW Scotland: implications for kinematic and structural evolution of the northernmost Moine thrust zone. *In*: LAW, R. D., BUTLER, R. W. H., HOLDSWORTH, R., KRABBENDAM, M. & STRACHAN, R. (eds) *Continental Tectonics and Mountain Building – The Legacy of Peach and Horne.* Geological Society, London, Special Publications, **335**, 623–662.

THIGPEN, J. R., LAW, R. D., LLOYD, G. E. & BROWN, S. J. 2010*b*. Deformation temperatures, vorticity of flow, and strain in the Moine thrust zone and Moine nappe: reassessing the tectonic evolution of the Scandian foreland-hinterland transition zone. *Journal of Structural Geology*, **32**, 920–940.

THIGPEN, J. R., LAW, R. D. *ET AL.* 2013. Thermal structure and tectonic evolution of the Scandian orogenic wedge, Scottish Caledonides: integrating geothermometry, deformation temperatures, and kinematic-thermal modelling. *Journal of Metamorphic Geology*, **31**, 813–842, http://dx.doi.org/10.1111/jmg.12046

THOMAS, J. B., WATSON, B. E., SPEAR, F. S., SHEMELLA, P. T., NAYAK, S. K. & LANZIROTTI, A. 2010. TitaniQ under pressure: the effect of pressure and temperature on the solubility of Ti in quartz. *Contributions to Mineralogy and Petrology*, **160**, 743–759.

VAZQUEZ, J. A., KYRIAZIS, S. F., REID, M. R., SEHLER, R. C. & RAMOS, F. C. 2009. Thermochemical evolution of young rhyolites at Yellowstone: evidence for a cooling but periodically replenished post-caldera magma reservoir. *Journal of Volcanology and Geothermal Research*, **188**, 186–196.

VOLL, G. 1991. The setting of the Ballachulish intrusive igneous complex in the Scottish Highlands. *In*: VOLL, G., TÖPEL, J., PATTISON, D. R. M. & SEIFERT, F. (eds) *Equilibrium and Kinetics in Contact Metamorphism: The Ballachulish Igneous Complex and Its Aureole*. Springer Verlag, Heidelberg, 3–17.

WARK, D. & WATSON, B. E. 2006. TitaniQ: a titanium-in-quartz geothermometer. *Contributions to Mineralogy and Petrology*, **152**, 743–754.

WARK, D., HILDRETH, W., SPEAR, F. S., CHERNIAK, D. J. & WATSON, B. E. 2007. Pre-eruption recharge of the Bishop magma system. *Geology*, **35**, 235–238.

WATT, G. 1997. Cathodoluminescence and trace element zoning in quartz phenocrysts and xenocrysts. *Geochimica et Cosmochimica Acta*, **61**, 4337–4348.

WEISS, S. & TROLL, G. 1989. The Ballachulish Igneous Complex, Scotland: petrography, mineral chemistry, and order of crystallization in the monzodiorite-quartz diorite suite and in the granite. *Journal of Petrology*, **30**, 1069–1115.

WEISS, S. & TROLL, G. 1991. Thermal conditions and crystallisation sequences as deduced from whole-rock and mineral chemistry. *In*: VOLL, G., TÖPEL, J., PATTISON, D. R. M. & SEIFERT, F. (eds) *Equilibrium and Kinetics in Contact Metamorphism: The Ballachulish Igneous Complex and Its Aureole*. Springer Verlag, Heidelberg, 67–98.

WIEBE, R. A., WARK, D. & HAWKINS, D. P. 2007. Insights from quartz cathodoluminescence zoning into crystallization of the Vinalhaven granite, coastal Maine. *Contributions to Mineralogy and Petrology*, **154**, 439–453.

WILSON, C., SEWARD, T., CHARLIER, B., ALLAN, A. & BELLO, L. 2012. An assessment of the validity of temperature and pressure estimates from Ti concentrations in magmatic and hydrothermal quartz. *Contributions to Mineralogy and Petrology*, **164**, 359–368.

WU, X.-Q., LIU, D.-L., LI, Z.-S. & YANG, Q. 2006. A new fractal method for the determination of deformation temperatures and strain rates – a case study of the Fuchashan tectonite in the Tanlu fault. *Geology in China*, **33**, 153–159.

Effect of strain geometry on the petrophysical properties of plastically deformed aggregates: experiments on Solnhofen limestone

SERGIO LLANA-FÚNEZ[1,2]* & ERNEST H. RUTTER[1]

[1]*School of Earth, Atmospheric and Environmental Sciences, University of Manchester, Oxford Road, Manchester M13 9PL, UK*

[2]*Present Address: Departamento d Geología, Universidad de Oviedo, calle Arias de Velasco s/n, 33005 Oviedo, Spain*

**Corresponding author (e-mail: slf@geol.uniovi.es)*

Abstract: Specimens of Solnhofen limestone were deformed under conditions where calcite deforms plastically using four experimental configurations: extension, torsion, direct shear and axisymmetric shortening. All experiments were run on dry specimens at the same temperature (600 °C), confining pressure (200 MPa) and comparable strain rates ($c.\ 10^{-4}\,\mathrm{s}^{-1}$). The different experimental settings and the heterogeneity of deformation within some of the specimens provided a large range of strain geometries. They allowed locally imposed strain geometries to be related to the crystallographic preferred orientation (CPO) patterns of calcite and the orientation of the shape fabric of calcite grains. CPO in calcite was measured using electron back-scattered diffraction (EBSD) in scanning electron microscopy. The development of CPO during deformation under the dominance of intracrystalline plasticity contains information about the strain geometry accumulated in rocks in 3D, although in nature the strain geometry can be modified by dynamic recrystallization that was not seen in the experiments. The different CPO patterns have a significant effect on the velocity structure of the deformed aggregates. Seismic properties inferred from CPO show that the orientation of the fastest V_p wave aligns with principal strain directions that are not equivalent in different strain geometries.

In rocks deforming at moderate temperatures, stress and strain-rate combinations may lead to the development of crystallographic preferred orientation (CPO) in relation to one predominant deformation mechanism, intracrystalline plasticity, or in relation to a combination of different deformation mechanisms involving not only intracrystalline plasticity but also diffusive mass transfer processes and grain boundary sliding (Poirier 1985). Owing to the strong velocity anisotropy in the propagation of seismic waves in most rock-forming minerals, the development of CPO will cause the rocks not to act as an aggregate of mineral grains with bulk isotropic properties but instead to show a seismic anisotropy, depending on the orientation pattern of minerals (Christensen 1984; Barruol & Mainprice 1993).

The activation of particular slip systems in any crystal is temperature dependent, as is the operation of diffusive mass transfer processes, the latter being more effective at smaller grain sizes. When a general deformation of a polycrystalline aggregate is to be accommodated geometrically in three dimensions (3D) only by intracrystalline plasticity (without grain boundary void opening and grain overlaps) the activation of five independent slip systems is required (Taylor 1938; Lister & Paterson 1979; Lister & Williams 1979; Lister & Hobbs 1980; Wenk *et al.* 1989). This geometrical constraint is relaxed when other deformation mechanisms are operating simultaneously or when deformation is heterogeneous at the grain scale, such as to avoid the opening of gaps and the 'overlap' of grain into one another. Some low symmetry minerals, such as plagioclase (Ji & Mainprice 1988) or omphacite (Mauler *et al.* 1998), have fewer independent slip systems than five and, in consequence, when deformed ductilely at moderate temperatures, intracrystalline plasticity is accompanied by additional deformation mechanisms to accommodate strain. On the other hand, some higher symmetry minerals with sufficient slip systems, for example olivine (Nicolas *et al.* 1973), seem to deform plastically by fewer independent slip systems than five because some systems are much harder than others. Thus, in consequence, geometrically they also require the operation of additional deformation processes.

From the geometrical point of view, the operation of slip along a slip plane in a constrained crystallite produces a rotation of the crystal lattice (and the slip system) with respect to an external reference frame until the crystallite reaches a steady orientation with respect to the stress or instantaneous strain rate imposed. This orientation in the geological literature is named as 'easy slip' based on the

From: LLANA-FÚNEZ, S., MARCOS, A. & BASTIDA, F. (eds) 2014. *Deformation Structures and Processes within the Continental Crust*. Geological Society, London, Special Publications, **394**, 167–187. First published online January 27, 2014, http://dx.doi.org/10.1144/SP394.12

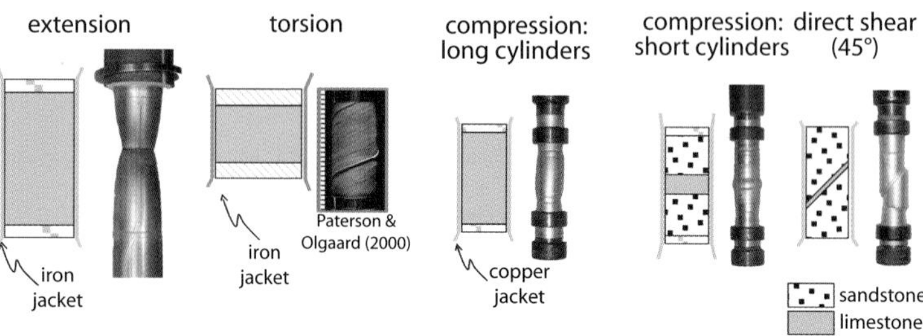

Fig. 1. Assemblage of the specimens in four experimental settings: axisymmetric shortening and extension, direct shear, and torsion. The dimensions of the specimens are given in Table 1.

observation in naturally deformed rocks that the easiest slip system tends to align in some way with the structural reference framework, usually the shear plane in a shear zone (e.g. Law 1990). The arrangement of a pair of conjugate slip planes in plane strain is sensitive to the sense of shear since only one will be in the 'easy' orientation, and the consequence is that asymmetry will be introduced in the CPO pattern. Over the scale of a polycrystalline aggregate, non-coaxial deformation will affect the angles of both the synthetic and antithetic slip planes with respect to the structural reference frame (external symmetry), and it will also affect the intensity in CPO patterns of both planes and slip directions, generating an internal symmetry (e.g. Passchier & Trouw 1996; Llana-Fúnez 2002). The CPO patterns can be used in these scenarios to infer the kinematics of the deformation, particularly when other potentially asymmetric microstructures such as pressure shadows, shear bands or delta porphyroblasts are absent. From the 3D distribution of slip planes and directions in a rock deformed by intracrystalline plasticity, it should then be possible to locate the orientation of the principal axes of strain rate in relation to the deformation accommodated by the rock.

The present paper attempts to relate experimentally the geometry of the imposed stress and accumulated strain to the development of the CPO pattern in a polycrystalline aggregate deforming predominantly by intracrystalline plasticity in a range of strain geometries (Figs 1 & 2). Calcite is a very accessible mineral from the experimental point of view in that it deforms at moderate temperatures and differential stresses and it has sufficient independent slip systems to accommodate plastic deformation (de Bresser & Spiers 1997). The CPO patterns in the deformed Solnhofen

limestone specimens will vary from sample to sample, depending on the strain path followed during the experiment, but also within the same specimen, as strain is distributed and accumulated heterogeneously.

Of course, CPO patterns with different geometries will also determine the petrophysical properties. In the discussion we explore how the orientation patterns created experimentally under different strain paths affect the propagation features of seismic waves. Whilst the experimental starting material, a micritic calcite limestone, may not be very common throughout the crust at depths where intracrystalline plasticity is expected to operate, quartz, on the other hand, is common, and has some very similar crystallographic features; in many cases quartz is the rock-forming phase that controls the mechanical behaviour of the middle and sometimes the lower continental crust. Our observations of the effects of 3D strain geometry by intracrystalline plasticity on the arrangement of crystallographic elements and the ultimate effects on the velocity structure of the aggregates are applicable to other rock-forming crustal minerals, such as quartz, in the continental crust.

Starting material and experimental approach

The starting material in the present experiments was Solnhofen limestone, a micritic calcite limestone, which has a mean grain size of 4 μm. This is sufficiently small to register strain gradients in deformed specimens over distances well below 1 mm. Secondary phases in the rock (less than 5 wt%) are mostly organic matter and clays, concentrated along grain boundaries. They have the effect

Fig. 2. Grain shape microstructures (polished sections, reflected light microscopy) of deformed specimens under: (**a**) axisymmetric extension, profile section (se05); (**b**) torsion, profile section (tor1); (**c**) direct shear, profile section (sf10); (**d**) axisymmetric shortening, top half profile section through the middle of the specimen (sf04).

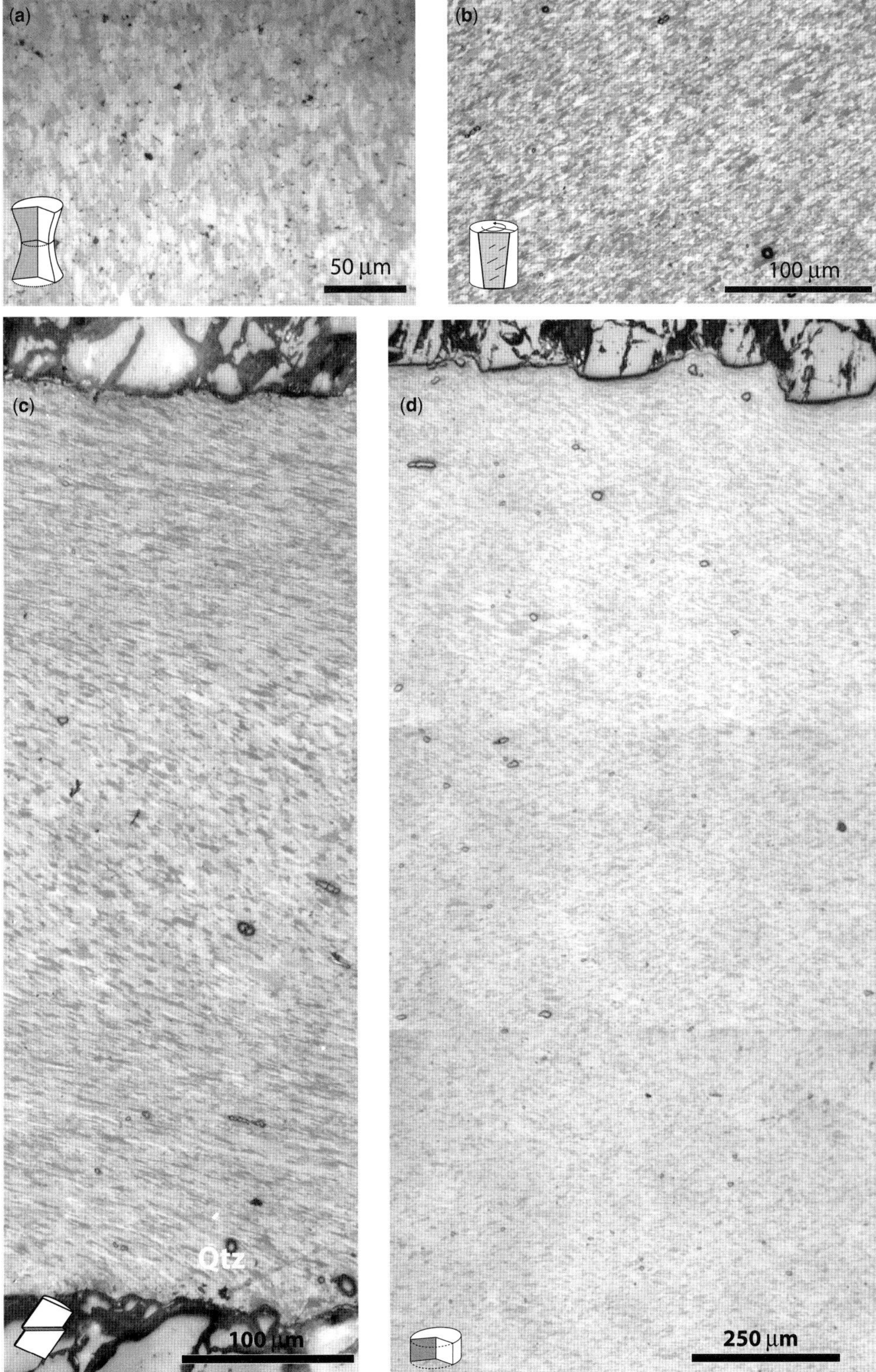
(a)
50 µm
(b)
100 µm
(c)
Qtz
100 µm
(d)
250 µm

Table 1. *Deformation tests on dry Solnhofen limestone at 200 MPa and 600 °C*

	Dimensional data		Bulk strain	
Test	Thickness/length Initial mm	Final mm	Thinning ε	Shear strain γ
Extension se05	31.09	41.03	−1.32	–
Torsion	10.0	10.0	–	1.55
Direct shear sf10	0.91	0.81	0.12*	2.66
Short cylinders sf04	4.11	2.1	0.49	–

*Consistent with a finite range in length from 13.42 to 14.32 mm.

of pinning the grain boundaries and reducing their mobility at high temperature (Walker *et al.* 1990). The microstructural arrangement of these secondary phases is inferred to be such that grain growth is prevented until temperatures above 700 °C, compared to pure, fine-grained calcite aggregates which grow at such temperatures (Schmid *et al.* 1987; Walker *et al.* 1990). This elevated minimum allowed the experiments to be run at higher temperatures than might otherwise be possible, with

the activation of slip systems and the possibility of flow becoming dominated by high-temperature grain boundary sliding. Solnhofen limestone has a weak initial CPO characterized by a distribution of c-axes preferentially within the bedding plane (Fig. 3b) (see also Casey *et al.* 1998). All experimental samples were cored from the same block of Solnhofen limestone in a direction normal to the bedding. Cores are from the same bed and in close proximity (the block is less than 30 cm long).

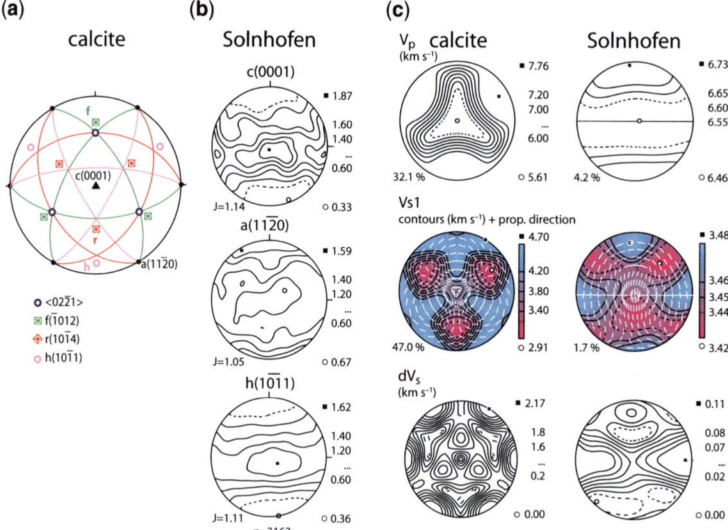

Fig. 3. Experimental starting material. (**a**) Equal area projection of a calcite crystal showing major crystallographic elements involved in intracrystalline plastic deformation. (**b**) CPO patterns in undeformed starting material: Solnhofen limestone (from Llana-Fúnez & Rutter 2005). Orientation is pole to bedding at the top of the primitive circle, contours are multiples of a uniform distribution (m.u.d.). (**c**) Seismic properties of single crystals of calcite and undeformed Solnhofen specimen from the same block used in the experiments (orientation as in (**b**)). All stereoplots are upper hemisphere equal area projections.

A wide range of strain geometries was achieved experimentally in two ways: firstly, by running experiments with different specimen configurations (extension, torsion, direct shear and axisymmetric shortening), and secondly, by taking advantage of strain heterogeneities that arise in certain specimen configurations. For example, friction at the specimen-piston/spacer interfaces when using short cylinders (discs) of rock (in axisymmetric shortening and in direct shear) can lead to substantial strain heterogeneity and variation in strain geometry (e.g. Llana-Fúnez & Rutter 2005). Therefore, the spatial variation of the CPO generated in this fine-grained material is accessible to study by using electron back-scattered diffraction (EBSD) in scanning electron microscopy (SEM), to measure systematically the lattice orientation at every point in a grid with 1 μm step size.

The experiments performed and the analytical technique used allowed changes in deformation geometry to be mapped and related either to strain accumulation or to the effect of stress heterogeneities inherent in some of the experimental conditions using shape fabric and CPO in calcite. Subsequent calculation of the seismic velocity from the orientation data showed that the geometry of the CPO has an effect on the seismic anisotropy, in particular, on the location of velocity maxima for the propagation of P-waves. This work is directly relevant to the understanding of the underlying processes of intracrystalline plasticity in rocks and also has direct application for the prediction of anisotropy of the propagation of seismic waves in rocks rich in quartz in crustal-scale shear zones.

Experimental procedures

Deformation apparatus

Two different deformation apparatus were used in the experiments: a Nimonic rig (e.g. Rutter *et al.* 1984; Covey-Crump 1992; and named for the Ni-alloy from which the pressure vessel is fabricated), and a Paterson rig (e.g. Paterson & Olgaard 2000). In both rigs, a cylindrical specimen, jacketed in an annealed copper sleeve with a wall thickness of 0.25 mm, was deformed between two loading pistons. The Nimonic rig uses an externally heated pressure vessel with water as the confining medium. The vessel hosts specimens of up to 10 mm in diameter and 25 mm in length. Differential loading capacity is 4 tonnes (equivalent to 500 MPa differential stress on the sample) at temperatures up to 700 °C and confining pressures up to 250 MPa. The displacement rate of the piston is kept constant at 0.3 mm s^{-1}, corresponding to a strain rate of the order of 10^{-4} s^{-1}. The Nimonic rig was used for standard shortening tests on cylinders with a

length to diameter ratio of 2:1 for axisymmetric compression tests using short cylinders (Llana-Fúnez & Rutter 2005), and for direct shear tests with saw-cut specimens positioned at an angle to the axial shortening direction (Llana-Fúnez & Rutter 2008) (Fig. 1).

The Paterson rig is an internally heated deformation apparatus, which uses argon gas as the confining medium. The specimens can be up to 15 mm in diameter and 30 mm in length. The differential loading capacity is 10 tonnes at temperatures up to 1200 °C, and confining pressures of up to 300 MPa can be applied. In this study the Paterson apparatus was used for extension and torsion tests. The extension test (Fig. 1) was run at the Rock Deformation Laboratory of the University of Manchester. The torsion test (Fig. 1) was run by J. Mecklenburgh at the Deformation Laboratory of the University of Bayreuth, using a similar deformation apparatus.

Experimental configurations

In the experimental configurations used in the deformation tests, the specimens have two types of surface: a free surface where the copper jackets are in contact with the confining medium (in our case always a fluid, therefore not supporting shear stress), and a solid–solid interface, which is the contact of the specimen either with a rock spacer (in direct shear) or directly against the loading pistons. All surfaces are subject to the confining pressure applied inside the pressure vessel and this is sufficient to hold the specimen assembly altogether, in torsion and in extension.

In extensional tests, the maximum principal stress is exerted by the confining pressure and the differential stress is obtained by withdrawing one of the axial pistons (Rutter 1998). To prevent separation of the specimen ends from the loading pistons, it is essential that the differential stress applied is less than the confining pressure. Depending on the rheology of the experimental material (Rutter 1998), deformation leads to development of a necking instability in the centre of the specimen (Figs 1 & 4) that can be either strongly localized or diffuse. A strong strain gradient develops on either side of the neck. In torsion tests, the torque is applied along the surface where the specimen ends contact the loading pistons, and the confining pressure must be sufficiently high to suppress frictional sliding across these surfaces. Because there is no shortening of the length of the specimen, this configuration produces a rather homogeneously deformed microstructure (Paterson & Olgaard 2000).

The orientation of the solid–solid interfaces with respect to the direction of axial loading determines normal and shear stresses across the interface. In axisymmetric shortening, there is nominally no interface-parallel shear stress because the applied axial load is normal to the interface. In this setup, the presence of unconstrained (jacketed) surfaces allows radial extrusion of the specimen with progressive shortening, thereby inducing shear strain in the specimen close to the contacts with the pistons if there is no slip between the specimen and the pistons. When samples are long with respect to the diameter, radial flow usually results in nucleation of conical shear zones; these zones isolate an internal, conical strain-free area (the strain shadow) from the rest of the sample (Jaeger & Cook 1976). At low bulk strains, these conical shear zones are broad and it is commonly assumed that outside the conical strain shadows, strain is fairly homogeneously distributed. As the pistons approach each other, or as the ratio of length to diameter decreases, the effect of the friction along the specimen ends becomes predominant in the system. The conical shear zones reduce their angle to the specimen ends, decreasing the size of the strain shadow, which starts to deform internally. The sideways extrusion of material (in the middle towards the sides) produces large shear strains parallel to the bases of the 'cylinders', passing into axisymmetric shortening in the original centre of the cylinder. With these boundary conditions, the lateral extrusion of rock is driven by the development of a stress gradient, reaching the maximum mean stress in the middle of the specimen (Jaeger 1962). In such axisymmetric shortening of a short cylinder of quartzite, the mean stresses in the middle of the specimen can be sufficiently high to produce the nucleation of coesite, the high-pressure polymorph of quartz (Tullis *et al.* 1973). Details of compression tests on short cylinders of Solnhofen limestone, conducted with the aim of inducing heterogeneity to produce a range of strain geometries, and the associated experimental results, were presented by Llana-Fúnez & Rutter (2005).

When the angle of the specimen–piston contact departs from $90°$ with respect to the shortening direction, the shear stress along this surface increases and bulk strain approaches simple shear parallel to the interface together with some coaxial component of interface-normal shortening. This widely employed specimen configuration is named 'direct shear' and a wide range of angles between the interface and the shortening direction can be employed. The most common angles are $45°$ and $30°$, both in high-temperature tests (Schmid *et al.* 1987; Dell'Angelo & Tullis 1989; Llana-Fúnez & Rutter 2008) and in experiments on fault gouge materials (Shimamoto & Logan 1981; Dresen 1991; Logan *et al.* 1992). There is also a $0°$ arrangement (double-direct shear) that has been used for high-temperature plastic flow and also for experiments on fault

gouge (e.g. Rutter & Rusbridge 1976; Saffer & Marone 2003). The propensity for lateral extrusion of the specimen decreases with decreasing the angle of the shortening direction with the orientation of the interface, because the interfacial normal stress decreases.

In all the experiments reported here, the samples were confined at 200 MPa and heated to 600 °C. Conventional strain rates in the different configurations were about $10^{-4}\,\mathrm{s}^{-1}$, at which conditions Solnhofen limestone predominantly deforms well within the intracrystalline plasticity regime (regime 2 in Schmid *et al.* 1987). Larger isolated grains do show the development of twins as a consequence of the experimental deformation (see fig. 9a in Llana-Fúnez & Rutter 2005); however, no twins were observed in smaller grains. The uniformity of grain size distribution and lack of original shape fabric in the original material make the specimen microstructure a perfect canvas on which to record the nature of finite strain accumulated during an experimental deformation event.

Preparation of deformed samples

Deformed specimens were cut parallel to the cylinder axis and mounted in a polyester resin block that was ground and polished using alumina and ultimately colloidal silica (Syton™). Microstructural observations on ultrapolished surfaces were made first using reflected light in a conventional optical microscope (see Bestmann *et al.* 2000). Further analysis in the SEM included the collection of EBSD patterns and simultaneous indexing by software Channel 5 (Oxford Instruments). Most analyses were done at the Material Science Centre at the University of Manchester using scan maps at 1 µm step size. The quality of the maps obtained was insufficient for a detailed grain size analysis. In the absence of large differences of grain size and considering the uniformity of grain size at the scale of every EBSD map, all measured points collected in the maps with sufficient quality were used for contouring. The work with experimentally deformed specimens and natural samples of gneisses, was completed at the Department of Earth and Ocean Sciences at the University of Liverpool. The CPO data were processed to calculate seismic velocities using software by D. Mainprice (Mainprice & Nicolas 1989; Mainprice 1990; Mainprice & Humbert 1994).

Seismic properties calculated from petrofabric data

Seismic properties of Solnhofen limestone deformed under various strain paths were calculated using the density of calcite together with the elastic properties of the deformed sample calculated from the orientation of calcite at every point in the EBSD maps collected. There are several averaging schemes for the elastic properties of the aggregate. Here, the arithmetic mean (Voigt-Reuss-Hill), between equal strain (Voigt) and equal stress (Reuss) averaging schemes, was used (Table 2). Velocities were calculated using software developed by D. Mainprice and co-workers and made freely available (Mainprice 1990; Mainprice & Humbert 1994). The elastic constants taken for calcite are those of Dandekar (1968). Anisotropy is defined as the difference between maximum and minimum velocity divided by the average velocity.

Calcite as a single crystal displays moderate anisotropy in the propagation of seismic waves, particularly P-waves (Fig. 3c). The most significant feature is that V_p is lowest parallel to the c-axis (Fig. 3c). The weak initial CPO in Solnhofen limestone with the arrangement of c-axes parallel to bedding implies that V_p will be fast at right angles to the bedding and slowest parallel to it (Fig. 3c). S-waves are always more difficult to interpret, but from Figure 3c it is obvious that the propagation direction for the fast wave will also be perpendicular to the bedding. The largest S-wave birefringence in Solnhofen limestone is found within the bedding plane and exceeds 0.1 km s^{-1}. Knowledge of the c-axis CPO patterns is necessary not only to interpret strain geometry in the rocks, but also to anticipate the seismic properties of P-waves because they propagate slowest along the c-axis.

Strain distribution recorded in the microstructure (shape and crystallographic preferred orientation)

Axisymmetric extension test

After the experiments, the originally cylindrical specimens lengthen and present a dog-bone external shape with a reduction of section towards the middle of the specimen (Fig. 4). In a section cut parallel to the axis of the specimen the rock develops a shape fabric defined by the stretching of the grains (Fig. 2a). This elongation fabric converges towards the middle of the specimen. The shape fabric is very weak close to the piston and increases its strength towards the middle and this is reflected in the fabric sketch in Figure 4b (bulk stretching strain is 125%).

The c-axis CPO is dominated by the symmetry of girdles and maxima with respect to the stretching direction. The c-axis pattern has two elements: a maximum parallel to the stretching direction and two small-circle girdles (or a double-cleft girdle)

Table 2. *Stiffness coefficients of the elastic tensor of studied aggregates*

EBSD map	Extension se05-m2	Torsion tor1-m1	Direct shear		Shortening		
			sf10-m1	sf10-m4	sf04-site2	sf04-m3	sf04-m7
C_{ij} GPa							
C11	115.64	116.33	120.29	113.04	115.12	119.62	120.87
C22	125.01	114.71	116.41	122.20	117.97	118.28	112.10
C33	115.69	120.51	113.79	119.94	114.77	116.77	117.43
C44	31.38	31.87	31.62	29.31	32.99	30.65	30.46
C55	31.24	32.15	32.24	31.26	31.19	31.58	33.12
C66	30.56	31.36	32.16	33.45	33.04	31.72	32.51
C12	52.50	52.71	54.10	55.13	54.87	53.61	54.22
C13	52.41	54.11	53.96	52.62	52.61	53.33	55.37
C14	0.58	0.48	0.69	−0.19	0.08	−0.50	0.07
C15	−1.47	−0.01	0.44	−1.77	0.04	−0.99	−1.83
C16	−0.46	−0.18	2.81	2.47	0.20	−2.60	−0.76
C23	53.45	53.65	52.97	51.06	54.78	52.16	51.69
C24	0.37	1.14	0.22	0.41	0.94	0.38	−0.24
C25	−1.00	−0.71	−0.34	0.27	−0.17	−1.25	−0.57
C26	−1.20	−0.06	−0.17	−0.89	0.49	−1.48	−0.02
C34	1.27	0.61	−0.64	1.82	0.58	−0.37	−0.56
C35	−1.35	−0.24	−1.61	−1.30	−0.10	−1.69	−0.35
C36	0.06	0.67	0.02	1.62	0.18	−0.61	−0.10
C45	0.11	0.63	−0.20	1.22	0.10	−0.23	−0.02
C46	−0.65	−0.64	−0.21	0.47	−0.15	−0.92	−0.33
C56	0.40	0.29	0.68	−0.39	−0.06	−0.47	0.14

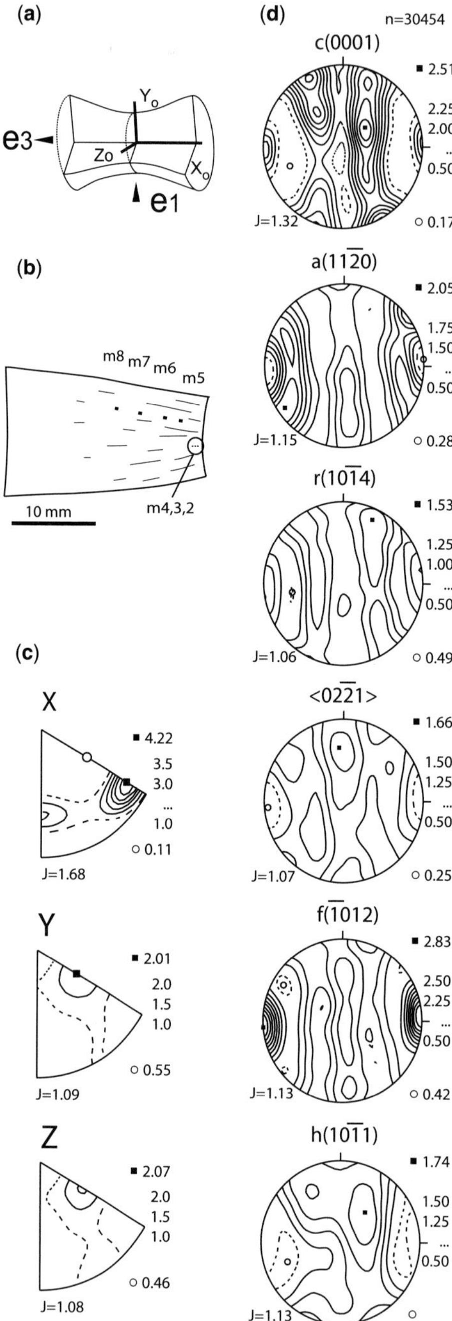

Fig. 4. CPO patterns in axisymmetric extension (specimen se05). (**a**) Sketch indicating principal deformation axes. (**b**) Sketch of the left half of the specimen (bulk stretching strain = 125%) showing the location of the EBSD map se05-m2 (in c and d) near the centre of the sample. (**c**) IPF for the directions x, y and z in (a). Contours in m.u.d. (**d**) pole figures (poles to c, a, r planes, <02–$21>$, poles to f and h planes).

also symmetric about the stretching direction. The inverse pole figures (IPF) taken for three orthogonal directions also reflect the axisymmetric nature of the strain, with greater intensity developed in the stretching direction, and a weaker orientation and similar patterns at right angles to it (y and z in Fig. 4c).

Torsion test

The specimen deformed in torsion (shear strain $\gamma = 1.55$, shear strain rate $= 4.47 \times 10^{-5}$ s^{-1} on outside of the specimen cylinder, Fig. 5) produced a homogeneous shape fabric in the calcite grains. Sections made tangential and parallel to the axis of the cylinder close to the edge incorporate areas of the specimen deformed to varying finite shear strain (it will be zero at the centre of the cylinder). Thus it is expected that foliation traces on the plane of observation will be curved. The bulk twist in the experiment was sufficiently low not to show an obvious change in foliation angle in an axis-parallel section of the cylinder close to the edge (Fig. 2b).

The CPO c-axis pattern collected is characterized by six submaxima, all located at the periphery of the pole figure, showing a hexagonal symmetry, with two of them parallel to the stretching direction (Fig. 5d). IPFs (Fig. 5c) show a classic pattern in the stretching and shortening directions but are undefined in the direction perpendicular to the section of observation.

Direct shear test

Tests in direct shear used pre-cut forcing blocks of sandstone with a 1 mm thick elliptical wafer of Solnhofen limestone placed between them (Fig. 6a, sample sf10, maximum shear offset = 2.2 mm, bulk shear strain $\gamma = 2.66$). The angle of cut was 45° to the cylinder axis and after some bulk shortening, strain accumulated heterogeneously in the specimens. A clear and well-developed shape fabric formed in the calcite grains, varying in intensity and angle with respect to the edges of the specimens. Shape fabric intensity was higher and the angle lower towards the contacts with the spacers and lower and higher respectively towards the middle of the specimen (Fig. 6b).

Because strain was strongly heterogeneous in intensity and geometry, two areas were selected for CPO collection to be representative of the two-deformation geometry end-members: high and low strain, at the edge and middle of the specimen respectively (Fig. 6b). CPO patterns of c-axis in the high strain area can be described as a double-cleft girdle and a maximum parallel to the lineation, rather similar to the pattern observed in extension

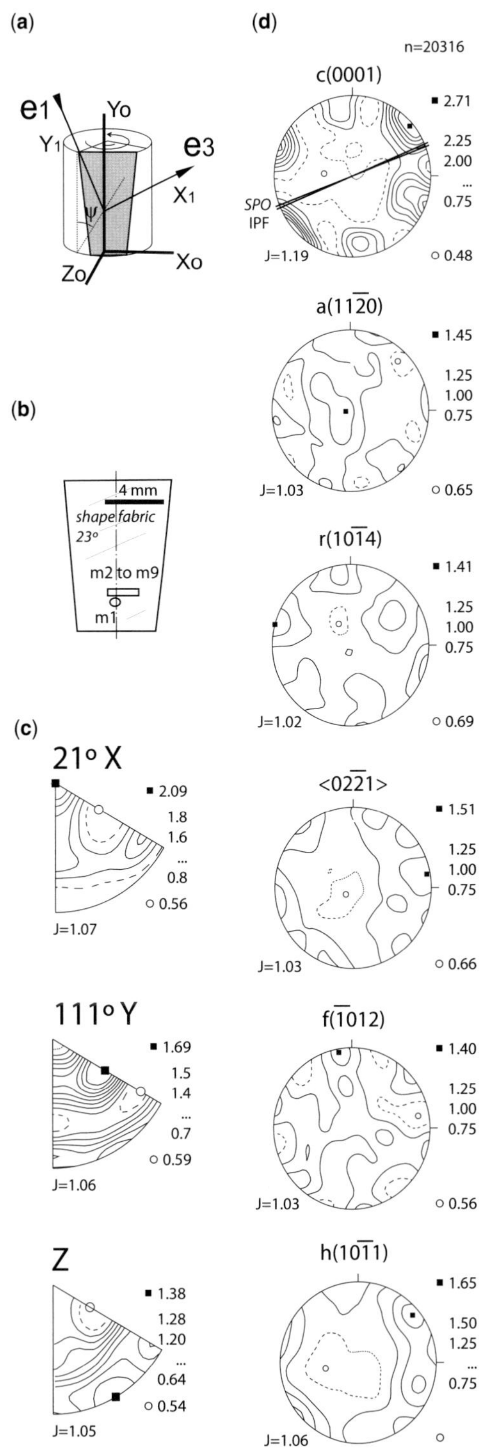

Fig. 5. CPO patterns in torsion (specimen tor1).
(**a**) Sketch indicating principal deformation axes. Finite
strain axes e_1 and e_3 lie in the tangent plane to the outer
surface of the cylinder. (**b**) The location of the EBSD
map m1 within the tangential thin section taken from the
sample (**c**) IPF for the extension, compression and
vorticity axes (indicated). (**d**) pole figures (poles to *c*, *a*, *r*
planes, <02–21>, poles to *f* and *h* planes).

(Fig. 6d). IPFs showing the preferred orientation of
the stretching (lineation) and shortening (pole to
flattening plane defined by the shape fabric) directions are typical of the type of patterns developed
along stretching and shortening (Schmid *et al.*
1981) but undefined in the direction perpendicular
to the other two, which coincides with the vorticity
axis (Fig. 6c). In the low strain area, c-axis CPO
patterns lost strength and varied slightly so that
the cleft girdles were further apart and the maximum of c-axes parallel to the lineation became a
great-circle girdle parallel to the foliation plane.

Axisymmetric shortening tests on discoid sample (short cylinders)

The most heterogeneously deformed samples were
produced when short cylinders (discs) of Solnhofen
limestone were deformed in compression (Llana-
Fúnez & Rutter 2005). The effect of the no-slip
condition at the piston–specimen interface was
maximized (Fig. 1), and led to radial extrusion of
the middle part of the specimen and introduction
of shear strain components at both interfaces
between the cylindrical sample and the loading
pistons (Figs 1 & 7a, b).

Optical micrographs show the development of
a shape fabric when strain is sufficiently high and
that this fabric changes orientation and intensity depending on the location within the sample
(Fig. 7c). The micrograph in Figure 2d shows a
traverse through one specimen deformed in this
configuration. Fabric data for specimen sf04 are presented in Figure 7c, d. The initial cylinder length
to diameter ratio was 0.43:1.

CPO patterns of c-axes in areas dominated by
flattening are characterized by the development of
a great-circle girdle parallel to the foliation plane
and two small-circle girdles centred on the location
of the foliation pole (Fig. 7d). In areas close to the
pistons, where the flow is dominated less by flattening and more by shearing with stronger contribution
of non-coaxial deformation, the CPO patterns show
partial cleft girdles and a small maximum parallel to
the stretching direction, resembling the patterns
observed in the direct shear test.

A third area of interest in this experimental
setup is the outer annulus of the specimen. Once
the sample material was extruded beyond the confines of the pistons the specimen strain became
dominated by circumferential extension. CPO in

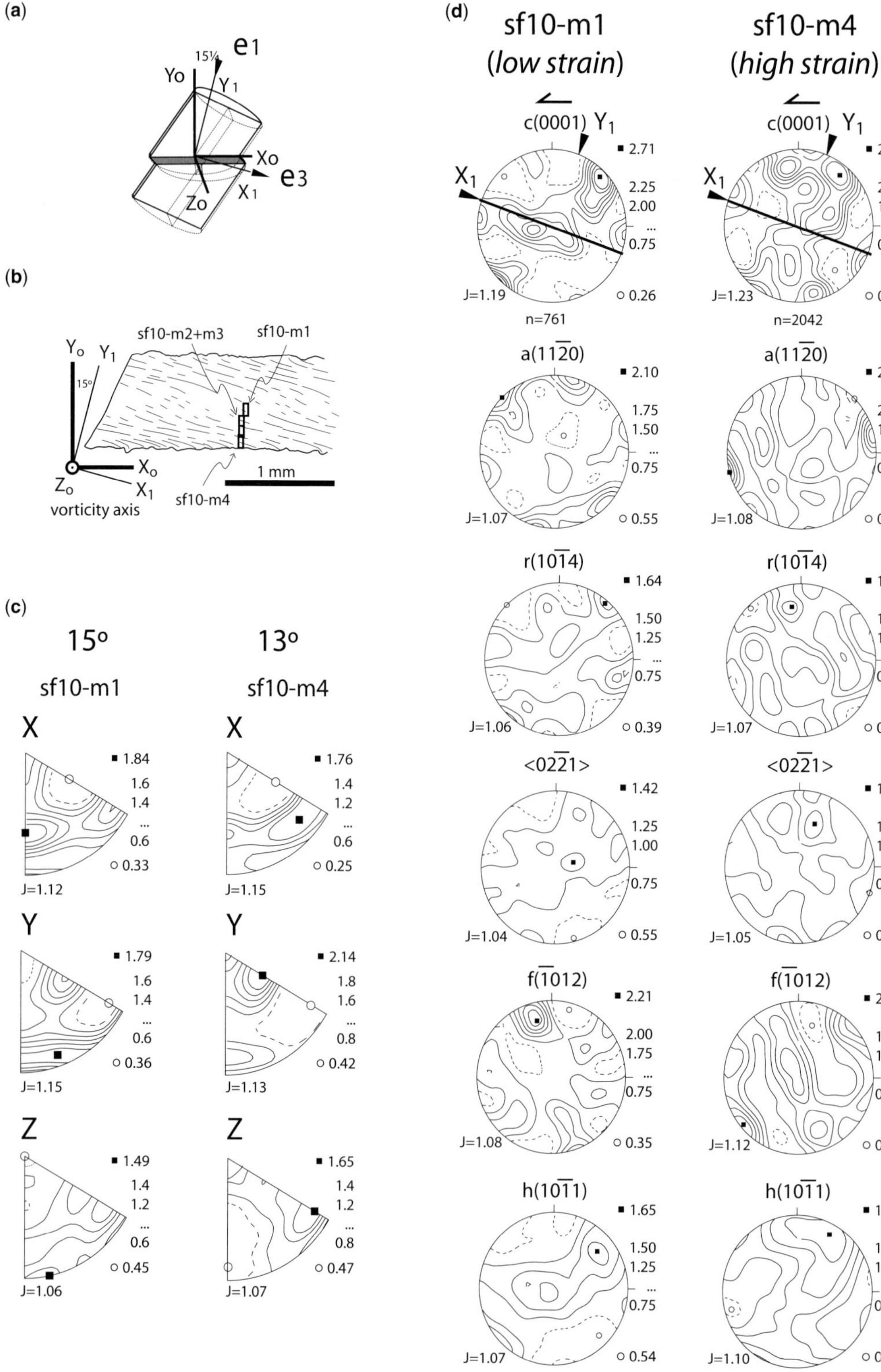

(a)
e1
15¼
Yo
Y1
Xo
e3
Zo
X1

(b)
Yo Y1
15°
sf10-m2+m3   sf10-m1
Xo
Zo X1
vorticity axis
sf10-m4
1 mm

(c)
15°   13°
sf10-m1   sf10-m4
X   X
1.84   1.76
1.6   1.4
1.4   1.2
...   ...
0.6   0.6
0.33   0.25
J=1.12   J=1.15
Y   Y
1.79   2.14
1.6   1.8
1.4   1.6
...   ...
0.6   0.8
0.36   0.42
J=1.15   J=1.13
Z   Z
1.49   1.65
1.4   1.4
1.2   1.2
...   ...
0.6   0.8
0.45   0.47
J=1.06   J=1.07

(d)
sf10-m1   sf10-m4
(low strain)   (high strain)
c(0001) Y1   c(0001) Y1
X1   X1
2.71   2
2.25   2
2.00   1
0.75   0
J=1.19   J=1.23
0.26   0
n=761   n=2042
a(11-20)   a(11-20)
2.10   2
1.75   2
1.50   1
0.75   0
J=1.07   J=1.08
0.55   0
r(10-14)   r(10-14)
1.64   1
1.50   1
1.25   1
0.75   0
J=1.06   J=1.07
0.39   0
<02-21>   <02-21>
1.42   1
1.25   1
1.00   1
0.75   0
J=1.04   J=1.05
0.55   0
f(-1012)   f(-1012)
2.21   2
2.00   1
1.75   1
0.75   0
J=1.08   J=1.12
0.35   0
h(10-11)   h(10-11)
1.65   1
1.50   1
1.25   1
0.75   0
J=1.07   J=1.10
0.54   0

this area is characterized by a great-circle girdle parallel to the foliation plane and double-cleft girdles centred on the stretching direction that here follows the perimeter of the specimen circumference (Fig. 7d).

Relation between finite strain and CPO intensity

Local finite strain in extension tests can be calculated from the reduction of the cross-sectional area and this has been done in Figure 8a, where strain is plotted with respect to distance from the middle (necked region) of the sample. A series of EBSD maps were collected from the poorly foliated part towards the centre of the specimen. The general features of c-axis CPO patterns is maintained, the main difference being the strength and definition of girdles within them (Fig. 8b). For CPO patterns with similar elements, we relate the strength of the CPO to strain using the intensity of two elements: the poles to the f-planes and the maximum of the IPF in the stretching direction. The maximum of the IPF in the stretching direction feeds from the intensity of the orientation of poles of f-planes and the direction [02−21]. Both crystallographic elements are expected to be active during intracrystalline plasticity of calcite at the experimental conditions (see de Bresser & Spiers 1997).

The strain in the torsion configuration is at a maximum at the edge of the specimen and decreases towards the axis of the specimen. The geometry permits the calculation of finite strain with distance from the edge, as shown in Figure 8c (which indicates the location of the EBSD maps collected). As mentioned earlier, the strain gradient across the section of observation is relatively low, and as a consequence no significant change is recorded in the intensity of CPO patterns (Fig. 8d).

In previous studies, we highlighted the heterogeneous character of the finite strain in terms of geometry, not just intensity, during deformation of discoid specimens, either in the direct shear configuration (Llana-Fúnez & Rutter 2008) or under axial compression (Llana-Fúnez & Rutter 2005). For the direct shear tests, we separated two 'components': (a) non-coaxial deformation dominating the volume within the specimen in contact with the spacers; and (b) the heterogeneous thinning that increases towards the middle of the specimen.

The strain geometry imposed upon the specimens under this experimental configuration is summarized in figure 10 in Llana-Fúnez & Rutter (2008) and shows the locations of the full set of EBSD maps collected. Here in Figure 8e we relate the intensity of IPFs in the shortening and stretching directions in all EBSD maps collected with respect to the strain calculated using the deflection of the shape fabric. The plot in Figure 8e shows that the intensity of IPF patterns in the shortening direction is more sensitive to strain than in the stretching direction, consistent with the fact that thinning of the specimen dominates over the stretching of the sample wafer.

Strain in compression of short cylinders can be estimated from the radius of the free surface, the contact between the sample jacket and the confining fluid, as proportional to heterogeneous thinning of the specimen (Fig. 8e). This approach provides a minimum strain and does not take into account shear localization at the bases of the disc and towards the corners of the specimen. Neither does this estimate consider the extrusion of the specimen beyond the confines of the pistons. Although the nature and arrangement of girdles seem to match the strain geometry expected in the various locations studied by EBSD, the intensity of the patterns in the locations studied does not increase systematically with shortening (Fig. 8f).

Seismic velocities in aggregates deformed by different strain geometries

Plastically deformed aggregates of Solnhofen limestone develop a moderate to high anisotropy to the propagation of seismic waves, in the range 2.5–5.6% for V_p and 1.9–8.5% for V_s (Fig. 9). These values in absolute terms are not very different from the seismic anisotropy calculated for the starting material (Fig. 3). However, there are significant differences in how the velocities are distributed with respect to the kinematic framework of each individual experiment, and within it with respect to the strain geometry recorded in the aggregate.

In general V_p is straightforward to interpret in terms of the structural framework of the studied aggregates. In axisymmetric extension, P-waves propagate faster along the extension direction (Fig. 9a). In the high strain area in the direct shear test,

---

Fig. 6. CPO patterns in direct shear (specimen sf10). (**a**) Sketch indicating principal deformation axes. (**b**) The location of the EBSD maps (low and high strain) within the sample thin section. The grain long axes were respectively at 15° and 13° to the shear direction (low and high strain). (**c**) IPF for extension, compression and vorticity axes (indicated). (**d**) Pole figures (poles to c, a, r planes, <02−21>, poles to f and h planes) for low and high strain cases.

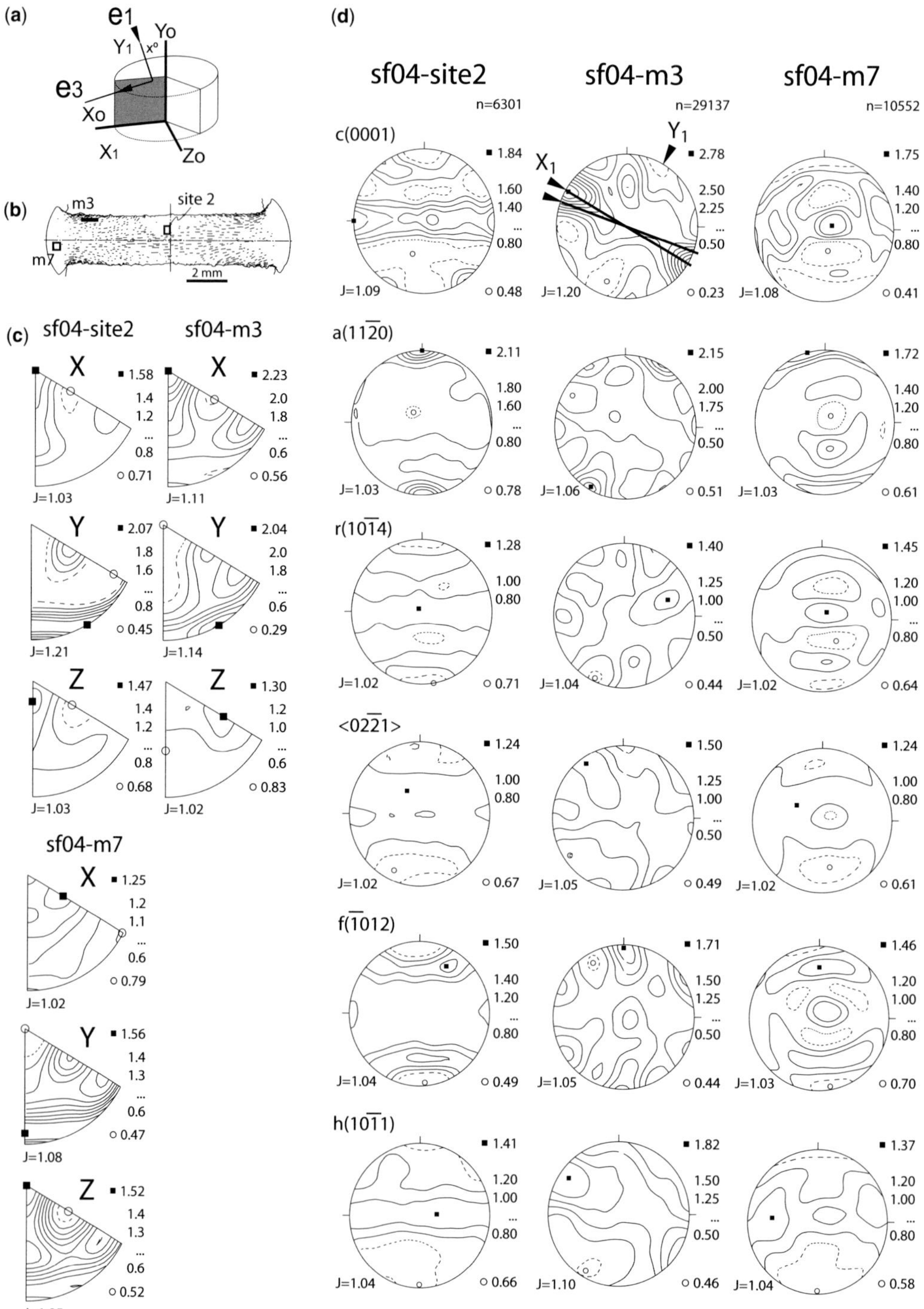

Fig. 7. CPO patterns in a shortened disc-shaped specimen (specimen sf04). (**a**) Sketch indicating principal deformation axes. Principal strain directions e_1 and e_2 lie in the shaded radial plane. (**b**) Location of the three EBSD maps within the sample section. These were chosen to demonstrate different strain histories. (**c**) IPF directions: for site 2, X and Z

V_p-max is also oriented sub-parallel to the extension direction. In the strain geometries dominated by the flattening component (Fig. 9c, d), V_p is fastest close to the perpendicular to the shortening direction. This pattern is observed not only in the axisymmetric shortening of a short disc, but also in the middle of the specimen deformed in direct shear (Fig. 9c). For the test run in a strain geometry closest to simple shear, in torsion, V_p is fastest along the vorticity axis of the flow. It is contained within the foliation plane but at right angles to the direction of movement (Fig. 9b).

Rather more difficult to interpret in detail is V_s, although the general pattern is similar to V_p in what amounts to similarities between similar strain geometries in different experimental configurations. The fast polarization direction for the fast V_s (or V_{s1}) is parallel to the stretching direction in axisymmetric extension and in the high strain area within the direct shear test; V_{s1} also develops a girdle of fast velocities perpendicular to the extension direction.

EBSD maps Site 2 and m7 in the shortening of disc-shaped specimen sf04 have remarkably similar patterns, in which V_{s1} is fastest in a small-circle girdle centred on the shortening direction (Fig. 9d). Areas dominated by shearing develop a V_{s1} pattern characterized by three fast directions, one perpendicular to the foliation and the other two contained within the foliation plane and perpendicular to each other (one following the vorticity axis). The test in torsion V_{s1} has only two maxima contained within the foliation plane, one parallel to the (finite) stretching direction, the other at right angles to it. Regarding the propagation direction, in all strain geometries where a planar foliation is developed, the propagation of the fast V_s is normal to it or highly oblique to it. Only in the axisymmetric extension does the fast V_s propagate along the stretching direction.

Birefringence patterns are similar in strain geometries dominated by extension (se05-m2 and sf10-m4) and those dominated by flattening (site 2 and m7).

Discussion

Comparison of CPO patterns with previous experimental studies

There have been a number of independent and separate studies reported that use either similar starting material or similar experimental configurations to those used here, allowing partial comparison. The most closely comparable set of tests was provided by Kern (1977, 1979), who successfully deformed in flattening and constriction Solnhofen limestone specimens to produce small- and great-circle girdles around dominant shortening or extension directions. In a later contribution, using a slightly different starting material and experimental configuration, specimens were deformed in plane strain, producing CPO patterns bearing similar features to our test in torsion (Kern & Wenk 1983). Either in pure or simple shear the c-axis pattern is characterized by the development of several point maxima at the periphery of the stereoplot. Modelling of CPO development by the same group of authors using the Taylor theory (homogeneous strain) reproduced well the CPO observed in experimentally deformed specimens, particularly when using the combination of slip systems that they regarded operative at high temperature (Wagner *et al.* 1982; Wenk *et al.* 1986). They showed that at high temperature a single girdle forms parallel to the foliation plane together with a small-circle girdle centred about the shortening axis.

With respect to the axisymmetric extension strain path, they modelled the development of a single point maximum along the stretching direction and a small-circle girdle centred in the extension direction, matching our experimental results in Figure 4.

The modelling study of Wenk *et al.* (1986) highlights the mechanical anisotropy that accumulation of strain by plasticity produces in the polycrystalline aggregate. Our experiments were run to relatively low strains (e.g. in axisymmetric extension to a bulk 125%, or in shortening to a bulk of 40%), therefore the comparison is relatively straightforward with numerical models simulating low strains. High strains are normally only achieved experimentally in the torsion configuration. More recent experimental data on calcite rocks at high strains in torsion do show some resemblance in c-axis patterns at low strain, particularly tests at 500 and 600 °C by Barnhoorn *et al.* (2004) in which they developed four-point maxima at the periphery of the stereoplot (with the primitive circle as the X–Z plane). Note that the angles between maxima are similar to those seen in our test, although we obtained two additional maxima to make a hexagonal symmetry that will be discussed below. At higher temperatures, Pieri *et al.* (2001) showed similar hexagonal symmetry at low strains, comparable to our test

Fig. 7. (*Continued*) correspond to compression and Y to extension; for map m3, X, Y and Z correspond to extension, compression and vorticity; for map m7, X and Y correspond to compression and Z to extension. (**d**) Pole figures (poles to c, a, r planes, <02–21>, pole to f and h planes) for each of the three EBSD sample sites.

S. LLANA-FÚNEZ & E. H. RUTTER

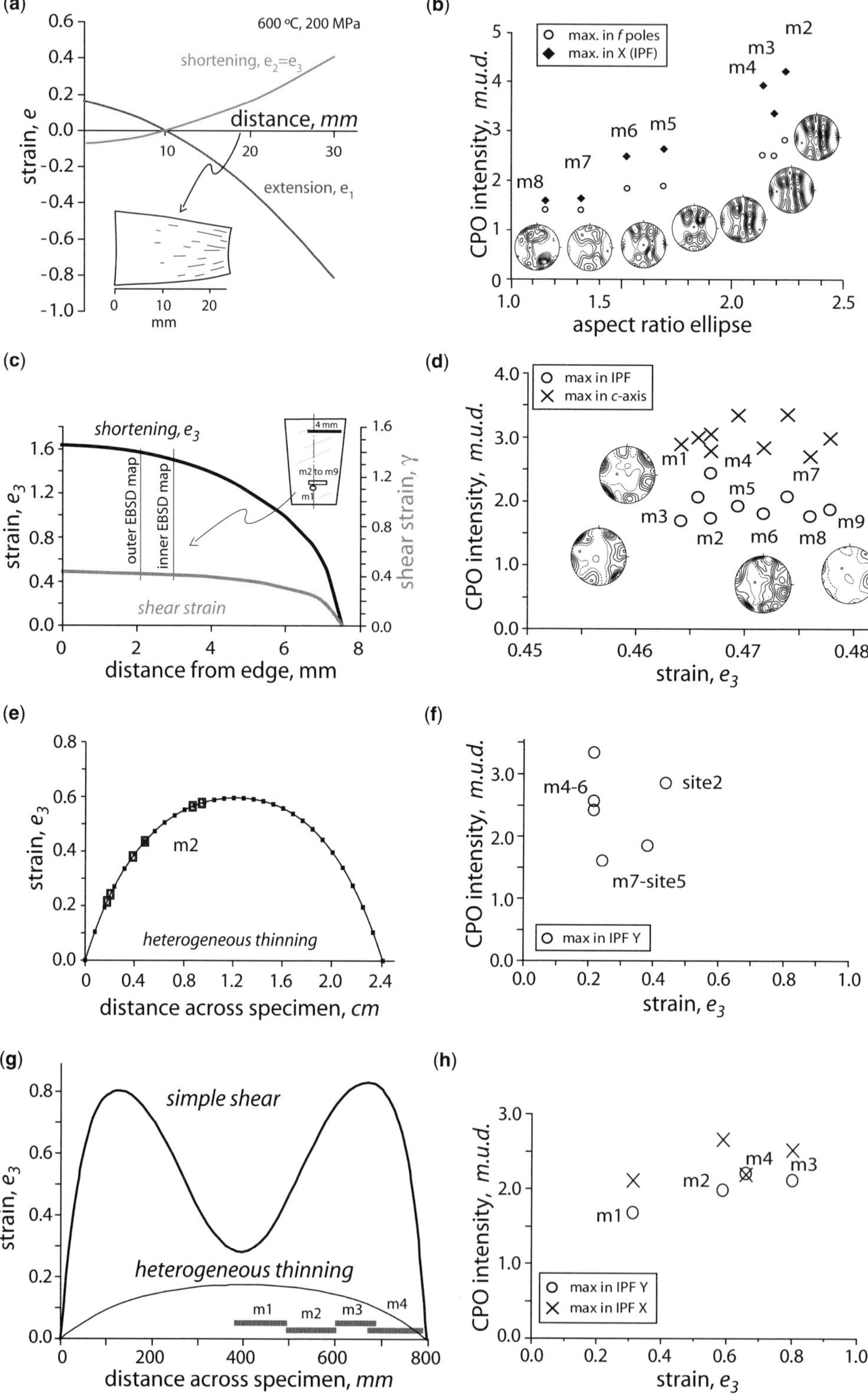

shown in Figure 5. Low strain patterns become wholly overprinted in large strain torsion tests.

CPO patterns in natural calcite tectonites

The orientation patterns in calcite deformed at moderate temperatures in our experiments differ substantially from natural calcite tectonites. However, it must be borne in mind that comparisons between natural and experimentally produced CPO patterns must take into account comparability of the strain path and the fact that relative activity on different slip systems varies with temperature. Therefore, exact correspondence may not necessarily be expected. In natural samples, girdles and point maxima distributions tend to be broad and less intense than in experimental samples, but the number of girdles and point maxima and their arrangement with respect to the structural framework, the foliation plane and the lineation, are also different. Figure 10 summarizes c-axis patterns most commonly reported in the literature for a range of deformation conditions, including low- and high-temperature deformations and different strain histories. Figure 10a resembles the CPO patterns in plane strain developed experimentally by Kern and co-workers (Kern 1977, 1979; Kern & Wenk 1983; Wagner *et al.* 1982) or Pieri *et al.* (2001) at moderate temperature. The data in figure 10b–d were reproduced experimentally by Casey *et al.* (1998) at high temperatures, close to the grain-size-sensitive creep field.

In all the above cases, recrystallization operated in nature during deformation, in contrast to our experimental dataset where it was absent. Dynamic recrystallization is achieved experimentally at relatively high strain rates at high temperatures (above 600 °C), but it is expected to occur even at low temperatures in nature, as low as 150 °C (Kennedy & White 2001). From the comparison between deformation of calcite polycrystalline aggregates in nature and in the laboratory, dynamic recrystallization may be a process that commonly modifies and influences the development of CPO patterns. Thus direct correspondence may be lost between numerical and experimental deformation studies compared to strains occurring naturally that imply accompanying recrystallization to overcome hardening during plastic flow.

Commonly in nature, at low temperatures and high stresses and particularly in coarser-grained aggregates, mechanical twinning is a significant contributor to CPO, usually leading to a strong c-axis maximum in uniaxial compression (or its corresponding types for other strain paths). This fabric is commonly observed both in naturally and experimentally deformed rocks (e.g. Rutter & Rusbridge 1976; Rowe & Rutter 1990; Rutter *et al.* 2007). In our present experiments, however, twinning was absent as a deformation mechanisms, owing to the combination of fine grain size and high temperature at which the Solnhofen limestone was deformed.

Effect of strain geometry on CPO

Figure 11 shows a summary sketch of characteristic c-axis patterns for some of the key strain geometries reproduced experimentally. Distribution elements of the c-axis, such as great- and small-circle girdles and point maxima, arrange in a similar fashion with respect to dominant strain directions (axisymmetric shortening and extension) as do numerical simulations of intracrystalline plasticity of quartz by Lister & Hobbs (1980), in which they modelled plasticity in quartz using the Taylor-Bishop-Hill analysis. In that seminal work, intracrystalline plasticity was neither accommodated nor accompanied by any other deformation mechanism and thus the orientation of girdles was strictly related to intracrystalline plasticity. The principal stress directions in the aggregate determine the orientation of favourable slip directions, which ultimately control the orientation of the c-axis. One of the conclusions of their work, that the location of the axis of maximum extension is generally associated with a pole-free area in c-axis patterns (in fact, along the axis of symmetry of small- and great-circle girdles), is confirmed for the first time by our experimentally deformed aggregates. In our experimental dataset we have not observed the effects of either recrystallization or any other deformation

---

Fig. 8. Relation between finite strain and CPO intensity. In axi-symmetric extension (**a**) shows the estimated strain based on the section reduction along specimen length and (**b**) compares the aspect ratio of strain ellipse with CPO intensity in multiples of a uniform distribution (m.u.d.). In specimens deformed under torsion, (**c**) shows the strain distribution with respect to the distance from the edge and (**d**) relates the calculated shortening, e_3, with CPO intensity in the EBSD maps collected. In shortening of specimens discs, (**e**) shows the estimated magnitude of shortening across the specimens, based on the geometry of the extruded specimen profile (from Llana-Fúnez 2005), and (**f**) the CPO intensity with shortening strain, e_3. In direct shear tests, (**g**) shows the distribution of strain across the specimen disc and the location of EBSD maps, based on Llana-Fúnez & Rutter (2008), and (**h**) the relation between CPO intensity with shortening, e_3.

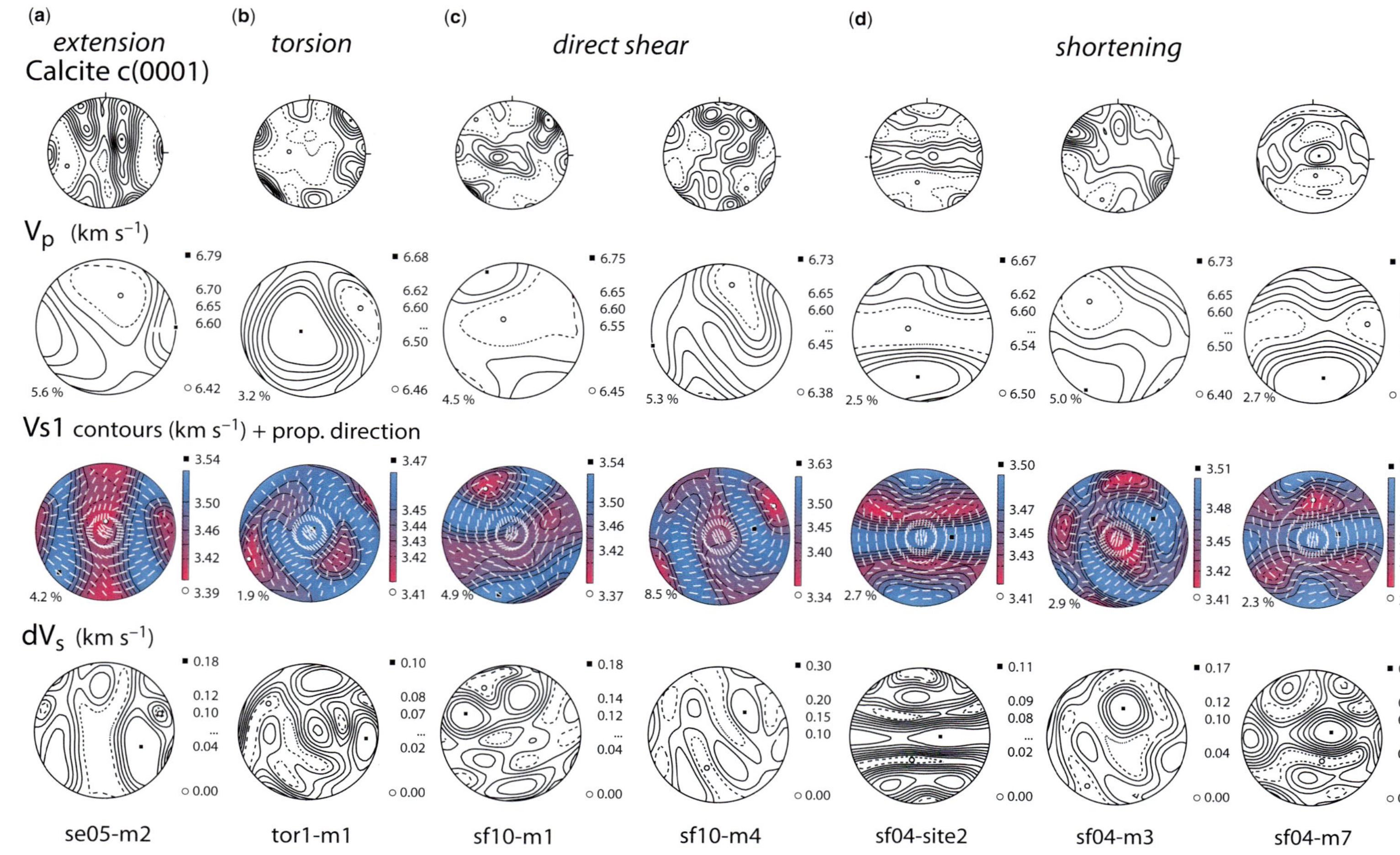

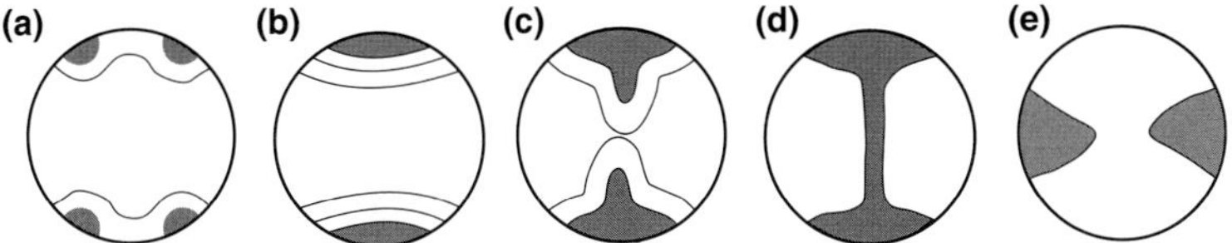

Fig. 10. Commonly observed CPO patterns for the c-axis in naturally deformed calcite rocks. In all cases the pole to foliation lies on the primitive circle at the top and bottom of the projection, and the lineation (stretching direction, if developed) lies on the primitive circle in the foliation plane. (**a**) Two point maxima with the shortening direction bisecting the acute angle between c-axis maxima (e.g. Leiss & Molli 2003; Oesterling *et al.* 2007). (**b**) Broad maximum axisymmetric with respect to the pole of foliation (e.g. Ratschbacher *et al.* 1991; Erskine *et al.* 1993; Bestmann *et al.* 2000; Molli *et al.* 2000; Herwegh & Kunze 2002; Trullenque *et al.* 2006; Ebert *et al.* 2007; Molli *et al.* 2011). (**c**) Intermediate pattern between a broad maximum about the pole of foliation and a partially developed girdle normal to the lineation (Schmid *et al.* 1987; de Bresser 1989; Leiss *et al.* 1999). (**d**) Broad maximum about the pole of foliation and a single great-circle girdle normal to the stretching direction (e.g. Dietrich & Song 1984; Busch & Van der Pluijm 1995; Bestmann *et al.* 2000; Kurz *et al.* 2000; Molli *et al.* 2000; Herwegh & Kunze 2002; Fernández *et al.* 2004; Romeo *et al.* 2007; Oesterling *et al.* 2007). (**e**) Broad maximum within the foliation plane and parallel to the stretching direction (Dietrich & Song 1984).

mechanism that may have modified or affected orientation patterns, so in that respect our approach is equivalent.

The summary of CPO patterns in a Flinn diagram in Figure 11 illustrates that the development of girdles relates to axisymmetric strain paths, whether in extension or in compression, and that the axis of strain symmetry is also the symmetry axis of the girdle. In contrast, the presence of several point maxima at the periphery of the stereoplot points to deformation conditions in plane strain. When interpreting orientation patterns in naturally deformed polycrystalline aggregates, the symmetry axis of girdle distributions should indicate that the orientation of principal finite strain axes is not necessarily parallel to the structural framework, particularly in triclinic 3D strains (e.g. Llana-Fúnez 2002).

In this study, we highlight the strong correlation that exists between strain intensity and the maximum in the pole of the *f*-plane (Fig. 8). The correlation was identified using IPF to characterize crystallographic elements that oriented along strain-rate principal directions. Despite the relatively higher critical resolved shear stress (CRSS) reported for slip along this family of planes (de Bresser & Spiers 1997) they seem to orient preferentially with respect to the finite strain geometry imposed upon the aggregate. From orientation data, we cannot distinguished between $f<10–11>\pm$ and $f<2–201>$, but according to de Bresser & Spiers

(1997), our temperature conditions will favour the operation of $f<2–201>$ in the experiments with Solnhofen limestone at 600 °C, 200 MPa and a strain rate of $10^{-4}\,s^{-1}$.

Expected seismic anisotropy in calcite rocks in geological settings

The starting material, Solnhofen limestone, is a micritic limestone with equigranular microstructure and no apparent shape fabric. It is characterized by a distribution of c-axis parallel to the bedding plane (Casey *et al.* 1998) (Fig. 3b) and, given that the V_p propagates slower along the c-axis, V_p velocities are expected to be lower along the bedding plane and faster at right angles to it (Fig. 3c). The largest birefringence to the propagation of S-waves is seen along bedding planes. In this respect, the starting material has a moderately developed anisotropy to the propagation of seismic waves along the bedding planes. The propagation direction of the fast wave (V_{s1}) is normal to bedding. Anisotropy for V_p is in excess of 4% and birefringence retardation is a maximum of 0.11 km s^{-1}.

The response to the propagation of seismic waves by Solnhofen limestone is in many respects opposite to what can be expected of a sedimentary rock with oriented phyllosilicates (Fig. 3). Phyllosilicates are strongly anisotropic materials in which V_p propagates faster along the orientation of sheets of mica (e.g. Naus-Thijssen *et al.* 2011).

Fig. 9. Petrophysical properties of Solnhofen limestone polycrystalline aggregates deformed experimentally at various strain geometries and at 600 °C, 200 MPa and $10^{-4}\,s^{-1}$ strain rate: (**a**) in axisymmetric extension, (**b**) in torsion, (**c**) direct shear and (**d**) shortening. In each column and from top to bottom are shown: c-axis orientation pattern, V_p distribution, V_s fast wave (contoured in colour) and propagation direction (superimposed, in white), and seismic birefringence (in km s^{-1}). Seismic velocities were calculated using Mainprice (1990) and Mainprice & Humbert (1994) software.

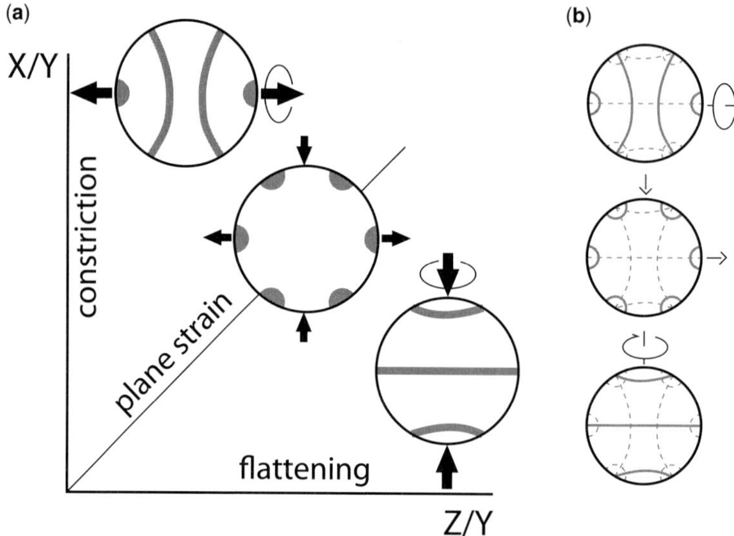

Fig. 11. Summary of CPO patterns produced experimentally related to strain using a Flinn diagram (**a**). Most test configurations produce patterns typical of the axisymmetric flattening regime, characterized by the development of a small-circle girdle centred about the shortening direction (that coincides with the axis of symmetry of the strain path) and a great-circle girdle parallel to the foliation in the rock. In extension, small-circle girdles are centred about the direction of extension. In the current dataset, only the test run in torsion produces a true plane-strain geometry, characterized by the development of several point maxima at the periphery of the stereoplot, and, more importantly, by the absence of girdles. Schematic c-axis patterns in (**b**) illustrate the effect of axisymmetric strain paths on the development of c-axis girdles and their orientation with respect to the dominant strain direction (whether the axis of symmetry corresponds to a compression or extension direction). In the case of plane strain, neither the extension nor the shortening direction are axisymmetric, therefore the development of point maxima is favoured over a girdle distribution.

Deformation processes cause the reorientation of mica to define a tectonic fabric, but it is feasible that sedimentation and diagenesis may also produce the orientation of micas parallel to bedding, albeit limited to a small fraction of small grains. Consequently, in equal volume fractions in a sedimentary sequence between micritic limestone and other sedimentary rocks, the effect of the micrites would be to reduce seismic anisotropy by reducing V_p along bedding and increasing V_p perpendicular to it, provided there is a sufficient fraction of mica sheets oriented parallel to bedding: 10% is already enough to control bulk velocity structure (e.g.Llana-Fúnez & Brown 2012).

From the experiments, and considering that dynamic recrystallization was avoided at the chosen temperature conditions, some of the first order features in the velocity structure of the starting material are maintained in the various strain geometries: for example, the propagation direction for the fast shear wave is mostly oriented perpendicular to the foliation plane. In addition to this, V_p shows fast and slow directions mostly oblique to the structural reference frame, the foliation plane and the lineation. In most scenarios, the foliation plane will represent a slow velocity for V_p, with one significant exception. This will be in simple shear, in which V_p will propagate faster along the vorticity axis. The latter effect of CPO on the velocity structure had not been shown before and opens up the possibility of using V_p patterns as a kinematic indicator, since it indicates indirectly the direction of tectonic transport in plane strain.

Concluding remarks

Experimental deformation of Solnhofen limestone specimens at high temperature and high pressure using different experimental configurations permits several non-plane-strain geometries to be produced. These may be common in some shear zones but also might be expected in terrains where deformation is widespread and not restricted to narrow, localized high-strain zones.

Strain geometries were produced dominated by:

(a) flattening but with varying degrees of stretching and shearing (shortening of specimen discs and direct shear testing);

(b) extension (axisymmetric extension and high strain in direct shear);

(c) plane-strain non-coaxial flow (torsion).

Shape orientation patterns of calcite grains within the deformed aggregates are in agreement with or match expected finite strain geometries. Crystallographic orientation patterns include small-circle girdles or maxima in accordance with strain geometry and vary in intensity according to the amount of finite strain accumulated. Despite following expected patterns of behaviour for deformation by intracrystalline plasticity, this has not previously been demonstrated experimentally to this extent. The CPO obtained from experimentally deformed specimens in the absence of dynamic recrystallization reproduces some of the models of quartz plasticity as developed by Taylor (1938) that also do not take into account dynamic recrystallization. Both numerical and experimental approaches coincide in indicating that dominant strain directions, such as shortening in flattening and extension in constriction, will produce small and/or great-circle girdles centred on those strain directions.

The characteristic orientation patterns observed in each deformation configuration give to the crystalline aggregates specific petrophysical properties. From each CPO pattern, together with the elastic properties and density of single crystals of calcite, the seismic properties of the deformed aggregates can be calculated. Although petrophysical properties and magnitudes cannot be directly related to strain intensity, it proved possible to characterize velocity patterns for seismic waves under the various deformation geometries. Axisymmetric deformation, such as flattening and constriction, develop characteristic patterns for both V_p and V_s. The former is fastest in a direction oblique to the shortening direction in a flattening strain path and follows the stretching direction in constriction. In torsion, the experimental strain geometry closest to simple shear and expected to dominate in discrete, parallel-sided shear zones, V_p is faster along the vorticity axis, perpendicular to transport direction. Patterns of V_s also relate to strain geometry, and here again axisymmetric flattening- and stretching-dominated regimes differ. In general, fastest propagation directions are perpendicular or highly oblique to the lineation within the foliation plane.

When deformed in nature, calcite provides a larger number of slip systems than most other rock-forming minerals, such as plagioclase, pyroxene or olivine. However, some of the observed effects of true 3D deformation geometries on calcite rock properties may be relevant to other rock-forming minerals deforming in natural conditions, such as quartz, provided that intracrystalline plasticity is the predominant deformation mechanism and that recrystallization processes are inhibited. This study demonstrates that for certain minerals strain geometry can have a measurable and significant impact on the petrophysical properties of rocks.

The experimental work was done at the University of Manchester while Sergio Llana-Fúnez held a Marie Curie Individual Fellowship of the European Commission (Marie Curie Fellowship contract Nr. HPMF-CT-2000-00778). Funding during the final writing at the University of Oviedo was provided by the Ramón y Cajal Programme, grant RYC2008-02067, funded by the Spanish Ministry of Science and Innovation (now Ministry of Economy and Competitiveness). EBSD data were collected at the Materials Science Centre, University of Manchester. The help of Ian Brough (data collection) and Walid Ben Ismail (data processing) is acknowledged. Dan Tatham (Department of Earth and Ocean Sciences, University of Liverpool) helped with the calculation of seismic properties of the aggregates. The authors thank Auke Barnhoorn and Nick Timms for feedback that aided revision of an earlier version of the manuscript.

References

BARNHOORN, A., BYSTRICKY, M., BURLINI, L. & KUNZE, K. 2004. The role of recrystallization on the deformation behaviour of calcite rocks: large strain torsion experiments on Carrara marble. *Journal of Structural Geology*, **26**, 885–903.

BARRUOL, G. & MAINPRICE, D. 1993. 3-D seismic velocities calculated from lattice-preferred orientation and reflectivity of a lower crustal section – examples of the Val Sesia Section (Ivrea Zone, Northern Italy). *Geophysical Journal International*, **115**, 1169–1188.

BESTMANN, M., KUNZE, K. & MATTHEWS, A. 2000. Evolution of a calcite marble shear zone complex on Thassos Island, Greece: microstructural and textural fabrics and their kinematic significance. *Journal of Structural Geology*, **22**, 1789–1807.

BUSCH, J. P. & VAN DER PLUIJM, B. A. 1995. Calcite textures, microstructures and rheological properties of marble mylonites in the Bancroft shear zone, Ontario, Canada. *Journal of Structural Geology*, **17**, 677–688.

CASEY, M., KUNZE, K. & OLGAARD, D. L. 1998. Texture of Solnhofen limestone deformed to high strains in torsion. *Journal of Structural Geology*, **20**, 255–267.

CHRISTENSEN, N. I. 1984. The magnitude, symmetry and origin of upper mantle anisotropy based on fabric analyses of ultramafic tectonites. *Geophysical Journal of the Royal Astronomical Society*, **76**, 89–111.

COVEY-CRUMP, S. J. 1992. *Application of a state variable description of inelastic deformation to geological materials*. PhD thesis, University College London.

DANDEKAR, D. P. 1968. Elastic constants of calcite. *Journal of Applied Physics*, **39**, 2971–2973.

DE BRESSER, J. H. P. 1989. Calcite c-axis textures along the Gavarnie thrust zone, central Pyrenees. *Geologie en Mijnbouw*, **68**, 367–375.

DE BRESSER, J. H. P. & SPIERS, C. J. 1997. Strength characteristics of the r, f, and c slip systems in calcite. *Tectonophysics*, **272**, 1–23.

DELL'ANGELO, L. N. & TULLIS, J. 1989. Fabric development in experimentally sheared quartzites. *Tectonophysics*, **169**, 1–21.

DIETRICH, D. & SONG, H. 1984. Calcite fabrics in a natural shear environment, the Helvetic nappes of western Switzerland. *Journal of Structural Geology*, **6**, 19–32.

DRESEN, G. 1991. Stress-distribution and the orientation of Riedel shears. *Tectonophysics*, **188**, 239–247.

EBERT, A., HERWEGH, M., EVANS, B., PFIFFNER, A., AUSTIN, N. & VENNEMANN, T. 2007. Microfabrics in carbonate mylonites along a large-scale shear zone (Helvetic Alps). *Tectonophysics*, **444**, 1–26.

ERSKINE, B. G., HEIDELBACH, F. & WENK, H.-R. 1993. Lattice preferred orientations and microstructures of deformed Cordilleran marbles: correlation of shear indicators and determination of strain path. *Journal of Structural Geology*, **15**, 1189–1205.

FERNÁNDEZ, F. J., BROWN, D., ÁLVAREZ-MARRÓN, J., PRIOR, D. J. & PÉREZ-ESTAÚN, A. 2004. Microstructure and lattice preferred orientation of calcite mylonites at the base of southern Urals accretionary prism. *Journal of the Geological Society, London*, **161**, 67–79.

HERWEGH, M. & KUNZE, K. 2002. The influence of nano-scale second-phase particles on deformation of fine grained calcite mylonites. *Journal of Structural Geology*, **24**, 1463–1478.

JAEGER, J. C. 1962. *Elasticity, Fracture and Flow*. Methuen & Co. Ltd and John Wiley & Sons Inc., London and New York.

JAEGER, J. C. & COOK, N. G. W. 1976. *Fundamentals of Rock Mechanics*. Science Paperback, London.

JI, S. C. & MAINPRICE, D. 1988. Natural deformation fabrics of plagioclase – implications for slip systems and seismic anisotropy. *Tectonophysics*, **147**, 145–163.

KENNEDY, L. A. & WHITE, J. C. 2001. Low-temperature recrystallization in calcite: mechanisms and consequences. *Geology*, **29**, 1027–1030.

KERN, H. 1977. Preferred orientation of experimentally deformed limestone marble, quartzite and rock salt at different temperatures and states of stress. *Tectonophysics*, **39**, 103–120.

KERN, H. 1979. Texture development in calcite and quartz rocks deformed at uniaxial and real triaxial states of strain. *Bulletin Mineralogie*, **102**, 290–300.

KERN, H. & WENK, H. R. 1983. Calcite texture development in experimentally induced ductile shear zones. *Contributions to Mineralogy and Petrology*, **83**, 231–236.

KURZ, W., NEUBAUER, F., UNZOG, W., GENSER, J. & WANG, X. 2000. Microstructural and textural development of calcite marbles during polyphase deformation of Penninic units within the Tauern Window (Eastern Alps). *Tectonophysics*, **316**, 327–342.

LAW, R. D. 1990. Crystallographic fabrics: a selective review of their applications to research in structural geology. *In*: KNIPE, R. J. & RUTTER, E. H. (eds) *Deformation Mechanisms, Rheology and Tectonics*. Geological Society, London, Special Publications, **54**, 335–352.

LEISS, B. & MOLLI, G. 2003. 'High-temperature' texture in naturally deformed Carrara marble from the Alpi Apuane, Italy. *Journal of Structural Geology*, **25**, 649–658.

LEISS, B., SIEGESMUND, S. & WEBER, K. 1999. Texture asymmetries as shear sense indicators in naturally deformed mono- and polyphase carbonate rocks. *Textures and Microstructures*, **33**, 61–74.

LISTER, G. S. & HOBBS, B. E. 1980. The simulation of fabric development during plastic deformation and its application to quartzite: the influence of deformation history. *Journal of Structural Geology*, **2**, 355–370.

LISTER, G. S. & PATERSON, M. S. 1979. The simulation of fabric development during plastic deformation and its application to quarzite: fabric transitions. *Journal of Structural Geology*, **1**, 99–115.

LISTER, G. S. & WILLIAMS, P. F. 1979. Fabric development in shear zones – theoretical controls and observed phenomena. *Journal of Structural Geology*, **1**, 283–297.

LLANA-FÚNEZ, S. 2002. Quartz c-axis texture mapping of a Variscan regional foliation (Malpica-Tui Unit, NW Spain). *Journal of Structural Geology*, **24**, 1299–1312.

LLANA-FÚNEZ, S. & BROWN, D. 2012. Contribution of crystallographic preferred orientation to seismic anisotropy across a surface analog of the continental Moho at Cabo Ortegal, Spain. *Geological Society of America Bulletin*, **124**, 1495–1513.

LLANA-FÚNEZ, S. & RUTTER, E. H. 2005. Distribution of non-plane strain in experimental compression of short cylinders of Solnhofen limestone. *Journal of Structural Geology*, **27**, 1205–1216.

LLANA-FÚNEZ, S. & RUTTER, E. H. 2008. Strain localization in direct shear experiments on Solnhofen limestone at high temperature – effects of transpression. *Journal of Structural Geology*, **30**, 1372–1382.

LOGAN, J., DENGO, C., HIGGS, N. & WANG, Z. 1992. Fabrics of experimental fault zones: their development and relationship to mechanical behaviour. *In*: EVANS, B. & WONG, T.-F. (eds) *Fault Mechanics and Transport Properties of Rocks*. Academic Press Ltd, London, International Geophysics Series, **51**, 33–67.

MAINPRICE, D. 1990. A Fortran program to calculate seismic anisotropy from the lattice preferred orientation of minerals. *Computers & Geosciences*, **16**, 385–393.

MAINPRICE, D. & HUMBERT, M. 1994. Methods of calculating petrophysical properties from lattice preferred orientation data. *Surveys in Geophysics*, **15**, 575–592.

MAINPRICE, D. & NICOLAS, A. 1989. Development of shape and lattice preferred orientations – application to the seismic anisotropy of the lower crust. *Journal of Structural Geology*, **11**, 175–189.

MAULER, A., KUNZE, K., BURG, J. P. & PHILIPPOT, P. 1998. Identification of electron backscatter diffraction patterns in a monoclinic solid-state solution series: example of omphacite. *Materials Science Forum*, **273/275**, 705–710.

MOLLI, G., CONTI, P., GIORGETTI, G., MECCHERI, M. & OESTERLING, N. 2000. Microfabric study on the deformational and thermal history of the Alpi Apuane marbles (Carrara marbles), Italy. *Journal of Structural Geology*, **22**, 1809–1825.

MOLLI, G., WHITE, J. C., KENNEDY, L. & TAINI, V. 2011. Low-temperature deformation of limestone, Isola Palmaria, northern Apennine, Italy – the role of primary textures, precursory veins and intracrystalline deformation in localization. *Journal of Structural Geology*, **33**, 255–270.

NAUS-THIJSSEN, F. M. J., GOUPEE, A. J., JOHNSON, S. E., VEL, S. S. & GERBI, C. 2011. The influence of crenulation cleavage development on the bulk elastic and seismic properties of phyllosilicate-rich rocks. *Earth and Planetary Science Letters*, **311**, 212–224.

NICOLAS, A., BOUDIER, F. & BOULLIER, A. M. 1973. Mechanisms of flow in naturally and experimentally deformed peridotites. *American Journal of Science*, **273**, 853–876.

OESTERLING, N., HEILBRONNER, R., STÜNITZ, H., BARNHOORN, A. & MOLLI, G. 2007. Strain dependent variation of microstructure and texture in naturally deformed Carrara marble. *Journal of Structural Geology*, **29**, 681–696.

PASSCHIER, C. W. & TROUW, R. A. J. 1996. *Microtectonics*. Springer-Verlag, Berlin.

PATERSON, M. S. & OLGAARD, D. L. 2000. Rock deformation tests to large shear strains in torsion. *Journal of Structural Geology*, **22**, 1341–1358.

PIERI, M., KUNZE, K., BURLINI, L., STRETTON, I., OLGAARD, D. L., BURG, J.-P. & WENK, H.-R. 2001. Texture development of calcite by deformation and dynamic recrystallization at 1000 K during torsion experiments of marble to large strains. *Tectonophysics*, **330**, 119–140.

POIRIER, J. P. 1985. *Creep of Crystals*. Cambridge University Press, Cambridge.

RATSCHBACHER, L., WENK, H.-R. & SINTUBIN, M. 1991. Calcite textures: examples from nappes with strain-path partitioning. *Journal of Structural Geology*, **13**, 369–384.

ROMEO, I., CAPOTE, R. & LUNAR, R. 2007. Crystallographic preferred orientations and microstructure of a Variscan marble mylonite in the Ossa-Morena Zone (SW Iberia). *Journal of Structural Geology*, **29**, 1353–1368.

ROWE, K. J. & RUTTER, E. H. 1990. Palaeostress estimation using calcite twinning: experimental calibration and application to nature. *Journal of Structural Geology*, **12**, 1–17.

RUTTER, E. H. 1998. Use of extension testing to investigate the influence of finite strain on the rheological behaviour of marble. *Journal of Structural Geology*, **20**, 243–254.

RUTTER, E. H. & RUSBRIDGE, M. 1976. The effect of noncoaxial strain paths on the development of preferred orientation in the experimental deformation of a marble. *Tectonophysics*, **39**, 75–86.

RUTTER, E. H., PEACH, C., WHITE, S. H. & JOHNSON, D. 1984. Experimental 'syntectonic' hydration of basalt. *Journal of Structural Geology*, **7**, 251–266, http://dx.doi.org/10.1016/0191-8141(85)90136-1

RUTTER, E. H., FAULKNER, D. R., BRODIE, K. H., PHILLIPS, R. J. & SEARLE, M. 2007. Deformation processes in the Karakoram fault zone, eastern Karakoram, Ladakh, NW India. *Journal of Structural Geology*, **29**, 1315–1326.

SAFFER, D. M. & MARONE, C. 2003. Comparison of smectite and illite frictional properties: application to the updip limit of the seismogenic zone along subduction megathrusts. *Earth and Planetary Science Letters*, **215**, 219–235.

SCHMID, S. M., CASEY, M. & STARKEY, J. 1981. The microfabric of calcite tectonites from the Helvetic Nappes (Swiss Alps). *In*: MCCLAY, K. R. & PRICE, N. J. (eds) *Thrust and Nappe Tectonics*. Geological Society, London, Special Publications, **9**, 151–158, http://dx.doi.org/10.1144/GSL.SP.1981.009.01.13

SCHMID, S. M., PANOZZO, R. & BAUER, S. 1987. Simple shear experiments on calcite rocks: rheology and microfabric. *Journal of Structural Geology*, **9**, 747–778.

SHIMAMOTO, T. & LOGAN, J. M. 1981. Effects of simulated clay gouges on the sliding behavior of Tennessee sandstone. *Tectonophysics*, **75**, 243–255.

TAYLOR, G. I. 1938. Plastic strain in metals. *Journal of the Institute of Metals*, **62**, 307–324.

TRULLENQUE, G., KUNZE, K., HEILBRONNER, R., STÜNITZ, H. & SCHMID, S. M. 2006. Microfabrics of calcite ultramylonites as records of coaxial and noncoaxial deformation kinematics: examples from the Rocher de l'Yret shear zone (Western Alps). *Tectonophysics*, **424**, 69–97.

TULLIS, J., CHRISTIE, J. M. & GRIGGS, D. T. 1973. Microstructures and preferred orientations of experimentally deformed quartzites. *Geological Society of America Bulletin*, **84**, 297–314.

WAGNER, F., WENK, H. R., KERN, H., VAN HOUTTE, P. & ESLING, C. 1982. Development of preferred orientation in plane strain deformed limestone: experiment and theory. *Contributions to Mineralogy and Petrology*, **80**, 132–139.

WALKER, A. N., RUTTER, E. H. & BRODIE, K. H. 1990. Experimental study of grain-size sensitive flow of synthetic, hot-pressed calcite rocks. *In*: KNIPE, R. J. & RUTTER, E. H. (eds) *Deformation Mechanisms, Rheology and Tectonics*. Geological Society, London, Special Publications, **54**, 259–284.

WENK, H. R., TAKESHITA, T., VAN HOUTTE, P. & WAGNER, F. 1986. Plastic anisotropy and texture development in calcite polycrystals. *Journal of Geophysical Research*, **91**, 3861–3869.

WENK, H. R., CANOVA, G., MOLINARI, A. & KOCKS, U. F. 1989. Viscoplastic modeling of texture development in quartzite. *Journal of Geophysical Research*, **94**, 17 895–17 906.

Sillimanite deformation mechanisms within a Grt-Sil-Bt gneiss: effect of pre-deformation grain orientations and characteristics on mechanism, slip-system activation and rheology

SANDRA PIAZOLO[1,2]* & PAULINA JACONELLI[1,3]

[1]*Department of Geology and Geochemistry, University of Stockholm, Sweden*

[2]*ARC Centre of Excellence for Core to Crust Fluid Systems and GEMOC National Key Centre Department of Earth and Planetary Sciences, Macquarie University, NSW 2109, Australia*

[3]*Villagatan 14, 931 62 Skellefteå, Sweden*

**Corresponding author (e-mail: sandra.piazolo@mq.edu.au)*

Abstract: Sillimanite-bearing metapelites are common in the middle to lower crust and play an important role in the rheology of these crustal levels. We report deformation microstructures in a metasedimentary gneiss (Greenland) produced during simple shear at *c.* 700 °C and *c.* 4.5 kbar. Electron backscatter diffraction analysis shows activation of the (001)[010] and (001)[100] slip systems in large grains. Additionally, (100)[001] and (010)[001] are activated in needles with their c-axes parallel to lineation. Medium-grained sillimanite aggregates mixed with biotite show a weak crystallographic preferred orientation (CPO) suggesting activation of (001)[010] and (001)[100] together with grain boundary sliding (GBS). GBS is the dominant deformation mechanism in the outer parts of fine-grained clusters of sillimanite stemming from andalusite replacement. Associated features are phase mixing, clockwise dispersion of c-axis orientations and change of misorientation axes from original c-axis rotation to rotation close to the kinematic Y axis. The latter features which result in strain localization around these clusters allow their 'preservation'. Rheologically, the medium-grained aggregates together with zones around clusters take up most of the deformation. Our study shows that deformation mechanism and slip-system activation, and CPOs are strongly dependent on the orientation relationship between stress axes and pre-deformation grain orientations and characteristics.

Sillimanite-bearing metapelites represent a significant part of the middle crust. As sillimanite is the high temperature, intermediate pressure phase of the three Al_2SiO_5 polymorphs, it is abundant in intermediate pressure granulite and upper amphibolite facies terrains (Fountain 1976; Condie 1982; Reid *et al.* 1989). In most cases, studies of such rocks focus on geothermobarometry due to their chemical suitability to constrain both pressure and temperature (e.g. Grambling 1981; Ghent *et al.* 1982; Kerrick & Speer 1988; Pant *et al.* 2012). Other studies investigate the phase transformations between the Al_2SiO_5 polymorphs, their kinetics, replacements and differential stress relationships (e.g. Carmichael 1969; Vernon 1987*a, b*; Cesare *et al.* 2002; Wheeler *et al.* 2004; Goergen *et al.* 2008). For example, deformation experiments show that sillimanite is the weakest of the Al_2SiO_5 polymorphs (Goergen *et al.* 2008). Several workers have concluded that high strain areas are nucleation and growth sites of sillimanite and that sillimanite networks allow deformation partitioning

accommodating high strain (Doukhan *et al.* 1985; Wintsch & Andrews 1988; Vernon 1987*a, b*; Cesare *et al.* 2002; Musumerci 2002; Kim & Bell 2005). As a consequence, sillimanite is expected to have a significant influence on the rheology of a deforming metapelitic rock. However, relatively little attention has been devoted to details of natural sillimanite deformation mechanisms and activation of slip systems, which govern their rheological response.

At middle to lower crustal conditions flow is either sensitive or insensitive to grain size. The so-called deformation mechanism maps (Ashby 1970; Frost & Ashby 1983) show for different minerals the grain-size-, temperature- and strain-rate-dependent fields of dominant deformation mechanisms. Each field is characterized by a flow law: Newtonian, linear creep for grain-size-sensitive flow and non-Newtonian, power-law creep with a stress exponent between 3 and 5 for grain-size-insensitive flow. The grain-size-insensitive deformation mechanism of dislocation creep occurs by

From: LLANA-FÚNEZ, S., MARCOS, A. & BASTIDA, F. (eds) 2014. *Deformation Structures and Processes within the Continental Crust*. Geological Society, London, Special Publications, **394**, 189–213.
First published online November 22, 2013, http://dx.doi.org/10.1144/SP394.10

generation and subsequent rearrangement of dislocations, such as their organization into subgrain boundaries within grains. The movement of dislocations is restricted to slip planes with a defined slip direction. Such slip systems are strongly governed by the mineral crystal structure. To activate a slip system the critical resolved shear stress (CRSS) for that specific system must be exceeded (Passchier & Trouw 2005), yet the CRSS changes with metamorphic and deformation conditions (e.g. Law 1990; Law *et al.* 1990). Furthermore, the active slip systems should be dependent on crystallographic orientation of the crystal with respect to the stress axes in addition to the aforementioned extrinsic parameters (Barrie *et al.* 2008 and references therein). The deformation of a large number of grains results in the development of crystallographic preferred orientation (CPO) observed in grain aggregates (e.g. Sander 1948; Turner & Weiss 1963; Hobbs 1985; Law 1990; Law *et al.* 1990). Traditionally, these CPOs have been used to derive the dominant slip systems and the conditions of deformation. Most experiments, detailed microstructural interpretation and modelling of CPOs have been performed on minerals such as quartz, olivine and calcite (e.g. Wenk 1985 and references therein; Heidelbach *et al.* 2000; Stipp *et al.* 2002; Bestmann & Prior 2003; Menegon *et al.* 2011). The relevance of using bulk CPOs for the interpretation of deformation temperatures has recently been questioned: Peternell *et al.* (2010) were able to demonstrate that the CPO can vary strongly between different areas within one sample. The reasons for this phenomenon are still unclear.

CPO data can be obtained by statistical diffraction analysis of a volume of rock using, for example, neutron and X-ray diffraction (e.g. Wenk 1985; Wenk *et al.* 1987). Until recently, slip systems were derived by either the aforementioned CPOs or transmission electron microscopy (TEM) (e.g. Boland *et al.* 1971) involving the study of very small sample volumes. Recently, a third technique has been developed that employs high resolution electron backscatter diffraction analysis (EBSD) to determine activated slip systems using the boundary analysis technique (e.g. Lloyd *et al.* 1997; Fliervoet *et al.* 1999; Prior *et al.* 2002; Reddy *et al.* 2007; Beane & Field 2007; Piazolo *et al.* 2008; Menegon *et al.* 2011).

At relatively small grain sizes and low stresses, grain-size-sensitive flow is dominant. Here, main mechanisms are diffusion creep and grain boundary sliding (GBS) accommodated by a variety of mechanisms including diffusion, dissolution − precipitation and dislocation glide (GBS accommodated by dislocation glide: DisGBS (Warren & Hirth 2006 and references therein)). Evidence for GBS has

been suggested to include randomization of misorientation axes and weakening of an existing CPO (Jiang *et al.* 2000; Bestmann & Prior 2003), low internal strain and equant grain shapes (Passchier & Trouw 2005), an increase of grain boundary angles going from low to high strain areas (Jiang *et al.* 2000) and phase mixing due to mechanical mixing or heterogeneous nucleation (e.g. Kruse & Stünitz 1999; Warren & Hirth 2006; Dimanov *et al.* 2007). In case of DisGBS, Prior & Hirth (2007) argue that misorientation axes between grains should be primarily controlled by deformation kinetics.

This study aims to use detailed microstructural analysis to determine dominant deformation mechanisms in sillimanite grains and clusters occurring in a typical Grt-Sil-Bt gneiss at mid-crustal conditions. We show that the same mineral deforms by different deformation mechanisms depending on its grain size and microstructural environment. In areas dominated by dislocation creep, detailed slip-system analysis allows for evaluation of the impact of the relationship between the mineral crystallographic orientation and the stress axes on the activation of slip systems. In other areas, grain sizes are significantly smaller and GBS accommodated by dislocation glide is inferred to be dominant. Finally, we discuss the effect of pre-deformation grain orientations and characteristics on the activation of deformation mechanisms and slip systems and resultant microstructural and rheological consequences.

Geological background

The studied sample is from one of the Archaean high-grade terranes from southern West Greenland. It was collected south of the Tummeralik area, SE of Nuuk, an area mapped during the 2006 GEUS (Geological Survey of Denmark and Greenland) mapping campaign, (Fig. 1). The Tummeralik area consists dominantly of tonalitic to granodioritic orthogneisses with subordinate supracrustal rock units. Amphibolite, metapelite and ultramafics rocks occur as boudins or linear bodies within the orthogneisses (Lee 2008). The investigated sample is from a Grt-Sil-Bt gneiss-bearing metapelitic unit representing a significant part of the supracrustal units of the area (UTM zone 22 (E 551448, N 7117040)). The unit is characterized by a well-developed foliation. Mineral and mineral aggregate lineations are defined by sillimanite needles and alignment of feldspar and quartz aggregates. Both foliation and lineation are parallel to the regional fabric seen in the orthogneisses. The general metamorphic evolution of the adjacent Tummeralik area has been defined as a clockwise Barrovian

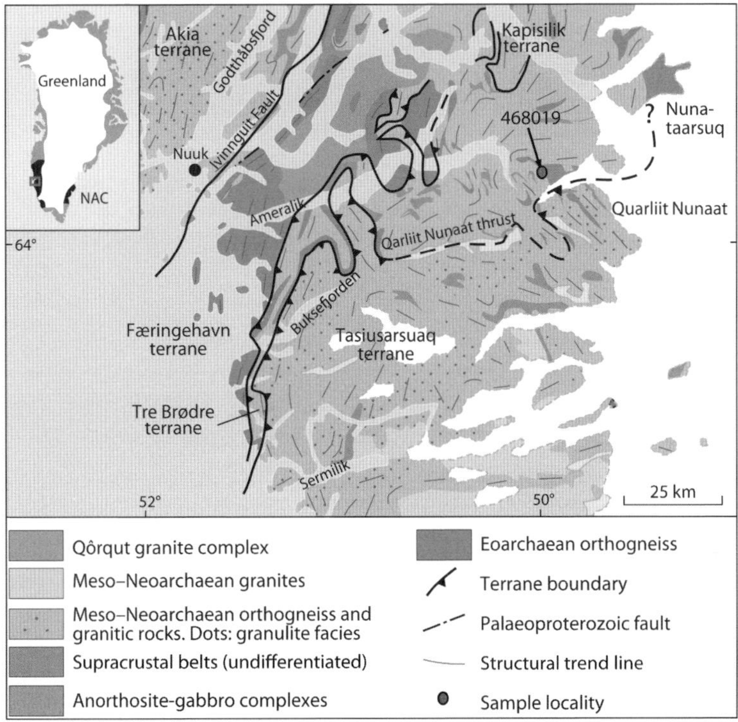

Fig. 1. Sample location shown on a map over the Nuuk region depicting the regional geology. Inset shows the location of the SE Nuuk region within Greenland; the black area is the Archaean North Atlantic Craton (NAC) (modified after Schersténn *et al.* 2008).

P–T path with peak conditions at upper amphibolite facies (*c.* 690 °C and *c.* 6.3 kbar, Lee 2008).

Sillimanite crystal structure and known slip systems

Similar to the crystal structure of andalusite, sillimanite contains chains of edge-shared aluminium-oxygen octahedra, parallel to the c-axis. Through corner sharing the polyhedra are cross-linked by double chains of tetrahedra containing alternately silica and aluminium (Fig. 2). This chain structure accounts for the elongated crystal habit parallel to the c-axis and possibly the perfect (010) cleavage in sillimanite crystals (for further details see Deer *et al.* 1997 and references therein). Although, there have been deformation studies on the Al_2SiO_5 polymorphs, studies on the internal deformation of sillimanite are relatively scarce (TEM studies: Doukhan & Christie 1982; Doukhan *et al.* 1985; Lambregts & van Roermund 1990; EBSD studies: Goergen *et al.* 2008). There are a number of suggested slip systems in sillimanite and the structurally closely related mineral mullite (Table 1).

Analytical methods

Chemical analysis and thermobarometry

Microprobe analysis was performed to obtain mineral chemistry data needed for the thermobarometric calculations. Analysis of biotite, plagioclase and garnet were obtained using the JEOL Superprobe JXL 8200 at the University of Copenhagen, Denmark. Natural and synthetic standards were used for the measurements of Si, Ti, Al, Fe, Mn, Mg, Ca, Na, K, F, Cl (wollastonite for Si and Ca, rutile for Ti, corundum for Al, hematite for Fe, $MnTiO_3$ for Mn, olivine for Mg, albite for Na, orthoclase for K, apatite for F and sodalite for Cl). Analytical conditions used during data acquisition were 15 kV acceleration voltage and 15 nA sample current with a spot size between 3 and 5 μm. Larger spot sizes were used to minimize the diffusion of Na in plagioclase. Counting time were 10 s on the peak and 10 s on the background. Analysed points were selected in accordance with the sequence of stable mineral assemblages determined by detailed optical microscopy.

Estimations of metamorphic pressure and temperature and their development through time were

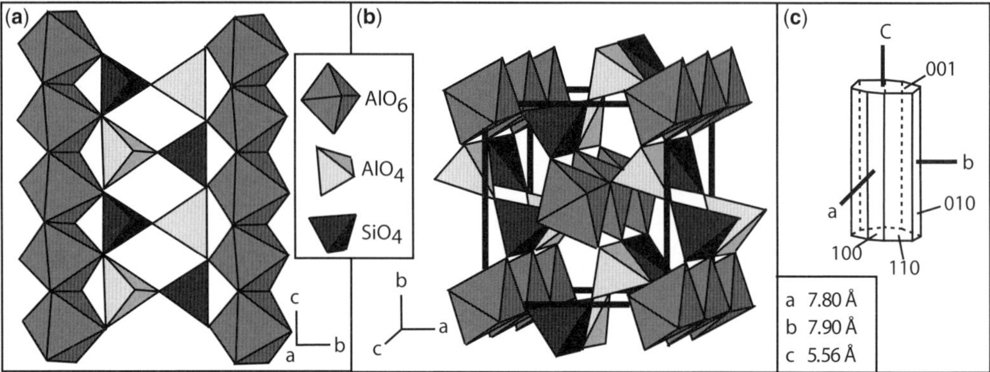

Fig. 2. Schematic representation of the crystal structure of sillimanite: (**a**) viewed down the a-axis and (**b**) at an oblique angle to c- and a-axes. The crystal shape is illustrated in (**c**) with the lattice parameters (modified after Burnham 1963).

done with the help of Spear and Kohn's program GTB (version GTB OSX August 2006 http://ees2.geo.rpi.edu/MetaPetaRen/metamorphicpetrologyindex.html). We chose to use Hodges & Spear's (1982) Bt − Grt thermometer and GASP barometer (Holdaway 2001). No ferric iron recalculations were used. Mineral abbreviations are used after Kretz (1983) and end member abbreviations after Holland & Powell (1998).

Electron backscatter diffraction (EBSD) analysis and orientation contrast imaging

Crystallographic orientation data were obtained from automatically indexed EBSD patterns collected on the in-house Philips XL-30-ESEM-FEG at the Department of Geology and Geochemistry, Stockholm University, and the Zeiss EVO at the Geochemical Analysis Unit, Macquarie University. EBSD analysis was carried out on two representative thin sections cut perpendicular to the foliation (XY plane of finite strain) and parallel to the stretching lineation (X-direction). Thin sections were SYTON®-polished (Fynn & Powell 1979) before the acquisition of EBSD patterns. Working conditions were: 15–20 kV accelerating voltage, c. 0.8 nA beam current, working distance of about 15 to 20 mm, 70 ° sample tilt, and low-vacuum mode (0.3–0.4 torr, Stockholm University) and high-vacuum mode (10^{-5}–10^{-6} torr, Macquarie University). For analysis at Macquarie University samples were carbon coated. Patterns were acquired

Table 1. *Summary of slip systems and microstructural features observed for sillimanite*

Observed microstructure	Burgers vectors	Active slip system	Deformation conditions	Material	Analytical technique	Reference
Heterogeneous dislocation densities and misoriented domains	[001] [100]	(100)[001] (010)[001] (010)[100]	Experimentally deformed at 11 kbar, from 500 °C to 900 °C compression along [110]	Gem quality single crystal	TEM	Doukhan & Christie (1982) Doukhan *et al.* (1985)
Sharp deformation-induced subgrain boundaries	[001] [100]	(100)[001] (001)[100]	Naturally deformed at 6 kbar and 750 °C	Prismatic sillimanite	TEM U-stage	Lambergts & van Roermund (1990)
Lattice preferred orientations relative to shear plane	n/a	(010)[001]	Experimentally deformed at 3 kbar and 1250 °C	Fine-grained aggregate of sillimanite	EBSD TEM	Goergen *et al.* (2008)
n/a	[100] [010] [110] [112]	n/a	Sintered in air at 1650 °C for 3 h	Commercially produced mullite	LACBED TEM	Gustafsson & Falk (2008)

LACBED, large-angle convergent beam electron diffraction.

on rectangular grids by moving the electron beam at a regular step size of 1 and 1.5 μm. Automatic EBSD indexing (Adams *et al.* 1993; Prior *et al.* 1999) was performed by detecting 6–7 Kikuchi bands using the Hough transform routine and comparing their orientation to 55 theoretical reflectors. The overall indexing in maps presented was between 76% and 92% for Sil excluding obvious scratches and holes. The raw EBSD data acquired were processed to remove false data and to enhance the continuity of data over the microstructures. Processing followed the procedure described in Prior *et al.* (2002) and Bestmann & Prior (2003). This involved first the removal of 'wild spikes', that is, orientation of very different orientation to their neighbouring analyses, then replacement of zero solutions with a nearest-neighbour algorithm (up to five neighbour-to-neighbour comparisons) and finally removal of systematic misindexing.

Crystallographic information is represented in a number of ways. Crystallographic axes and misorientation axes orientations without spatial information are presented in pole figures in the structural (XYZ) reference frame using equal-area, upper-hemisphere projections. We present microstructural data in a number of ways. Orientation maps show for each data point a colour according to the inverse pole figure (IPF) colouring scheme. Textural component maps show the difference in orientation of each data point relative to a chosen reference orientation (marked with a cross). Based on EBSD data we define grains as areas fully enclosed by grain boundaries with a misorientation of >10°. Boundaries with misorientation between 1.5 and 10° are defined as subgrain boundaries. Among all the combinations of symmetrically equivalent axis/angle pairs, we follow the general convention described by Wheeler *et al.* (2001) to select the pair with the minimum misorientation angle. Crystallographic misorientations are investigated on misorientation profiles and misorientation axis/angle pairs and their orientation. It should be noted that misorientation axes for small misorientation angles (<5°) are relatively poorly constrained (Prior 1999). We use the boundary trace analysis to determine the geometry of low-angle boundary and the active slip system(s) from EBSD data. When the 3D orientation of a boundary is unknown the boundary trace analysis provides a crystallographically consistent solution for the boundary geometry, if ideal tilt and twist boundary models are assumed. This method considers the dispersion of the orientation data around a rotation axis for an area sampled across a selected 2D trace of a low-angle boundary on EBSD maps. The rotation axis is identified on the pole figures as the direction with little or no dispersion (Lloyd & Freeman 1994). In the case of a tilt boundary, the boundary plane

must contain the 2D boundary trace and the rotation axis. A plane at high angle (ideally at 90°) to the boundary plane and containing the rotation axis represents the most likely active slip plane and must contain the slip direction. In the case of a twist boundary, the rotation axis is perpendicular to the boundary plane. Boundary trace analysis has been used successfully in EBSD studies of various geological materials: quartz (e.g. Lloyd *et al.* 1997; Menegon *et al.* 2011), halite (Borthwick &Piazolo 2010), calcite (Bestmann & Prior 2003), garnet (Prior *et al.* 2002), kyanite (Beane & Field 2007), zircon (e.g. Reddy *et al.* 2007; Piazolo *et al.* 2012) and pyrite (Barrie *et al.* 2008). The method has also been used on ice (Piazolo *et al.* 2008).

Results

Microstructure and phase relationships: general

The investigated sample is dominated by Qtz − Pl − Bt-rich, 3–5 mm-wide ribbons alternating with Sil − Bt-rich, 2–5 mm-wide bands commonly associated with up to 3 mm large, pale pink Grt porphyroblasts (Fig. 3a, b). Grt exhibits two growth stages with an inclusion-rich core (Grt1) mantled by a relatively inclusion-free outer rim (Grt2) (Fig. 3c, d). Inclusions in Grt consist of Qtz, Pl (Pl1), Bt, Rt, Mnz and Sil (Sil1). Bt inclusions show two groups Bt1 and Bt2, where they occur in the core and rim, respectively. In the matrix, Pl occurs as euhedral porphyroclasts up to 5 mm in size (Pl2) with recrystallized, 100–200 μm large grains (Pl3) at the porphyroclasts' borders. Recrystallized 50–200 μm anhedral Pl grains are also a part of the Qtz−Pl ribbons. Qtz grains (50–200 μm) are anhedral, moderately elongated showing undulose extinction. Symplectitic intergrowths of Pl and Qtz are found in pressure shadows in stable contact with the Sil−Bt layers and/or Grt. Bt (Bt3) and Sil (Sil2) are the principal foliation-forming minerals wrapping Grt porphyroblasts and Pl porphyroclasts (Fig. 3a, b). Bt3 is additionally seen in strain shadows next to Grt (Fig. 3b). Rutile occurs in the Qtz−Pl−Bt ribbons, the Sil−Bt foliation and as inclusions in Grt1. Muscovite is not present in the rock. No retrograde alteration or significant Grt resorption are observed.

The foliation-forming mineral Sil2 occurs as four different types, where each type is defined by a specific crystallographic orientation to the structural framework XYZ and/or shape (Fig. 4a, b). The area (%) relative to the whole thin section varies from *c.* 3% for both type 1 and type 2, up to 6–8% for type 3 and 8–10% for type 4. Type

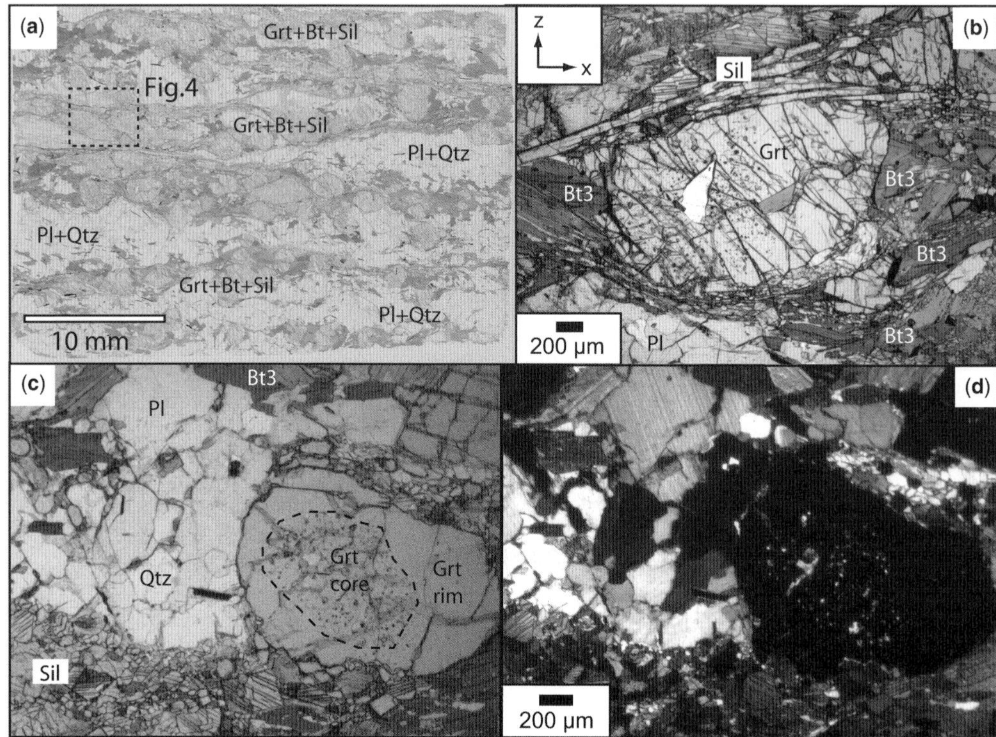

Fig. 3. Photomicrographs showing the typical microstructures of the investigated samples, here thin section 468019 cut parallel to lineation (X) and perpendicular to foliation (X–Y plane). (**a**) Overview showing the domains of Grt + Bt + Sil and those of Pl + Qtz, which define the foliation, plain light; stippled box outlines the location of microphotographs shown in Figure 4a, b. (**b**) Close-up of Sil and Bt wrapping around a Grt – note that Bt grows in the strain shadow of the Grt. (**c, d**) Close-up of typical microstructure within and surrounding Grt – note the inclusion-rich core and inclusion-poor Grt rims: (c) plain light, (d) cross-polars.

1 consists of euhedral rhombic to blocky crystals, 150–300 μm in diameter, with visible perfect {010} cleavage and c-axis orientation parallel to Y (Fig. 4). Close to these grains the orientation of Sil grains of type 3 or 4 may be disturbed and/or folded (Fig. 4d, f). Type 2 is characterized by up to 5 mm-long needles elongated with the x-direction (lineation); thus c-axis orientation is parallel to X, and a- and b-axes are perpendicular to X (Fig. 4a, e). Type 3 grains occur as grain aggregates commonly associated with Bt. These aggregates anastomose around Grt, Pl–Qtz lenses and type 4 clusters (for details on the latter, see below). Individual grains are anhedral, round to elongate in shape (Fig. 4a–c) and commonly between 30 and 100 μm in size (Fig. 4c). Within type 3 aggregates, small grains commonly exhibit a foam texture with 120° triple junctions. In type 3 aggregates a few individual, larger grained type 2 and type 1 grains are present (Fig. 4a, e, f). Anhedral to subhedral type 4 grains form grain aggregates occurring as rhomb to oval shaped, 2 by 5 mm large clusters

(Fig. 4a, b, d). These fine-grained clusters are commonly bordered by relatively bigger anhedral (type 3 matrix grains) and/or type 1 or 2 grains. Optical analysis shows that within an individual cluster crystallographic orientations are similar but may change towards the cluster borders. However, the orientation of individual clusters differs. Two main crystallographic orientations dominate, where the c-axes are either sub-parallel or sub-perpendicular to X (Fig. 4a, b). Depending on the c-axis orientation of these clusters, grain sizes determined on the XZ section vary. If Sil grains are oriented with their c-axis sub-perpendicular to X, their grain size is fine (3–10 μm). If Sil grains are oriented with their c-axis sub-parallel to X, the measured grain size is larger (20–50 μm) and grains are slightly elongated parallel to X (Fig. 4a, b). Within type 4 clusters, individual type 2 and type 1 grains are common (Fig. 4a, e, f), where type 1 grains are predominantly seen in those clusters with c-axis sub-parallel to Y, and type 2 grains in clusters with c-axes sub-parallel to X.

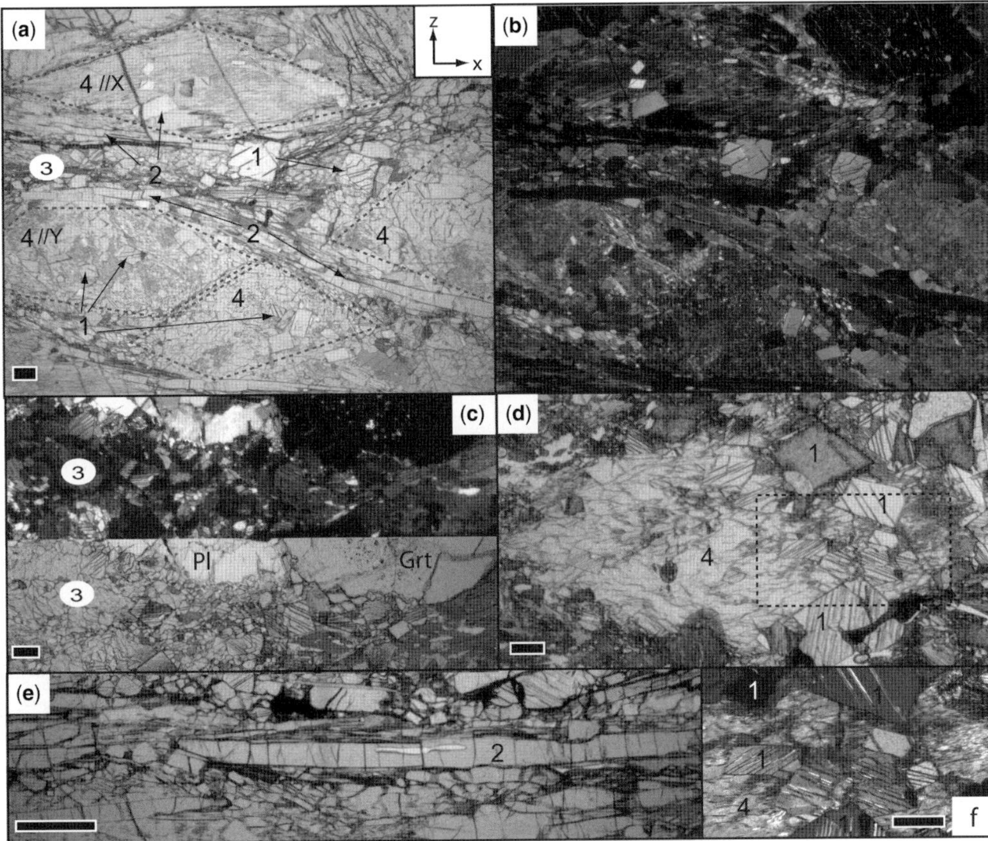

Fig. 4. Photomicrographs of sillimanite microstructures. (**a, b**) Relationship between type 1–4 Sil grains marked with numbers 1–4. Type 1 and 2 represent the euhedral large grains. They occur within type 3 and type 4 domains. Type 4 anhedral masses are outlined with a stippled line, note the similarity of crystallographic orientation within an individual cluster, while different clusters exhibit different general orientations of Sil grains, approximate orientation is indicated, for example //X. The local reorientation of type 4 domains is well illustrated by the change in grey in (b) towards the edge of the type 4 cluster. For location within large-scale microstructure, see stippled box in Figure 3. Type 3 represents the anhedral matrix anastomosing around the Grt porphyroclasts and type 4 domains: (a) plain light; (b) cross-polars. (**c**) Type 3 mixed with Bt and Pl, cross-polars (top), plain light (bottom). (**d**) Example of group 4 mass with a few type 1 Sil grains; close-up of area marked with the stippled box is shown in (f), plain light. (**e**) Type 2 elongate Sil grain within type 3 matrix. (**f**) Close-up of type 1 Sil grain embedded in type 4 domain shown in (d) – note the disturbance of the preferred orientation of type 4 grains close to type 1 grains; cross-nicols. Scale bar on all micrographs is 200 μm.

No And or kyanite remnants were found in the sample by optical or EBSD analysis. However, we suggest that individual type 4 domains represent paramorphic replacement of individual, And grains, where a large number of similarly oriented Sil replaces one individual And grain. This interpretation is based on several lines of evidence. The fact that type 4 domains are monomineralic and commonly rhomb or oval shaped (Fig. 4a) is consistent with paramorphism of And. Furthermore, the size of the clusters between 2 and 5 mm is in the range of the typical grain size of metamorphic And in

pelitic rocks (e.g. Vernon 1987a, b). In addition, type 4 clusters occur as well-defined individual domains with similar crystallographic orientation within one individual cluster (Fig. 4a, b). According to Vernon (1987a, b) intergrown Sil and And shows parallelism of their c-axes. If a grain of And is replaced by Sil, it would be expected that the area of the pre-existing And is replaced with clusters of Sil grains exhibiting similar c-axis orientations. Our interpretation is consistent with the observation that paramorphic replacements after And by Sil are common in low pressure metamorphic areas

(Vernon 1987*b*; Kerrick 1990; Cesare *et al.* 2002; Kim & Bell 2005 and references therein). This may be due to the fact that the Sil−And paramorphism is a near-perfect volume-to-volume replacement reaction (Kerrick 1990) allowing this transformation to take place once P−T conditions change from the And to the Sil stability field during prograde metamorphism.

According to the microstructural relationships described above, we derived the following succession of mineral assemblages, all of which were stable with Qtz and Pl (Fig. 5). Initially, And was stable (assemblage A_1), then the And was replaced by Sil clusters (now type 4 Sil), some Sil grains and Bt (Bt1) grew in the matrix as well as Grt which is now seen as the core (Grt1) forming a Grt−Sil−Bt−rt assemblage (A_2) (Fig. 3b, c). In the next stage the general assemblage stayed the same (Grt−Bt−Sil−Rt); however, a largely inclusion-free Grt rim formed (Grt2) which rarely includes Bt (Bt2). As the inclusion-free Grt rim is not resorbed, we suggest that the Sil grains in the matrix (Sil2) and Bt (Bt3), which are now aligned with the main foliation, form a stable assemblage (Grt2 − Bt2/ Bt3−Sil2) (A_3) at peak metamorphic conditions.

Microstructures show that deformation only occurred after the first growth of Grt as there is no obvious relict foliation preserved in the Grt1 cores in the form of aligned inclusions (Figs 3c, d & 5). During the second growth phase of Grt (Grt 2) no pronounced foliation was present either, as the Bt included in the outer rim of the Grt shows no preferred orientation. However, Bt, Sil and Qtz−Plag domains are now seen to be aligned parallel to each other suggesting deformation occurred while

these phases were stable. Foliation wraps around Grt (Fig. 3b). In the first stage of deformation (D_1) Sil2 and Bt3 and Pl + Qtz clusters aligned to form a foliation. During progressive deformation (D_2) shearing intensified, Bt3 rotated successively into parallelism with the foliation and new Bt3 has grown in strain shadows. In the vicinity of Grt, both Bt3 and Sil2 deformed. For example, D_2 resulted in the bending of the elongated Sil needles (type 2), giving the Sil2 a bent lenticular shape (Fig. 3b) Dextral shear is indicated through asymmetry of type 4 Sil clusters and strain shadows around Grt-forming σ-clasts. D_1 and D_2 are part of the same deformation event (D_{1+2}) and are distinguished here to emphasize progression of deformation. D_{1+2} occurs in the stability field of Grt2, Sil2 and Bt3, consequently the Sil deformation of Sil2 and the resultant generation of Sil microstructures discussed below occurred during the D_{1+2} at high-grade metamorphic conditions.

Mineral chemistry, thermobarometry and deformation conditions

Representative chemical data of the main phases are presented in Table 2. Garnets show compositional prograde zonation between the cores (Grt1) and the rims (Grt2). Cores exhibit a slightly higher X_{pyp} and X_{grs} than rims (Fig. 6a, Table 2). In general, the dominant Pl composition is X_{ab} *c.* 0.7 (Fig. 6b, Table 2); however, feldspar porphyroclasts (Pl2) exhibit some diffuse variation in X_{ab} and X_{an} from centre to rim. Biotite shows a general trend with decreasing Fe^T from Bt1 to Bt2 and Bt3

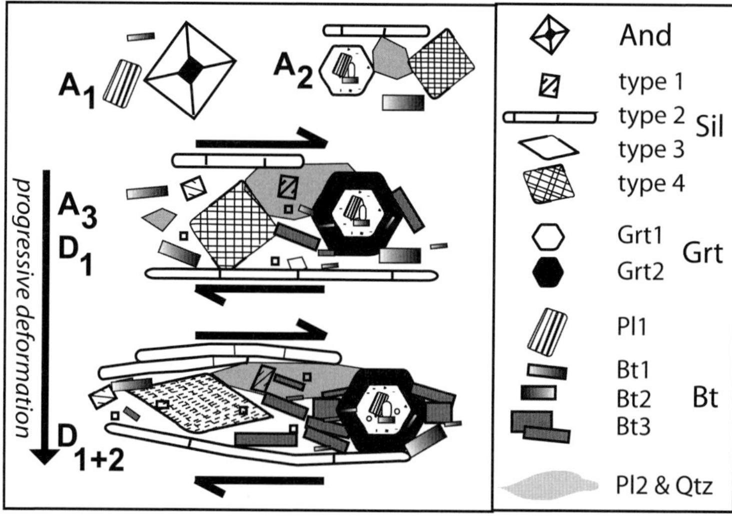

Fig. 5. Schematic illustration of the textural (D_1–D_{1+2}) and mineralogical (A_1–A_3) evolution of the studied sample.

Table 2. *Representative microprobe analyses of main phases*

Mineral	Core								Rim							
	Bt1a	Grt1a	Bt1b	Grt1b	Grt1c	Pl1a	Grt1d	Pl1b	Bt2a	Grt2a	Bt2b	Grt2b	Grt2c	Pl2a	Grt2d	Pl2b
Wt%tot	99.32	102.556	100.222	103.04	101.412	100.733	102.364	99.986	99.252	103.056	98.348	103.601	103.589	100.214	102.93	100.32
Si	2.846	2.984	2.783	2.958	2.965	1.024	2.986	1.02	2.85	2.963	2.839	2.957	2.965	1.027	2.981	1.024
Al	1.523	1.974	1.571	1.996	1.964	0.472	2.012	0.468	1.566	1.998	1.567	2.002	1.995	0.472	1.998	0.474
Ti	0.183	–	0.181	0.001	0.011	–	0.002	–	0.189	0.002	0.183	0.001	–	–	–	–
Fe^{3+}	–	0.057	–	0.083	0.089	–	0.013	–	–	0.071	–	0.081	0.077	–	0.038	–
Mg	1.916	1.139	1.943	1.089	1.228	–	1.157	–	1.74	0.952	1.776	1.006	0.996	–	0.98	–
Fe^{2+}	0.609	1.722	0.603	1.74	1.626	0.006	1.714	0.003	0.743	1.868	0.738	1.82	1.836	0.001	1.876	–
Mn	0.003	0.027	0.003	0.032	0.029	–	0.025	–	–	0.041	0.001	0.029	0.032	0	0.032	0
Ca	0.001	0.091	–	0.097	0.083	0.11	0.092	0.107	–	0.101	–	0.098	0.091	0.105	0.091	0.109
Na	0.038	0.002	0.038	–	0.002	0.274	–	0.274	0.027		0.035	0.002	0.004	0.272	–	0.273
K	0.881	–	0.879	–	0.003	0.001	–	0.002	0.884	0.001	0.862	–	–	0.002	–	0.002
H_2O	1.999	–	1.998	–	–	–	–	–	1.996	–	1.999	–	–	–	–	–
F	–	–	0.001	–	–	–	–	–	–	–	0.001	–	–	–	–	–
Cl	0.001	–	0.001	–	–	–	–	–	0.001	–	0.003	–	–	–	–	–

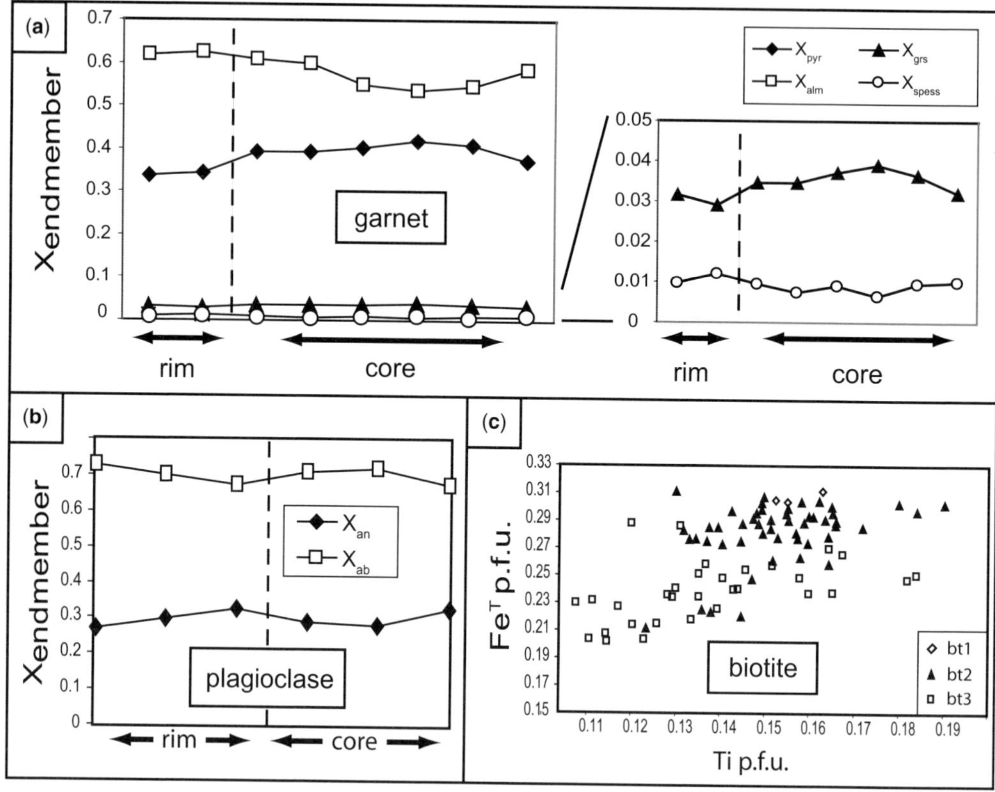

Fig. 6. Mineral chemistry of main phases: (**a**) profile across a representative garnet from core to rim (see Table 2 for details, Grt1-2); (**b**) profile across representative Pl porphyroclast; and (**c**) Bt1–b3 composition depicted on Fe_t^T p.f.u. (per formula unit) against Ti p.f.u. graph. Fe^T represents the total amount of iron, that is Fe^{2+} and Fe^{3+}.

(Fig. 6c, Table 2) while Bt2 and Bt3 composition largely overlaps.

Estimated pressure and temperature for Grt cores (Grt1–Bt1) range between 3–4.5 kbar and 600–650 °C, whereas those for the Grt rims (Grt2–Bt2/Bt3) range between 3.5–4.5 kbar and 675–700 °C, (Fig. 7). Both Grt core (A_2) and rim (A_3) are lying in the Sil stability field showing a possible prograde path in the mid- to upper-amphibolite facies. Metamorphism therefore is characterized by low pressures. This is consistent with the reported P–T conditions for peak metamorphism (Lee 2008). Deformation conditions for the Sil2 grains studied in detail here are the peak metamorphic conditions (3.5–4.5 kbar and 675–700 °C) (Fig. 7) as there is no significant deformation recorded before or after peak metamorphism.

Detailed crystallographic and microstructural characterization

The detailed crystallographic and microstructural analysis focused on the intracrystalline deformation in the euhedral type 1 and 2 grains providing data to determine the most likely slip system(s) activated during their deformation. For type 3, in addition, pole figures depicting CPO and misorientation axes are presented. Type 4 analysis concentrated on CPO data, and misorientation axes and angle distributions.

Intracrystalline orientation characteristics of type 1 sillimanite grains. Type 1 Sil grains are oriented with their c-axes parallel to the Y direction, perpendicular to the lineation (X) (Fig. 8). Internal structures of the grains show low-angle boundaries with dominantly 3–10° misorientation angles propagating through the grains (Fig. 8a, b). In type 1 Sil, three main subgrain boundary trace directions could be detected: (*i*) oriented *c.* 45° oblique to X, subsequently referred to as diagonal; (*ii*) sub-perpendicular to X, subsequently referred to as vertical; and (*iii*) sub-parallel to X, subsequently referred to as horizontal (Fig. 8a). Texture component maps and linear misorientation profiles, within a grain but not crossing subgrain boundaries,

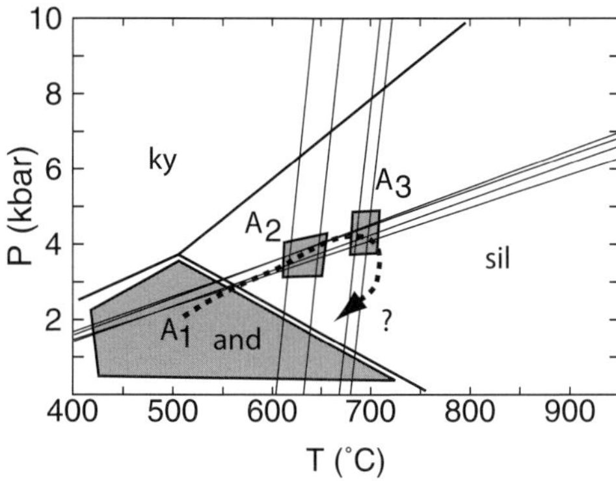

Fig. 7. P–T diagram showing the PT conditions for the three assemblages A_1–A_3 (cf. Fig. 5) using microprobe data from garnet cores (A_2) and rims (A_3) results (cf. Table 2). The inferred clockwise low pressure P–T path is shown as a stippled line.

show internal bending up to 1.5–$2°$ per 100 μm (Fig. 8b, c). Diagonal boundaries exhibit a rotation axis $<100>$, shown by the small circle dispersion around this axis in the pole figure plots. Vertical and horizontal boundaries have a rotation axis of $<110>$ (Fig. 8).

Intracrystalline orientation characteristics of type 2 sillimanite. Crystallographic data of type 2 Sil grains show that these grains primarily have two distinct orientations. Therefore, we subdivide them into type 2a and type 2b (Figs 9 & 10). For type 2a c-axes are oriented parallel to X with the a-axes sub-parallel to Y and b-axes sub-parallel to Z (Fig. 9a, c). In these grains internal structures show dominantly $1.5°$ to $6°$ low-angle boundaries propagating through the needle. Two main orientations of these boundaries are seen. Boundaries are either sub-perpendicular to X (vertical) (boundary (*i*) and (*ii*) in Fig. 9) or sub-parallel to X (horizontal) (boundary (*iii*) in Fig. 9). Texture component maps and misorientation profiles, within a grain but not crossing subgrain boundaries, show significant continuous crystal lattice bending of up to $5°$ per 100 μm (Fig. 9b, d). Vertical boundaries have rotation axes (*i*) $<100>$ or (*ii*) $<310>$ shown by the dispersion on the other axes in the pole figure plots (Fig. 9e). The discontinuous horizontal boundary (*iii*) has a rotation axis of $<001>$ (Fig. 9e). Type 2b grains are characterized by c-axes oriented parallel to X with the a-axes roughly sub-parallel to Z and b-axes roughly sub-parallel to Y (Fig. 10a, b, c). Internal structures show dominantly $1.5°$ to $10°$ low-angle boundaries propagating through the needle with two main directions: (*i*) and (*ii*) oriented sub-perpendicular to X (vertical); and a less frequent and discontinuous $1.5°$ low-angle

boundary, (*iii*) oriented sub-parallel to X (horizontal) (Fig. 10a, b). Texture component maps and misorientation profiles, within a grain but not crossing subgrain boundaries, show continuous bending of the crystal up to $7.5°$ per 100 μm (Fig. 10b, d). Vertical boundaries have rotation axes (*i*) $<010>$ or (*ii*) $<110>$ shown by the dispersion of the other axes in the pole figure plots (Fig. 10e). Discontinuous horizontal boundaries (*iii*) have rotation axis $<130>$ (Fig. 10e).

Orientation characteristics of type 3 sillimanite. Type 3 grains occur together with Bt and form highly elongate aggregates anastomosing between Grt porphyroclasts and type 4 Sil clusters (Figs 4a, b & 11). Pole figures show a weak girdle distribution sub-parallel to the (001) plane and weak maxima for both [010] and [100] directions parallel to Z (Fig. 11b), while exhibiting a significant influence of the adjacent type 4 preferred orientation (see below for details). Misorientation axes plotted in crystal coordinates show a dominance of c-axis rotations and rotations in the plane between [100] and [010] (Fig. 11c). Subgrain boundaries are rare in type 3 grains; only a few larger grains allow boundary trace analysis where systematic crystal bending is observed (Fig. 11d). These dominantly show rotation around [010] and [100] for differently oriented grains (Fig. 11e).

Orientation and grain characteristics of type 4 sillimanite. Type 4 clusters are dominated by anhedral grains with curved boundaries. Some grains may show some elongation. However, no statistically significant shape-preferred orientation is detected within an individual domain (Figs 12a & 13). Grains exhibit irregular subgrain boundaries. In smaller grained areas, subgrain boundaries are less abundant. Pole figures show an exceptionally strong clustering of the c-axes (Figs 12d & 13c). A- and b-axes form large circle distributions (Figs 12 & 13c). There is a marked dispersion of the strong clustering of c-axis and with that a- and b-axis-girdle distributions in the outer border zones of the type 4 domain. This change of orientation pattern is very abrupt (indicated as dashed lines in Figs 12b & 13b) and is commonly associated with the presence of small Bt grains dispersed between type 4 grains and simultaneous grain-size reduction (Fig. 13). In cases when the change from strong c-axis maxima to a c-axis dispersion is not associated with Bt, grain sizes increase slightly (Fig. 12). Rotation axes between individual type 4 grains are dominated by a rotation around the c-axis; however, in the border zones this dominance is less distinct (Fig. 12c). In c-axis pole figures a clockwise rotation around the kinematic Y axis at type 4 domain border zones is seen (Fig. 12d). Misorientation

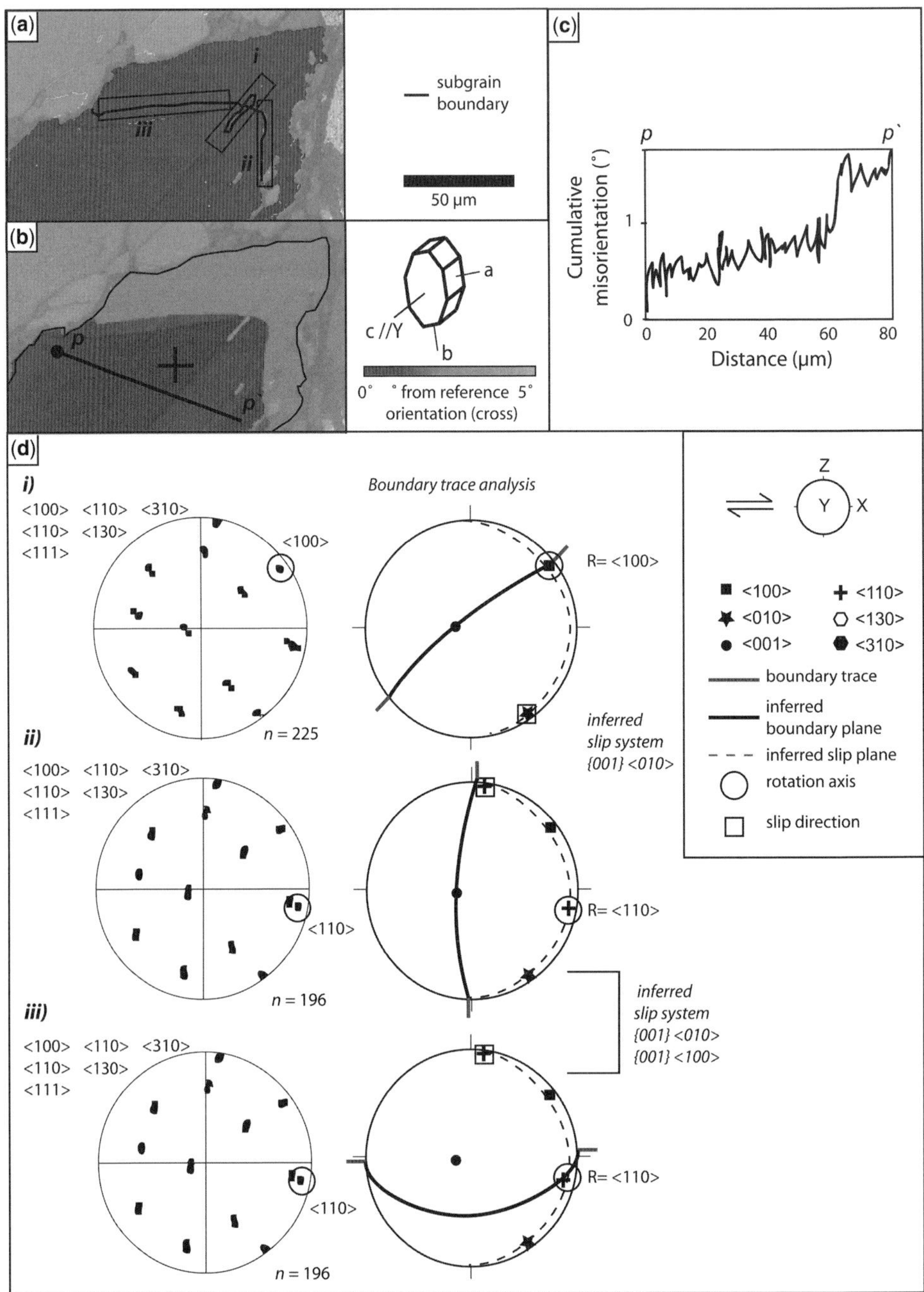

Fig. 8. Orientation characteristics of the internal microstructure of a representative type 1 euhedral grain. (**a**) Grain with subgrain boundaries of 1.5–6° misorientation across the boundary in black. Characteristic subgrain boundaries (*i*)–(*iii*) are outlined by black boxes. (**b**) Texture component map with greyscale variations indicating the progressive internal misorientation 0–5° (black to light grey) relative to the orientation at the black cross. *p–p'* outlines misorientation profile shown in (c); inset shows schematically crystal orientation in three dimensions. (**c**) Misorientation profile within grain. (**d**) Orientation data for subgrain boundaries (*i*), (*ii*) and (*iii*). Pole figures (left) show the orientation of main crystallographic axes across the respective subgrain boundary; note the small circle dispersion. Pole figure (right) depicts related boundary trace analysis. *Italic* = slip system(s) consistent with data; *n* = number of analysed points; R = rotation axis.

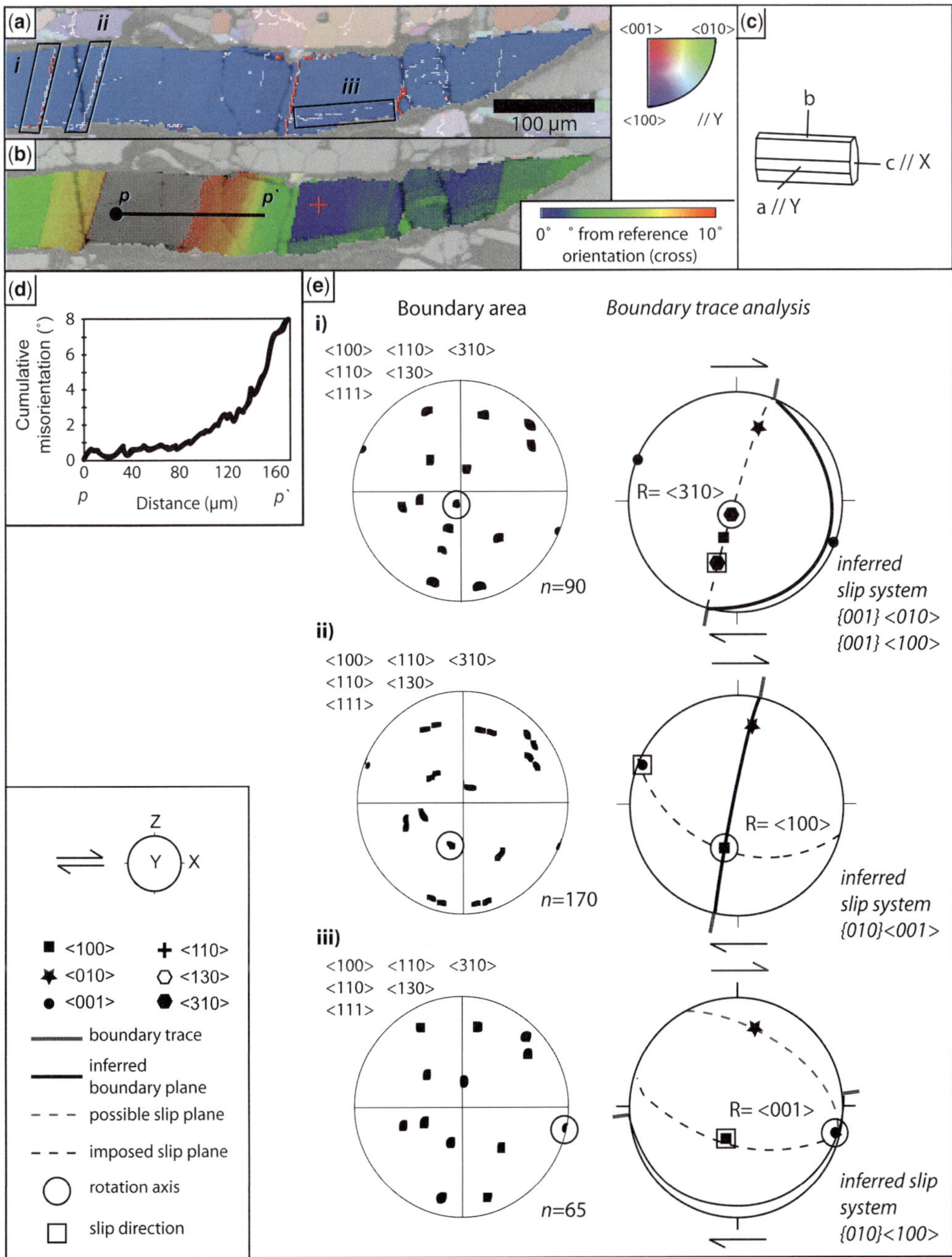

Fig. 9. Orientation characteristics of the internal microstructure of type 2a needles. (**a**) Orientation map with IPF colouring scheme and defined subgrain boundaries (see Fig. 8 caption). Characteristic subgrain boundaries (*i*)–(*iii*) are outlined by black boxes. (**b**) Texture component map; colour variations indicate the progressive internal misorientation of 0–10° (blue to red) with respect to the orientation at the red cross. *p*–*p′* outlines misorientation profile shown in (d). (**c**) Schematic crystal orientation shown in three dimensions. (**d**) Misorientation profile within grain. (**e**) Orientation data for subgrain boundaries (*i*), (*ii*) and (*iii*). Pole figures (left) show the orientation of the main crystallographic axes across the respective subgrain boundary; note the small circle dispersion. Pole figure (right) depicts related boundary trace analysis. *Italic* = slip system(s) consistent with data; *n*, number of analysed points; R, rotation axis.

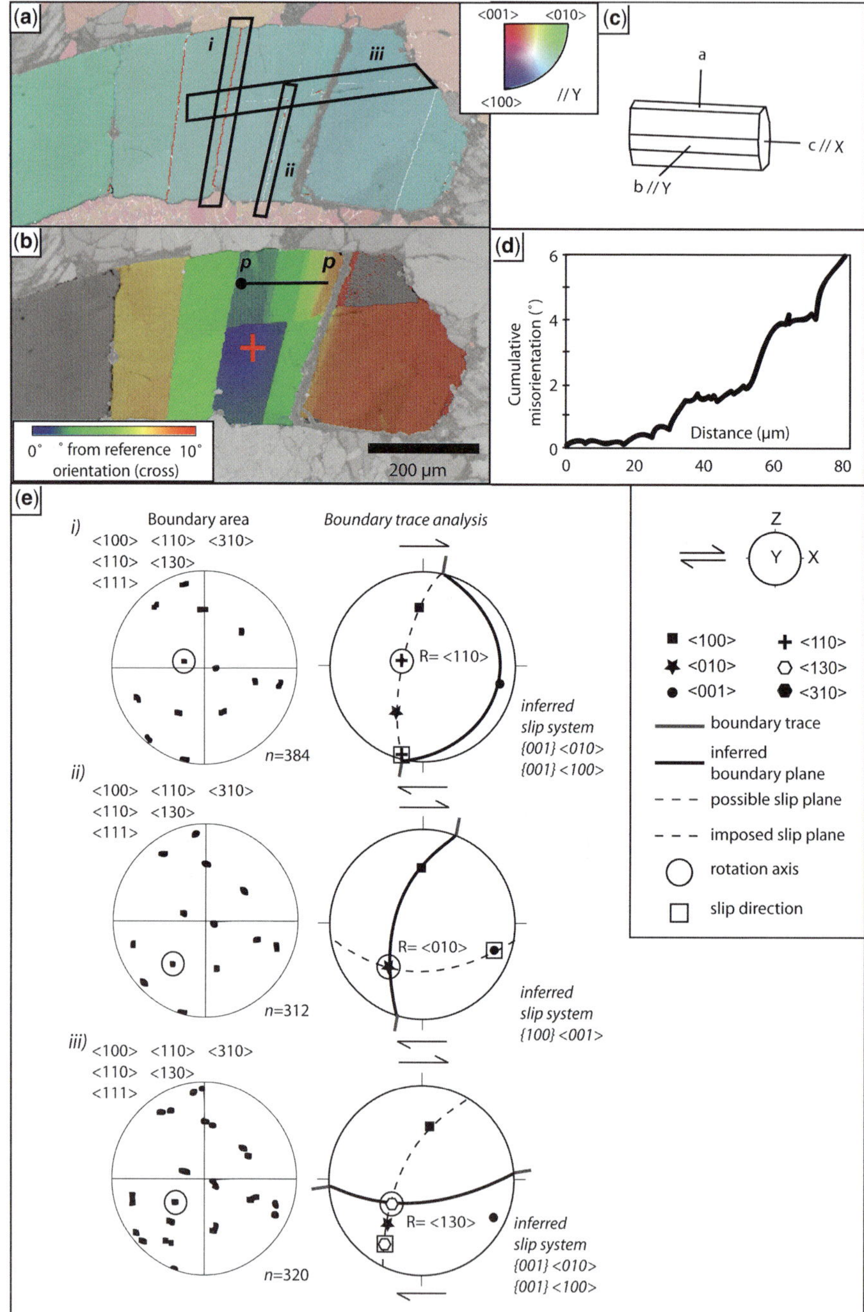

Fig. 10. Orientation characteristics of the internal microstructure of type 2b needles. (**a**) Orientation map with IPF colouring scheme and defined subgrain boundaries (see Fig. 8 caption). Characteristic subgrain boundaries (*i*), (*ii*) and (*iii*) are outlined by black boxes. (**b**) Texture component map, colour variations indicate the progressive internal misorientation 0–10° (blue to red) from the orientation at the red cross. p–p' outline misorientation profile shown in (d). (**c**) Schematic crystal orientation in three dimensions. (**d**) Misorientation profile within grain. (**e**) Orientation data for subgrain boundaries (*i*), (*ii*) and (*iii*). Pole figures (left) show the orientation of main crystallographic axes across the respective subgrain boundary; note the small circle dispersion. Pole figure (right) depicts related boundary trace analysis. *Italic* = slip system(s) consistent with data; *n*, number of analysed points; R, rotation axis.

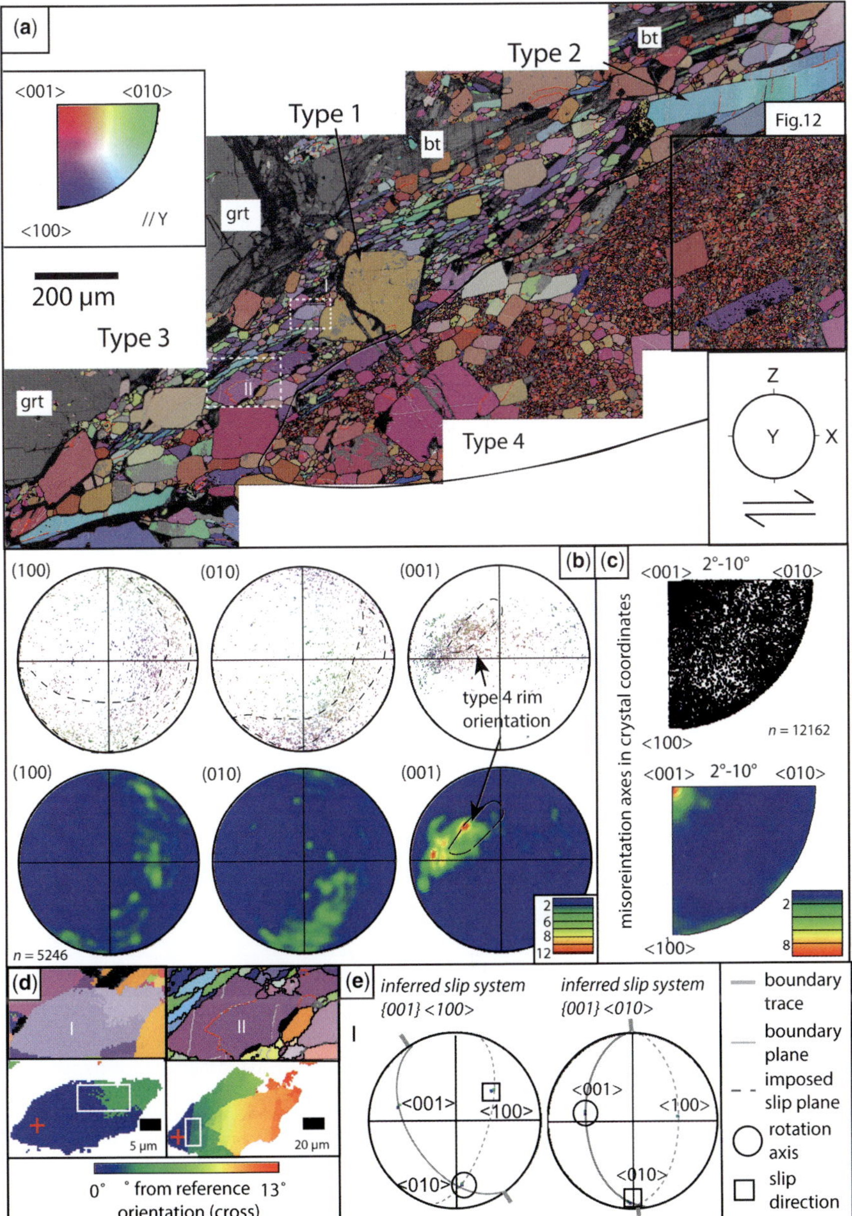

Fig. 11. Orientation characteristics of type 3 matrix grains. (**a**) IPF colour-coded orientation map showing the spatial relationship of type 1–4 grains; note presence of Grt and Bt; only Sil is colour coded. Type 4 domain is not completely covered by data but its boundaries are outlined in black; arrows point at type 1 and 2. White dashed boxes show areas I and II of type 3 grains analysed in detail in (d and e); black box shows area of data shown in Figure 12. (**b**) Pole figures showing orientation of all type 3 grains as one point per grain; top – point data, bottom – contoured plot with half space of 10°; the orientation of type 4 grains close to the type 3 domain is shown as stippled areas. (**c**) Misorientation axes in crystal coordinates, point data and contoured plot (half-width of 5°). (**d**) Characteristics of grain I (left) and II (right); top map represents IPF coloured orientation map (for colour scheme see inset in (a)); lower map shows texture component maps, colour variations indicate the progressive internal misorientation 0–13° (blue to red). (**e**) Orientation data for subgrain boundaries of grain I and II. Pole figures show the orientation of main crystallographic axes across the respective subgrain boundary; note the small circle dispersion; also depicted on pole figure is the related boundary trace analysis. *Italic* = slip system(s) consistent with data.

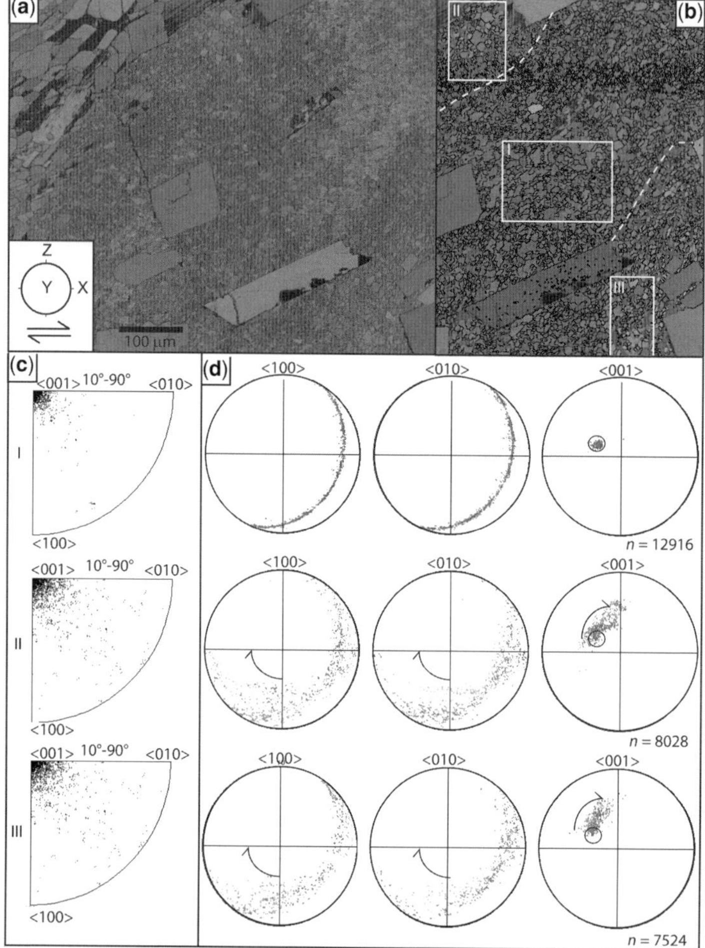

Fig. 12. Characteristics of representative type 4 domain I; same area as shown in black box in Figure 11a. (**a**) Band contrast map showing the grain microstructure; inset indicates the kinematic framework. (**b**) Greyscale orientation map of left part of (a); 3–10° and >10° boundaries are shown in grey and black, respectively. Different areas analysed in (c, d) are shown as white boxes; white dashed lines outline significant orientation changes from centre to border zone. There is an area with an increased amount of high-angle boundaries in the upper third of the orientation map, which is an analytical artefact. (**c**) Misorientation axes shown in crystal coordinates; note dispersion at rims (area II, area III) relative to domain centre (area I). (**d**) Pole figures from areas I–II. The main c-axis orientation from area I (centre of type 4 domain) is shown in all c-axis pole figures for comparison. Note the increased dispersion of orientations in areas II and III as well as the clockwise rotation of c-axes around Y.

axes plotted relative to the kinematic framework show a pronounced shift: in the domain centre a perfect c-axis rotation is seen while at the domain rim rotations are close to the kinematic Y axis (Fig. 13c). In addition, misorientation angle distributions show a decrease in low-angle boundaries and an increase in very high-angle boundaries (Fig. 13d). Orientation relationships of type 3 domains adjacent to type 4 domains show a significant influence of the strong preferred orientation of type 4 domains (Fig. 11c).

Discussion

Activation of slip systems: the influence of stress axis orientation relative to individual grain orientation: type 1 and type 2 sillimanite

Deformation of type 1 and type 2 Sil grains is characterized by both continuous and systematic lattice bending and the presence of distinct sub-grain boundaries (Figs 8–10). These characteristics are indicative of crystal plastic behaviour through

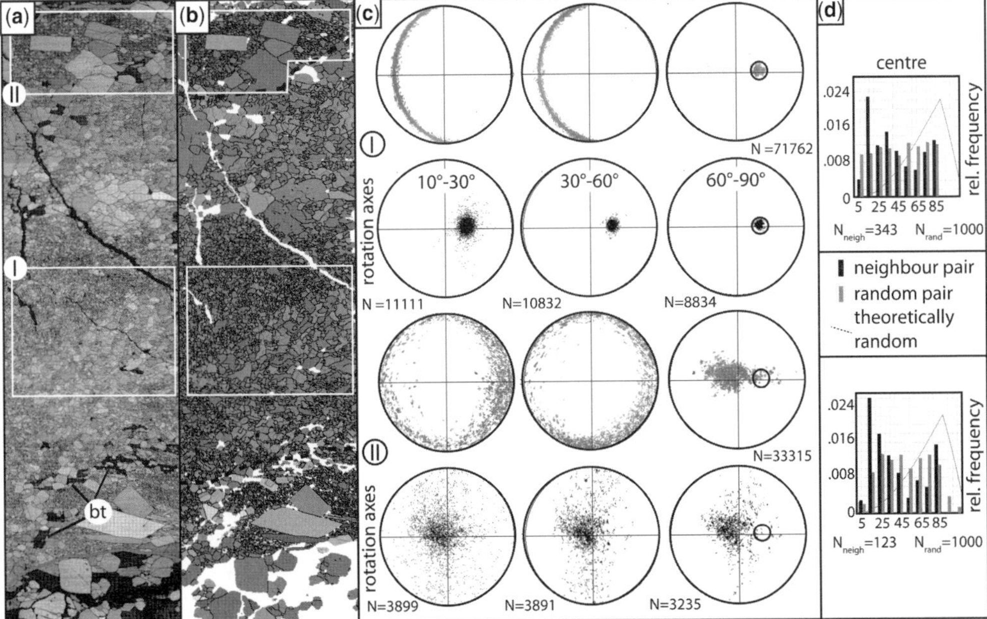

Fig. 13. Characteristics of representative type 4 domain II. (**a**) Band contrast map showing the grain microstructure of type 4 domains; the kinematic framework is the same as shown in Figure 12a. Red lines at top and bottom show the border of the type 4 domain with type 3 matrix grains and Bt grains. (**b**) Greyscale orientation map of left part of (a); 3–10° and >10° boundaries are shown in grey and black, respectively. Black dashed lines show the abrupt change from a strong c-axis maximum to a dispersed c-axis distribution from centre to border zone; note this is strongly associated with a higher abundance of Bt. Areas I and II analysed in detail in (c) are shown as white boxes in (a) and black boxes in (b). (**c**) Pole figure (top) and rotation axes in sample coordinates (bottom) shown for areas I and II. In area II the main cluster seen in area I is depicted for comparison. (**d**) Misorientation angle distribution for centre and rim of type 4 domain, neighbour pair and random pair distribution in black and grey, respectively.

dislocation glide and/or creep (e.g. Prior *et al.* 1999). In-depth analysis shows that: (a) no cross slip is activated as no T-shaped patterns are seen (Reddy *et al.* 2007; Figs 8–10); (b) there are a number of directions which act as rotation axes, suggesting a variety of active slip systems; and (c) rotation axes differ depending on crystal orientation relative to the kinematic framework, i.e. foliation and lineation. While [100] & [001] rotation axes are consistent with previously reported slip systems (Table 1 and references therein), directions [110], [310], [130] are not reported in the literature. In order to explain the latter rotation axis without involving cross slip, we consider the activation of two rotation axes in the same plane, whether or not in the same proportions (Lloyd 2004). For example, the observed rotation around [110] can be explained by equal rotation around both [100] and [010] axes in the (001) plane. This concept is illustrated in Figure 14 using vectors. The net of two perpendicular vectors is a third drawn diagonally through them. Slip direction is perpendicular to the rotation axis and can only be sinistral or

dextral, as a result either sinistral or dextral slip direction gives the net slip direction for rotation axis [110] consistent with <110> direction. Therefore, unequal proportions of (001)[100] and (001) [010] give rotation axes [310] and [130]. Table 3 summarizes the possible combinations of slip systems in this scenario.

Before discussing the dependence of slip-system activation and crystal orientation relative to stress axes, let us review the observed slip systems for the three crystal orientations examined (types 1, 2a & 2b). For type 1 data are consistent with {001} <010> slip for the low-angle boundary oriented (*i*) diagonal and with rotation axis [100] (Fig. 8). Moreover the illustrated probable orientation of the boundary plane relative to the rotation axis is consistent with a tilt boundary as the rotation axis lies within the boundary plane. For the (*ii*) vertical and (*iii*) horizontal, low-angle boundaries with a rotation axis [110] suggest both {001} <100> and {001} <010> slip systems in equal proportions (Fig. 14). Illustrated probable orientation of the boundary planes, supported by optical microscopy

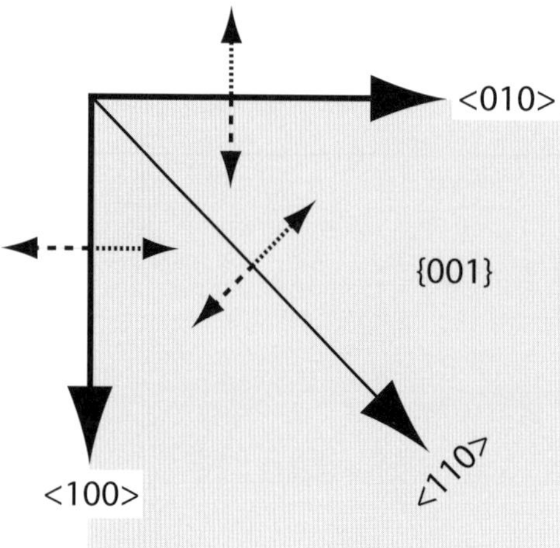

Fig. 14. Schematic illustration of the combined rotation of two axes, <010> and <100>, in a plane, {001}, with vectors. The net of the two perpendicular vectors is a third drawn diagonal through them, <110>. The small stippled secondary vectors illustrate the slip directions perpendicular to the rotation axes. Only the combination of either dextral or sinistral rotation is possible (small arrows: stippled line denotes sinistral, dashed line denotes dextral).

relative to the rotation axis is consistent with a twist boundary for (*ii*) and a tilt boundary for (*iii*) (Fig. 8).

For type 2a, data are consistent with activation of both {001} <100> and {001} <010> slip systems in unequal proportions for the low-angle boundary oriented (*i*) vertical with rotation axis [310] (Fig. 9). Data from the vertical boundary (*ii*) with rotation axis [100] suggest {010} <001>. Illustrated probable orientation of the vertical boundary planes, supported by optical microscopy, relative to the rotation axes are consistent with a twist boundary for (*i*) and a tilt boundary for (*ii*) (Fig. 9). The discontinuous horizontal boundary (*iii*) with rotation axis [001] can have either {010} <100> or {100}

<010> as the activated slip system (Fig. 9). However, with respect to the imposed stress axes the activation of {010} <100> is more plausible.

Type 2b data suggest the activation of both {001} <100> and {001} <010> occurs in equal proportions for the low-angle boundary oriented (*i*) vertical and rotation axis [110] (Fig. 10). For the vertical boundary with rotation axis [010] (*ii*) though, {100} <001> activation is most consistent with the data. The most probable orientation of the boundary planes, supported by optical microscopy, is consistent with a twist boundary for (*i*) and a tilt boundary for (*ii*). The discontinuous horizontal boundary (*iii*) with rotation axis [130] can be explained by activation of {001} <100> and {001} <010> in unequal proportions (Fig. 10).

Table 4 summarizes the above results. For all types {001} <010> and {001} <100> are activated. Additional activated slip systems are governed by the orientation of the Sil crystal. In general, the activated slip plane is sub-perpendicular to oblique against Z. Type 2a needles lie oriented so that activation of slip system {010} <001> is promoted, whereas type 2b needles are oriented to promote {100} <001> slip. For type 1, slip on {001} is dominant. However, the combination of slip directions observed is not likely to be one with lowest critical resolved sheer stress (CRSS). The development of the vertical (*ii*) and horizontal (*iii*) boundaries relative to the c-axis are interpreted to be related to the specific crystal shape and local stress/strain environment. The rhomb shape results in sharp corners causing local high stress concentrations (Fig. 8), which can result in activation of high CRSS slip systems (Montagnat *et al.* 2011).

Type 1 and 2 grains are seen in type 3 and 4 domains. The respective response of type 1 and 2 grains v. type 4 domains helps to evaluate if further deformation mechanisms are active. For example, next to type 1 grains, type 4 grains are folded and/or disturbed (Fig. 4d, f). This suggests that type 1 grains are seen rheologically to be

Table 3. *Relationship between slip systems and rotation axes*

Slip system Rotation axis	{001} <100> [010]	{100} <001> [010]	{010} <001> [100]
{001} <010> [100]	**{001} <110>*** **[110]**	–	–
{100} <010> [001]	–	**{100} <011>*** **[011]**	–
{010} <100> [001]	–	–	**{010}<101>*** **[101]**

*Combinations of simultaneously active slip systems that operate equally are shown in bold.

Table 4. *Summary of type 1 and 2 microstructural features including derived slip systems*

	Observed microstructure					
	Boundary angle	Internal misorientation*	Boundary orientation	Rotation axes	Interpreted active slip system	Boundary type
Type 1 (Euhedral) *c-axis ‖ Y*						
	1.5–6°	*c.* 1.5°	**(i)** Diagonal[†]	**[100]**	*{001}<010>*	tilt
	1.5–6°		**(ii)** Vertical[‡]	*[110]*	*{001}<010>*	twist
	1.5–6°		**(iii)** Horizontal[§]	[110]	*{001}<100>* {001}<010> {001}<100>	tilt
Type 2 (Needle) *c-axis ‖ X*						
Type 2a *a-axis ‖ Y*	1.5–6° 1.5–6°	*c.* 6°	**(i)** Vertical[‡]	*[310]*	*{001}<010>* *{001}<100>*	twist
			(ii) Vertical[‡]	**[100]**	**{010}<001>**	tilt
	1.5°		**(iii)** Discontinuous horizontal[§]	[001]	{010}<100>	tilt
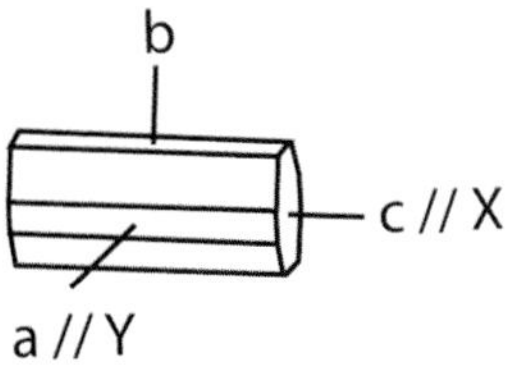 **Type 2b** *b-axis ‖ Y*	3–10° 3–10°	*c.* 8°	**(i)** Vertical[‡]	*[110]*	*{001}<010>* *{001}<100>*	twist
			(ii) Vertical[‡]	**[010]**	**{100}<001>**	tilt
	1.5°		**(iii)** Discontinuous horizontal[§]	[130]	{001}<010> {001}<100>	tilt

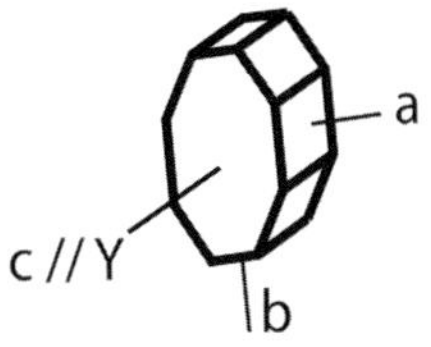

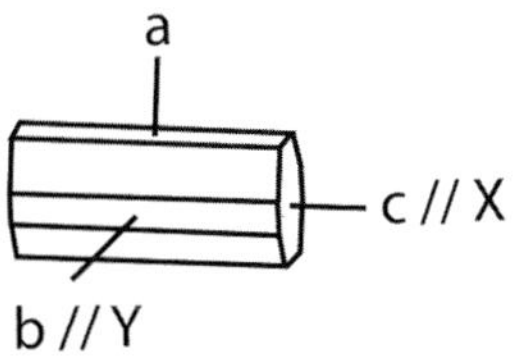

*Maximum systematic change in orientation in areas between distinct subgrain boundaries in a grain.
[†]Subgrain boundary oriented oblique *c.* 45° to X.
[‡]Subgrain boundary‖Z.
[§]Subgrain boundary‖X.
Bold Dominant slip systems and related rotation axes.
Bold & Italic Dominant slip system seen in all types.
Italic Orientation of elementary axes in relation to sample orientation.

harder than type 4 domains. They deform to a minor extent by dislocation creep but mainly rotate clockwise by rigid body rotation around the kinematic Y axis while remaining with their c-axes sub-parallel to Y (Fig. 4d, f). The low aspect ratio of type 1 grains results in significant rotation (Jeffery 1922; Griera *et al.* 2013 and references therein). In contrast, the high aspect ratio of type 2 grains allows these grains to remain parallel to the fabric attractor (Passchier & Trouw 2005) without further rotation, resulting in little to no rigid body rotation.

Deformation mechanisms and activation of slip systems within grain aggregates: type 3 matrix

Analysis of type 3 matrix grains shows that the adjacent type 4 grains have been incorporated into the type 3 matrix domains. Accordingly, type 3 pole figures show an inherited type 4 strong c-axis rotation. However, there is an additional component of crystal plasticity, even though only some individual grains show significant internal deformation

(Fig. 11). The crystallographic orientation data viewed in pole figures suggest the activation of the two slip systems {001}<010> and {001}<100>, which is the same slip-system combination inferred to be responsible for a large part of crystal plastic deformation within individual type 1 and 2 grains. However, the preferred orientation is not very well defined, suggesting the activation of another deformation mechanism. We suggest that in the type 3 matrix domains, GBS contributes significantly to the deformation of these domains. This is supported by the phase mixture with Bt and the lack of internal deformation in small grains. The commonly observed near-120° triple junctions in small grain areas suggest in these areas that GBS is accommodated by grain boundary migration whereas in larger grained parts GBS is accommodated by dislocation glide, as suggested by the discussed internal crystal lattice distortions (Fig. 11d, e).

Comparison of derived activated slip systems with data from the literature

The slip systems determined in this study (Table 4) are consistent with microstructural features previously observed in Sil (Table 1) with the exception of the activation of {001} slip plane and <010> slip direction. We infer that the {001} <010> and {001} <100> slip systems are activated to accommodate the resulting [110] rotation axes. Nevertheless, a 'pure' {001} slip plane with a 'pure' <010> slip direction is also documented in the investigated sample. It is generally assumed that slip in the {001} plane is unlikely due to the necessity to cut or distort the strong AlO_6 octahedral chains in the Sil crystal structure (Fig. 2). However, a TEM study on the structurally related mineral mullite ($3Al_2O_3 \cdot 2SiO_2$ to $2Al_2O_3 \cdot SiO_2$, orthorhombic, space group Pbam) identified [010], [110] and [112] slip directions (Gustafsson & Falk 2009). The authors' conclusion was that vacancies, impurities and distortions in the crystal structure could facilitate and allow slip on {001}. Therefore these three factors (vacancies, impurities and distortions) are possibly facilitating slip on {001}. The proposed activation of two rotation axes in the same plane and the suggested (001)[010] cannot be proven with EBSD data alone; hence, this needs to be addressed in future studies using supplementary TEM analysis.

Deformation of type 4 domains: an example of GBS accommodated by dislocation glide and grain boundary migration

Type 4 domains offer the unique opportunity to investigate how a fine-grained aggregate with a known pre-existing microstructure, grain–grain and crystal orientation relationships behaves in response to simple shear deformation. As opposed to previous studies that have investigated the deformation of fine-grained aggregates (e.g. Bestmann & Prior 2003; Warren & Hirth 2006; Svahnberg & Piazolo 2010, 2013; Menegon et al. 2011), in our study no significant grain size reduction in response to recrystallization processes has been observed to have occurred as part of deformation. We suggest that the main deformation mechanism seen in type 4 border zones is GBS. This interpretation is based on: (a) dispersion crystallographic orientations with shift in crystallographic preferred orientation (CPO); (b) rotation of main axes around the kinematic axis; (c) shift of rotation axes towards the kinematic axis; (d) increase of higher angle boundaries; and (e) some phase mixing. The observed dispersion around the kinematic Y axis of the initially very strong preferred orientation of all crystallographic axes with deformation is expected to be one of the key features of GBS (e.g. Jiang et al. 2000; Bestmann & Prior 2003; Warren & Hirth 2006 and references therein). In addition, rotation axes which were initially dominated by rotation around <001> become less clustered in the border zones of the type 4 domains (Fig. 12c) as expected during activation of GBS. Rotation axes rotate away from the c-axis rotation dominance to the kinematic Y axis (Fig. 13c). Towards border zones, low-angle boundaries decrease and higher misorientation angles are observed (Fig. 13d). The direct spatial correlation of phase mixing with crystallographic changes as described above further supports our interpretation.

Due to space problems, deformation by GBS must be accommodated by other deformation mechanisms. These may include dislocation glide, grain boundary migration or diffusion. In our case, diffusion is unlikely to accommodate GBS due to the relatively low diffusion rates for Al at the derived deformation temperature (675–700 °C). DisGBS (Warren & Hirth 2006) is predicted to result in a weak CPO and boundaries between grains that show higher misorientation angles. In addition, Prior & Hirth (2007) suggested that GBS accommodated by dislocation glide should result in a rotation of grains around the kinematic rotation axis and Menegon et al. (2011) showed this to be true for recrystallized Qtz grains within a narrow high strain zone. In our case, a clockwise rotation around the Y axis (i.e. synthetic with the dextral sense of shear) is indeed observed (Fig. 12d). Even though we observe the predicted change in misorientation angles and shift of rotation axes, it is difficult to determine if a weak CPO developed. We need to resort to type 3 characteristics where we still see a strong influence of type 4 orientations.

In this case, a weak preferred orientation develops consistent with DisGBS. In type 4 domains (Fig. 12) a slight grain size increase is observed in the border zone. This can be explained by grain boundary migration acting as an accommodating process for GBS. However, where we see a significant amount of second phase particles such as fine-grained Bt, no grain size increase is observed; hence grain boundary migration could not accommodate GBS (Fig. 13).

Interestingly, the switch from very strong preferred orientation originating from the paramorph replacement of And by Sil to a dispersed crystallographic orientation pattern is abrupt. This implies that once deformation starts at the border zone of the type 4 aggregates, deformation is strongly localized and does not continue to incorporate larger areas of the type 4 aggregates. This then results in the microstructural 'preservation' of the large central parts of the clusters. This is consistent with a deformation mechanism requiring relatively low stress to operate, that is, a grain-size-sensitive flow law such as GBS resulting in Newtonian flow.

Effect of pre-deformation grain orientations and characteristics on deformation mechanisms and slip-system activation, and resultant rheology

The example studied here shows a large range of grain sizes and different well-defined individual grain orientations for one mineral (Sil). Much of the grain size variation is inferred to be due to variations in pre-existing grain characteristics. These pre-existing microstructural differences have a marked effect on the activation of dominant deformation mechanism(s), slip systems and resulting CPOs and rheological responses. In the following, we put results discussed above in the context of these relationships.

The initially large type 1 and 2 grains show that the difference in habit and crystal orientation with regard to stress axes leads to significant differences in strain accommodation in terms of the dominance of deformation mechanism and slip systems (Figs 4, 8, 9 & 10). Type 1 euhedral grains, with their rhomb shape within the XZ plane deform significantly by rigid body rotation around Y whereas the type 2 needles with extreme aspect ratios and alignment of the long axis (c-axis) with X could accommodate little to no strain by rigid body rotation. Hence, the latter deform to a large extent by intracrystalline deformation through slip. This leads to higher dislocation input in type 2 needles than for type 1 grains resulting in higher intracrystalline bending in the former. At the same time, differences in the orientation of crystal axes lead to differences in slip-system activation. In general, the activated slip plane is sub-perpendicular to oblique to Z. For type 1 grains slip on (001) is dominant, while for type 2 slip systems (010) [001] and (100)[001] are activated depending on b-axis orientation.

Even though, type 1 and 2 grains deform, within the rock deformation is mainly localized in type 3 aggregates and type 4 border zones. Microstructurally, this is supported by the anastomosing nature of the type 3 domains around the harder $grt-pl-qtz$ and type 4 domains. Such anastomosing patterns are classic examples where strain partitions into the anastomosing weak zones (e.g. Ramsay 1980; Carreras & Casas 1987; Rutter 1999; Carreras *et al.* 2010). We suggest that the partitioning is facilitated by the presence of initially small- to medium-grained aggregates. While for type 4 it can be shown that the pre-existing small grain size is a direct consequence of the paramorphic replacement of And, for type 3 aggregates the original microstructure is less well defined. Two possible scenarios for pre-existing microstructures can be envisaged: medium-grained $Sil-Bt$ aggregates with near-random orientation which grew during the A_2/D_1 event (scenario I); or And-replacement clusters with c-axis orientation at high angle to X and Y(scenario II). In scenario I the phase mixture is not an effect of GBS, and deformation is initially accommodated to a large extent by dislocation creep. With increasing deformation of type 4 cluster border zones, type 3 aggregates and type 4 cluster border zones mix. Consequently, the average grain size decreases and with that deformation by GBS becomes increasingly important. In scenario II, initially small-grained Sil clusters deform by GBS and successively rearrange completely due to GBS accommodated by dislocation glide and grain boundary migration. Phase mixing through GBS results in the observed type 3 $Sil-Bt$ aggregates. The fact that the c-axis orientation of 'preserved' And clusters, that are still microstructurally distinct, is dominated by two main orientations (c-axes $\pm//$ X, c-axes $\pm//$ Y) suggests that, assuming random orientation of original And grains, clusters with a different c-axes orientation deformed beyond recognition. We suggest that only clusters with a specific orientation are 'preserved'. Clusters with c-axes sub-parallel to Y have for geometric reasons the smallest possible grain size on the XZ plane assuming the paramorphs after And have the same c-axes while a- and b-axes are randomly oriented. In this case, GBS is most effective and strain localization at the border of the cluster results in 'preservation' of the central parts of the clusters. On the other hand, clusters with c-axes orientation parallel to X will be unable to rotate due to their long aspect ratio and parallelism of their long axes with X. Thus, rigid body rotation and GBS will be very limited and deformation is accommodated by limited dislocation creep only, resulting in the observed slight modification of the original cluster shape (see Fig. 4). However, clusters with initially c-axes orientation non-parallel to X or Y, will deform by GBS combined with dislocation creep depending on the grain size and will consequently localize deformation. However, the observed localization is not as extreme as in the type 4 clusters investigated, therefore deformation results in a successive incorporation of the cluster leading to the 'obliteration' of the cluster. We prefer scenario II as our data suggest selective preservation of type 4 clusters. Consequently, dominance of both deformation mechanisms and partitioning is highly dependent on the pre-existing orientation of grain aggregates.

The presented influence of crystallographic orientation relative to the stress axes together with the 'preservation' of some microstructure through strain localization could explain the different CPOs in different domains of a thin section as reported by Peternell *et al.* (2010). In domains that originally exhibited similar orientations, for example, a Qtz vein in a quartzofeldspathic rock, different slips systems would be activated from those in the bulk of the rock purely due to the relationship between crystal orientation and stress axes. Furthermore, initial, pre-existing grain size differences may help to localize deformation and 'preserve' pre-existing CPOs within specific domains.

In our case, even though, type 3 aggregates make up a low area fraction of the thin section, the interconnectivity of these together with type 4 border zones suggests that deformation was dominated by the grain-size-dependent deformation mechanism GBS. Therefore, we infer that during metamorphism and deformation this rock was deformed by near-linear Newtonian flow. As a consequence, within metasedimentary sequences, including metapsammites, mafic metasediments (amphibolites) and quartzite, metapelitic rocks are expected to be one of the weaker components, if not the weakest.

Conclusion

Detailed analysis of sillimanite microstructures shows that during dislocation creep at amphibolite facies conditions, sillimanite deforms mainly by a combination of both (001)[010] and (001)[100]. Additionally, (100)[001] and (010)[001] are activated in needles with their c-axes parallel to lineation. The same main combination of slip systems is seen in medium-grained sillimanite aggregates mixed with biotite. However, in these aggregates grain boundary sliding accommodates a significant

part of the deformation resulting in phase mixture and weak crystallographic preferred orientation. Grain boundary sliding is the dominant deformation mechanism in the outer parts of fine-grained clusters originating from sillimanite replacing andalusite. Associated features are phase mixing, clockwise dispersion of c-axis orientations and change of misorientation axes from original c-axis rotation to rotation close to the kinematic Y axis. This results in pronounced strain localization around these clusters allowing for their 'preservation'. The 'preservation' of such clusters is highly dependent on the original orientation; only clusters with c-axes orientations parallel or perpendicular to lineation are preserved, while clusters oriented differently develop into medium-grained sillimanite aggregates. These aggregates form an anastomosing and interconnected network where deformation localizes.

Our study on sillimanite microstructures in a naturally deformed amphibolite facies metapelite shows that pre-existing grain orientations and grain-size differences markedly govern the microstructural development during high-grade deformation and consequently deformation partitioning and bulk rheology.

P. Jaconelli and S. Piazolo thank H. Svahnberg (Stockholm University) for the sample, maps and background information, A. Berger (University of Copenhagen) for microprobe assistance, and V. Borthwick and D. Prior for constructive criticism which significantly improved this manuscript. Thorough reviews by Santanu Misra and Robert A. Hunter, and the handling editor S. Llana-Funez helped to improve the manuscript further and are greatly appreciated. P. Jaconelli and S. Piazolo acknowledge the financial support by the Swedish Research Council (VR 621-2004-5330), the Knut and Alice Wallenberg Foundation (financing of equipment in Stockholm). This contribution has been financially supported by the European Science Foundation (ESF), EUROCORES Programme EuroMinScl, FP6. S. Piazolo acknowledges financial support by the Australian Research Council through DP120102060 and FT1101100070. Analytical work at Macquarie University was undertaken using instrumentation funded by DEST Systemic Infrastructure Grants, ARC LIEF, NCRIS, industry partners and Macquarie University. This is contribution 323 from the ARC Centre of Excellence for Core to Crust Fluid Systems (http://www.ccfs.mq.edu.au) and 888 in the GEMOC Key Centre (http://www.gemoc.mq.edu.au)

References

ADAMS, B. L., WRIGHT, S. I. & KUNZE, K. 1993. Orientation imaging: the emergence of a new microscopy. *Metallurgical Transactions*, **24A**, 819–831.

ASHBY, M. F. 1970. Deformation of plastically non-homogeneous materials. *Philosophical Magazine*, **21**, 399–424.

BARRIE, C. D., BOYLE, A. P., COX, S. F. & PRIOR, D. J. 2008. Slip systems and critical resolved shear stress in pyrite: an electron backscatter diffraction (EBSD) investigation. *Mineralogical Magazine*, **72**, 1181–1199.

BEANE, R. J. & FIELD, C. K. 2007. Kyanite deformation in whiteschist of the ultrahigh-pressure metamorphic Kokchetav Massif, Kazakhstan. *Journal of Metamorphic Geology*, **25**, 117–128.

BESTMANN, M. & PRIOR, D. J. 2003. Intragranular dynamic recrystallization in naturally deformed calcite marble: diffusion accommodated grain boundary sliding as a result of subgrain rotation recrystallization. *Journal of Structural Geology*, **25**, 1597–1613.

BOLAND, J. N., MCLAREN, A. C. & HOBBS, B. E. 1971. Dislocations associated with optical features in naturally deformed olivines. *Contributions to Mineralogy and Petrology*, **30**, 53–63.

BORTHWICK, V. E. & PIAZOLO, S. 2010. Complex temperature dependent behaviour revealed by in-situ heating experiments on single crystals of deformed halite: new ways to recognize and evaluate annealing in geological materials. *Journal of Structural Geology*, **32**, 982–996.

BURNHAM, C. W. 1963. Refinement of the crystal structure of sillimanite. *Zeitschrift für Kristallographie*, **118**, 127–148.

CARMICHAEL, D. M. 1969. On the mechanism of prograde metamorphic reactions in quartz-bearing pelitic rocks. *Contributions to Mineralogy and Petrology*, **20**, 244–267.

CARRERAS, J. & CASAS, J. M. 1987. On folding and shear zone development: a mesoscale structural study on the transition between two different tectonic styles. *Tectonophysics*, **135**, 87–98.

CARRERAS, J., CZECK, D. M., DRUGUET, E. & HUDLESTON, P. J. 2010. Structure and development of an anastomosing network of ductile shear zones. *Journal of Structural Geology*, **32**, 656–666.

CESARE, B., GÓMEZ-PUGNAIRE, M. T., SÁNCHEZ-NAVAS, A. & GROBERTY, B. 2002. Andalusite–sillimanite replacement (Mazarrón, SE Spain): a microstructural and TEM study. *American Mineralogist*, **87**, 433–444.

CONDIE, K. C. 1982. *Plate Tectonics and Crustal Evolution*. Pergamon, New York.

DEER, W. A., HOWIE, R. A. & ZUSSMAN, J. 1997. *Rock-forming Minerals: Orthosilicates*. Geological Society, London, **1a**.

DIMANOV, A., RYBACKI, E., WIRTH, R. & DRESEN, G. 2007. Creep and strain-dependent microstructures of synthetic anorthite–diopside aggregates. *Journal of Structural Geology*, **29**, 1049–1069.

DOUKHAN, J.-C. & CHRISTIE, J. M. 1982. Plastic deformation of sillimanite Al_2SiO_5 single crystals under confining pressure and TEM investigation of the induced defect structure. *Bulletin de Minéralogie*, **105**, 583–589.

DOUKHAN, J.-C., DOUKHAN, N., KOCH, P. S. & CHRISTIE, J. M. 1985. Transmission electron microscopy investigation of lattice defects in Al_2SiO_5 polymorphs and plasticity induced polymorphic transformations. *Bulletin de Minéralogie*, **108**, 81–96.

FLIERVOET, T. F., DRURY, M. R. & CHOPRA, P. N. 1999. Crystallographic preferred orientations and misorientations in some olivine rocks deformed by diffusion or dislocation creep. *Tectonophysics*, **303**, 1–27.

FOUNTAIN, D. M. 1976. The Ivrea-Verbano and Strona-Ceneri zones, northern Italy: a cross section of the continental crust – new evidence from seismic velocities of rock samples. *Tectonophysics*, **33**, 145–165.

FROST, H. J. & ASHBY, M. 1983. *Deformation-Mechanism Maps: The Plasticity and Creep of Metals and Ceramics*. Pergamon Press, Oxford.

FYNN, G. W. & POWELL, W. J. A. 1979. *The Cutting and Polishing of Electro-Optic Materials*. Wiley, New York.

GHENT, E. D., KNITTER, C. C., RAESIDE, R. P. & STOUT, M. Z. 1982. Geothermometry and geobarometery of pelitic rocks, upper kyanite and sillimanite zones, Mica Creek area, British Columbia. *The Canadian Mineralogist*, **20**, 295–305.

GOERGEN, E. T., WHITNEY, D. L., ZIMMERMAN, M. E. & HIRAGA, T. 2008. Deformation-induced polymorphic transformation: experimental deformation of kyanite, andalusite, and sillimanite. *Tectonophysics*, **454**, 23–35.

GRAMBLING, J. A. 1981. Kyanite, andalusite, sillimanite, and related mineral assemblages in the Truchas Peaks region, New Mexico. *American Mineralogist*, **66**, 702–722.

GRIERA, A., LLORENS, M-G., GOMEZ-RIVAS, E., BONS, P. D., JESSELL, M. W., EVANS, L. A. & LEBENSOHN, R. 2013. Numerical modelling of porphyroclast and porphyroblast rotation in anisotropic rocks. *Tectonophysics*, **587**, 4–29.

GUSTAFSSON, S. & FALK, L. K. L. 2009. Defocus convergent beam electron diffraction analysis of dislocation in mullite. *Journal of Microscopy*, **233**, 346–351.

HEIDELBACH, F., KUNZE, K. & WENK, H.-R. 2000. Texture analysis of a recrystallized quartzite using electron diffraction in the scanning electron microscope. *Journal of Structural Geology*, **22**, 91–104.

HOBBS, B. E. 1985. The geological significance of microfabric analyses. *In*: WENK, H.-R. (ed.) *Preferred Orientation in Deformed Metals and Rocks*. Academic Press, Orlando, 463–484.

HODGES, K. V. & SPEAR, F. S. 1982. Geothermometry, geobarometry and the Al_2SiO_5 triple point at Mt. Moosilauke, New Hampshire. *American Mineralogist*, **67**, 1118–1134.

HOLDAWAY, M. J. 2001. Recalibration of the GASP geobarometer in light of recent garnet and plagioclase activity models and versions of the garnet–biotite geothermometer. *American Mineralogist*, **86**, 1117–1129.

HOLLAND, T. J. B. & POWELL, R. 1998. An internally-consistent thermodynamic dataset for phases of petrological interest. *Journal of Metamorphic Geology*, **16**, 309–344.

JEFFERY, G. B. 1922. The motion of ellipsoidal particles in a viscous fluid. *Proceedings of the Royal Society of London, Series A*, **102**, 161–179.

JIANG, Z., PRIOR, D. J. & WHEELER, J. 2000. Albite crystallographic preferred orientation and grain misorientation distribution in a low-grade mylonite: implications for granular flow. *Journal of Structural Geology*, **22**, 1663–1674.

KERRICK, D. M. 1990. The Al$_2$SiO$_5$ polymorphs. *Reviews in Mineralogy, Mineralogical Society of America*, **22**.

KERRICK, D. M. & SPEER, J. A. 1988. The role of minor element solid solution on the andalusite–sillimanite equilibrium in metapelites and peraluminous granitoids. *American Journal of Science*, **288**, 152–192.

KIM, H. S. & BELL, T. H. 2005. Multiple foliations defined by different sillimanite habits, partial melting and the late metamorphic development of the Cannington Ag–Pb–Zn deposit, Northeast Australia. *Gondwana Research*, **8**, 496–509.

KRETZ, R. 1983. Symbols of rock-forming minerals. *American Mineralogist*, **68**, 277–279.

KRUSE, R. & STÜNITZ, H. 1999. Deformation mechanisms and phase distribution in mafic high-temperature mylonites from the Jotun Nappe, southern Norway. *Tectonophysics*, **303**, 223–249.

LAMBREGTS, P. J. & VAN ROERMUND, H. L. M. 1990. Deformation and recrystallization mechanisms in naturally deformed sillimanites. *Tectonophysics*, **179**, 371–378.

LAW, R. D. 1990. Crystallographic fabrics: a selective review of their applications to research in structural geology. *In*: KNIPE, R. J. & RUTTER, E. H. (eds) *Deformation Mechanisms, Rheology and Tectonics*. Geological Society, London, Special Publications, **54**, 335–352.

LAW, R. D., SCHMID, S. M. & WHEELER, J. 1990. Simple shear deformation, quartz crystallographic fabrics: a possible natural example from the Torridon area of NW Scotland. *Journal of Structural Geology*, **12**, 29–45.

LEE, N. R. 2008. *The Neoarchaean tectonothermal evolution of the SE Nuuk region, southern West Greenland*. PhD thesis. University of Edinburgh, UK.

LLOYD, G. E. 2004. Microstructural evolution in a mylonitic quartz simple shear zone: the significant roles of dauphine twinning and misorientation. *In*: ALSOP, G. I., HOLDSWORTH, R. E., McCAFFREY, K. J. W. & HAND, M. (eds) *Flow Processes in Faults and Shear Zones*. Geological Society, London, Special Publications, **224**, 39–61.

LLOYD, G. E. & FREEMAN, B. 1994. SEM electron channeling analysis of dynamic recrystallization in a quartz grain. *Journal of Structural Geology*, **13**, 945–953.

LLOYD, G. E., FARMER, A. B. & MAINPRICE, D. 1997. Misorientation analysis and the formation and orientation of subgrain and grain boundaries. *Tectonophysics*, **279**, 55–78.

MENEGON, L., PIAZOLO, S. & PENNACCHIONI, G. 2011. The effect of Dauphiné twinning on plastic strain in quartz. *Contributions to Mineralogy and Petrology*, **161**, 635–652, http://dx.doi.org/10.1007/s00410-010-0554-7

MONTAGNAT, M., BLACKFORD, J. R., PIAZOLO, S., ARNAUD, L. & LEBENSOHN, R. A. 2011. Measurement and full-field predictions of deformation heterogeneities in ice. *Earth and Planetary Science Letters*, **305**, 153–160.

MUSUMERCI, G. 2002. Sillimanite bearing shear zones in syntectonic leucogranite: fluid assisted brittle–ductile deformation under amphibolite facies conditions. *Journal of Structural Geology*, **24**, 1491–1505.

PANT, N. C., KUNDU, A., D'SOUZA, M. J. & SAIKIA, A. 2012. Petrology of the Neoproterozoic granulites from Central Dronning Maud Land, East Antarctica – implications for southward extension of East African Orogen (EAO). *Precambrian Research*, **227**, 389–408.

PASSCHIER, C. W. & TROUW, R. A. J. 2005. *Microtectonics*, 2nd edn. Springer Verlag, Berlin.

PETERNELL, M., HASALOVA, P., WILSON, C. J., PIAZOLO, S. & SCHULMANN, K. 2010. A comparative study of quartz EBSD and fabric analyser: crystallographic preferred orientation from the Thaya region, Czeck Republic. *Journal of Structural Geology*, **32**, 803–817.

PIAZOLO, S., MONTAGNAT, M. & BLACKFORD, J. R. 2008. Sub-structure characterization of experimentally and naturally deformed ice using cryo-EBSD. *Journal of Microscopy*, **230**, 509–519.

PIAZOLO, S., AUSTRHEIM, H. & WHITEHOUSE, M. 2012. Brittle–ductile microfabrics in naturally deformed zircon: deformation mechanisms and consequences for U–Pb dating. *American Mineralogist*, **97**, 1544–1563.

PRIOR, D. J. 1999. Problems in determining the misorientation axes, for small angular misorientations, using electron backscatter diffraction in the SEM. *Journal of Microscopy*, **195**, 217–225.

PRIOR, D. J. & HIRTH, G. 2007. Microstructural recognition of grain boundary sliding and its rheological implications. DRT conference abstract. *Rendiconti della Societa Geologica Italiana*, **5**, 187.

PRIOR, D. J., BOYLE, A. P. *ET AL.* 1999. The application of electron backscatter diffraction and orientation contrast imaging in the SEM to textural problems in rocks. *American Mineralogist*, **84**, 1741–1759.

PRIOR, D. J., WHEELER, J., PERUZZO, L., SPIESS, R. & STOREY, C. 2002. Some garnet microstructures: an illustration of the potential of orientation maps and misorientation analysis in microstructural studies. *Journal of Structural Geology*, **24**, 999–1011.

RAMSAY, J. G. 1980. Shear zone geometry: a review. *Journal of Structural Geology*, **2**, 83–99.

REDDY, S. M., TIMMS, N. E., PANTLEON, W. & TRIMBY, P. 2007. Quantitative characterization of plastic deformation of zircon and geological implications. *Contributions to Mineralogy and Petrology*, **153**, 625–645.

REID, M. R., HART, S. R., PADOVANI, E. R. & WANDLESS, G. A. 1989. Contribution of metapelitic sediments to the composition, heat production, and seismic velocity of the lower crust of southern New Mexico, USA. *Earth and Planetary Science Letters*, **95**, 367–381.

RUTTER, E. H. 1999. On the relationship between the formation of shear zones and the form of the flow law for rocks undergoing dynamic recrystallization. *Tectonophysics*, **303**, 147–158.

SANDER, B. 1948. *Einführung in die Gefügekunde der geologischen Körper*. Bd. I, Springer Verlag, Wien.

SCHERSTÉN, A., STENDAL, H. & NÆRAA, T. 2008. Geochemistry of greenstones in the Tasiusarsuaq terrane, southern West Greenland. *Geological Survey of Denmark and Greenland Bulletin*, **15**, 69–72.

STIPP, M., STUNITZ, H., HEILBRONNER, R. & SCHMID, S. M. 2002. The eastern Tonale fault zone: a 'natural laboratory' for crystal plastic deformation of quartz over a temperature range from 250 to700 °C. *Journal of Structural Geology*, **24**, 1861–1884.

SVAHNBERG, H. & PIAZOLO, S. 2010. The initiation of strain localisation in plagioclase-rich rocks: Insights from detailed microstructural analyses. *Journal of Structural Geology*, **32**, 1404–1416.

SVAHNBERG, H. & PIAZOLO, S. 2013. Interaction of chemical and physical processes during deformation at fluid-present conditions: a case study from an anorthosite–leucogabbro deformed at amphibolite facies conditions. *Contributions to Mineralogy and Petrology*, **165**, 543–562, http://dx.doi.org/10.1007/s00410-012-0822-9

TURNER, F. J. & WEISS, L. E. 1963. *Structural Analysis of Metamorphic Tectonites*. McGraw-Hill, New York.

VERNON, R. H. 1987a. Oriented growth of sillimanite in andalusite, Placitas – Juan Tabo area, New Mexico, U.S.A. *Canadian Journal of Earth Sciences*, **24**, 580–590.

VERNON, R. H. 1987b. Growth and concentration of fibrous sillimanite related to heterogeneous deformation in K-feldspar-sillimanite metapelites. *Journal of Metamorphic Geology*, **5**, 51–68.

WARREN, J. C. & HIRTH, G. 2006. Grain size sensitive deformation mechanisms in naturally deformed peridotites. *Earth and Planetary Science Letters*, **248**, 438–450.

WENK, H.-R. (ed.) 1985. *Preferred Orientation in Deformed Metals and Rocks: An Introduction to Modern Texture Analysis*. Academic Press, London.

WENK, H.-R., TAKESHITA, T., BECHLER, E., ERSKINE, B. G. & MATTHIES, S. 1987. Pure and simple shear calcite textures. Comparison of experimental, theoretical and natural data. *Journal of Structural Geology*, **9**, 731–745.

WHEELER, J., PRIOR, D. J., JIANG, Z., SPIESS, R. & TRIMBY, P. J. 2001. The petrological significance of misorientations between grains. *Contributions to Mineralogy and Petrology*, **141**, 109–124.

WHEELER, J., MANGAN, L. S. & PRIOR, D. J. 2004. Disequilibrium in the Ross of Mull Contact Metamorphic Aureole, Scotland: a consequence of polymetamorphism. *Journal of Petrology*, **45**, 835–858.

WINTSCH, R. P. & ANDREWS, M. S. 1988. Deformation induced growth of sillimanite: 'stress' minerals revisited. *Journal of Geology*, **96**, 109–262.

What happens to deformed rocks after deformation? A refined model for recovery based on numerical simulations

V. E. BORTHWICK[1]*, S. PIAZOLO[1,2], L. EVANS[3,2], A. GRIERA[4] & P. D. BONS[5]

[1]*Department of Geological Sciences, Stockholm University, Svante Arrhenius väg 8C, Stockholm 10691, Sweden*

[2]*Department of Earth and Planetary Sciences, Australian Research Council Centre of Excellence for Core to Crust Fluid Systems/GEMOC, Macquarie University, NSW 2109, Australia*

[3]*School of Geosciences, Monash University, 3800, Victoria, Australia*

[4]*Departament de Geologia, Universitat Autònoma de Barcelona, 08193 Bellaterra (Cerdanyola del Vallès), Spain*

[5]*Institut für Geowissenschaften, Eberhard Karls Universität Tübingen, Wilhelmstr. 56, D-72074 Tübingen Germany*

**Corresponding author (e-mail: verity.borthwick@gmail.com)*

Abstract: Deformation, in large parts of the middle crust, results in strained rocks consisting of grains with variable dislocation densities and microstructures which are characterized by gradual distortion and subgrain structures. Post-deformation residence of these rocks at elevated temperatures results in microstructural adjustments through static recovery and recrystallization. Here, we employ a numerical technique to simulate intragrain recovery at temperatures at or below the deformation temperature. The simulation is based on minimization of the stored energy, related to misorientation through local rotation of physical material points relative to their immediate environment. Three temperature- and/or deformation-geometry-dependent parameters were systematically varied: (1) deformation-induced dislocation types, (2) dislocation mobility and (3) size of dislocation interaction volume. Comparison with previously published *in situ* experiments shows consistency of numerical and experimental results. They show temperature- and dislocation-type-dependent small-scale fluctuations in subgrain-boundary misorientations and orientation variation within subgrains. These can be explained by the combined effect of increase in dislocation interaction volume and activation of climb. Our work shows microstructure can be significantly modified even if the post-deformational temperature is at or below the deformation temperature: a scenario relevant for most deformed rocks.

Samples of rocks deformed at mid to lower crustal levels now found at the Earth's surface may have remained at medium to high temperatures for extended time spans after deformation (e.g. Fairbairn 1949; Turner & Weiss 1963; Hobbs 1968; Molli & Heilbronner 1999; Stoeckert & Duyster 1999; Molli *et al.* 2000; Park *et al.* 2001; Bergman & Piazolo 2011). By gaining an in-depth understanding of the post-deformational changes that have occurred in rocks, geologists can develop tools to (a) establish a 'window' to the original deformation characteristics and (b) elucidate annealing conditions occurring after deformation (e.g. Heilbronner & Tullis 2002; Bestmann *et al.* 2005; Piazolo *et al.* 2006). Studying such behaviour in rocks is limited as we can see only the final structure, the endpoint of a long series of processes active at different temperatures.

In a most general sense, post-deformational annealing within a polycrystal alters microstructures by two main processes: recovery and recrystallization (Turner & Weiss 1963; Urai *et al.* 1986; Humphreys & Hatherly 2004 and references therein). Post-deformational annealing reorganizes the intracrystalline substructure, which encompasses all structures present at the subgrain scale, and can alter grain and subgrain boundaries to produce entirely new, strain-free grains (Wilson 1982; Ree & Park 1997; Heilbronner & Tullis 2002; Barnhoorn *et al.* 2005). The use of advanced analytical techniques, such as electron backscatter diffraction analysis (EBSD), in geology (e.g. Venables & Harland 1973; Dingley 1984; Dingley & Randle 1992; Prior *et al.* 1999) has enabled simplified elucidation of intragranular characteristics such as subgrain-boundary orientations and misorientations, rotation

From: LLANA-FÚNEZ, S., MARCOS, A. & BASTIDA, F. (eds) 2014. *Deformation Structures and Processes within the Continental Crust*. Geological Society, London, Special Publications, **394**, 215–234.
First published online December 9, 2013, http://dx.doi.org/10.1144/SP394.11

axes as well as intracrystalline distortions for the interpretation of mineral deformation (e.g. Prior *et al.* 1999; Trimby *et al.* 2000; Reddy *et al.* 2007; Barrie *et al.* 2008; Piazolo *et al.* 2008, 2012*a*, *b*). However, EBSD studies investigating the changes during post-deformational annealing at the intra-grain scale are sparse (e.g. Borthwick & Piazolo 2010; Borthwick *et al.* 2012). In the study presented here, we aim to close this gap. The study focuses on the intracrystalline changes commonly termed recovery. Recovery encompasses the reduction of stored energy in a deformed crystal structure by annihilation of dislocations and their rearrangement into lower-energy arrays (Sellars 1978; Urai *et al.* 1986; McQueen & Evangelista 1988; Humphreys & Hatherly 2004). Recovery thus affects the intra-granular *substructure*, where, in a general sense, a bent lattice with distributed dislocations evolves to one with subgrains separated by sharp tilt walls (Urai *et al.* 1986; Bons *et al.* 2001). These processes are facilitated by dislocation movement, which is thermally activated (e.g. Ranalli 1995). In particular, climb allows dislocations in a boundary to rearrange to decrease dislocation spacing (e.g. Hu 1963; Kocks 1976, 1985; Hull & Bacon 2001; Gottstein & Shvindlerman 2009).

Changes during recovery are dependent on (a) non-conservative dislocation mobility, which is strongly influenced by both the temperature and type of dislocation in terms of screw or edge type, Burgers vector and the related slip system, and (b) long-range interaction between dislocations, which is thought to be temperature dependent (Hull & Bacon 2001 and references therein). Hence, to model recovery, the following parameters need to be considered: (a) density, type and energy of dislo-cations in the system and (b) the reduction in stored elastic energy through removal and rearrangement of dislocations in space and time.

Experiments and numerical simulations are valuable tools in the study of microstructural evolu-tion. *In situ* annealing experiments conducted inside the scanning electron microscope, coupled with EBSD, allow us to follow the evolution in 'real-time' (Le Gall *et al.* 1999; Prior *et al.* 1999; Hum-phreys 2001; Seward *et al.* 2002; Piazolo *et al.* 2005). There have been a number of studies on a variety of different materials such as Ni (e.g. Zhang *et al.* 2009) and Al (e.g. Vandermeer & Juul Jensen 1998), but most relevant for this study are those on halite (Bestmann *et al.* 2005; Piazolo *et al.* 2006; Borthwick & Piazolo 2010). This method has drawbacks, such as (a) the possibility of surface effects influencing results (Frost *et al.* 1990), (b) the necessity of examining materials in which changes are fast enough for observation on labora-tory timescales and (c) the limited range of tempera-tures or pressures that are currently achievable in a laboratory. Numerical simulation has been used to model grain and subgrain growth using front-tracking models (Bons *et al.* 2001; Weygand *et al.* 2001; Moldovan *et al.* 2002; Jessell *et al.* 2003; Roessiger *et al.* 2011; Bergman & Piazolo 2011), Monte Carlo algorithms (Jessell *et al.* 2001; Holm *et al.* 2003, 2004; Gruber *et al.* 2009), phase field models (Sreekala & Haataja 2007) and cellular auto-mata models (Raabe & Becker 2000; Miodownik 2002). The strength of numerical modelling is that it enables the study of the effects of different par-ameters and process algorithms on model results. The latter allows testing of the theoretical basis of underlying processes (Bons *et al.* 2008). The quality of simulations is limited by the relevance of the fundamental laws and theories applied as well as the availability and accuracy of physical material constants (Piazolo *et al.* 2004).

By combining *in situ* experiments and numerical simulations, we seek to overcome the limitations of each technique and explore the parameters con-trolling lattice distortion and low-angle boundary (LAB – boundary with a misorientation between adjacent analysis points of less than 15°) behaviour within individual grains during post-deformational annealing. The numerical simulation is based on theoretical consideration and first interpretation of *in situ* experiments put forward by Borthwick & Piazolo (2010). Numerical results not only provide a method to investigate the parameter space of recovery but also can be tested for consistency with results from *in situ* experiments (Borthwick & Piazolo 2010). Experimental results emphasize the close link between pre-annealing deformation conditions, annealing behaviour itself and tempera-ture of annealing. They also highlight the effect of deformation geometry and presence of dominant slip systems. This is particularly important for many geological materials as they tend to have a low-symmetry crystallography that involves one or more preferred slip systems: quartz (e.g. Hobbs 1968; Law *et al.* 1990 and references therein), olivine (e.g. Ave Lallemant *et al.* 1970; Chopra & Paterson 1981), feldspar (e.g. Olsen & Kohlstedt 1984), ice (e.g. Wakahama 1964; Hondoh 2000 and references therein). We present here a refined model of recovery behaviour at temperatures at and below that of deformation and discuss its importance for interpretation of deformed rocks that have undergone some post-deformational annealing.

Experimental background

In situ experiments were performed on a *c.* 7 × 10 × 15 mm single halite crystal, pre-deformed by uniaxial shortening in the [001] direction, at a homologous temperature (T_h) of 0.68 (453 °C). Final strain was 0.65 (i.e. 35% shortening), using a strain rate of 6.9×10^{-6} s^{-1} (for further details see Borthwick & Piazolo 2010). Our main observations on post-deformational annealing, at temperatures at and below deformation temperatures, were (Fig. 1):

(1) Post- deformation geometry resulted in activation of two perpendicular sets of slip systems.

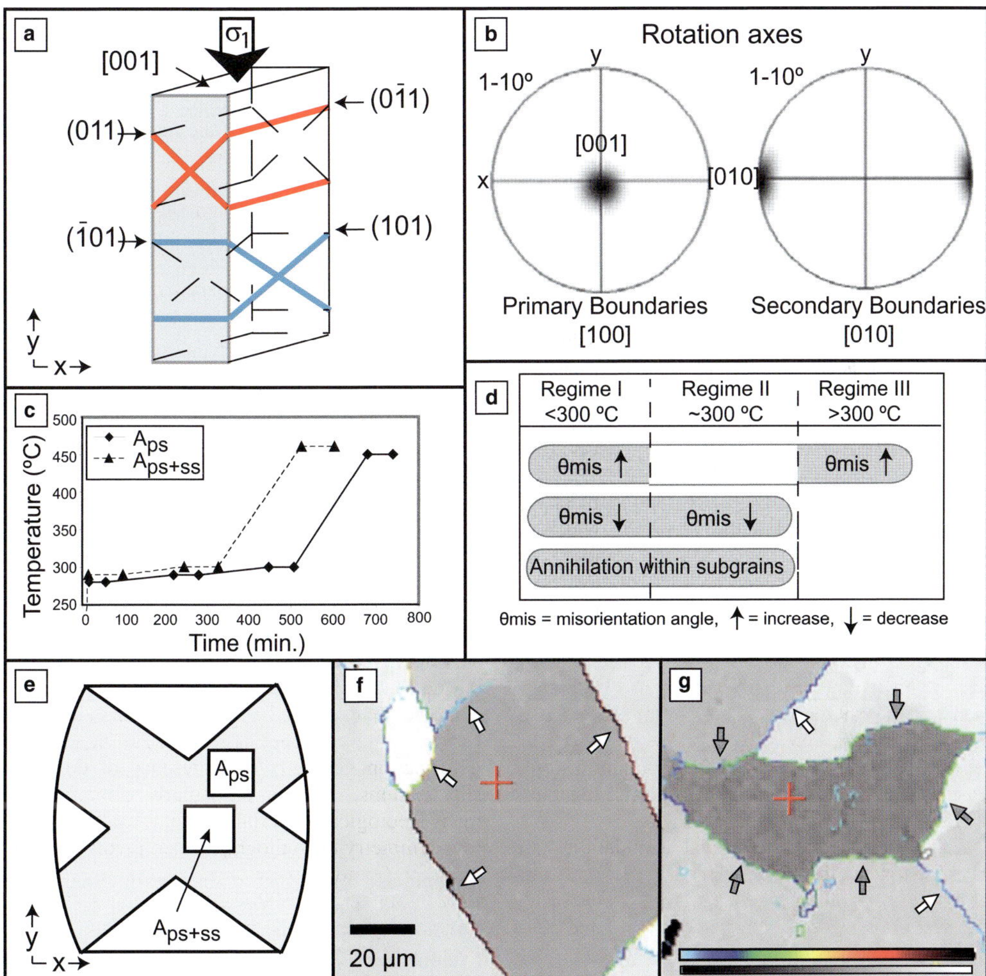

Fig. 1. Summary of experimental method and results from the 2D *in situ* annealing study (Borthwick & Piazolo 2010). (**a**) Deformation geometry of the halite crystal, showing the observation surface as a grey surface, the main compression axis as a thick black arrow, the primary boundaries in red and the secondary boundaries in blue. (**b**) Equal-area, lower-hemisphere contoured pole figures showing the dominant rotation axes for primary boundaries and secondary boundaries. (**c**) Heating method used for annealing experiments. Temperature was decreased to *c.* 25 °C after each heating stage for electron backscatter diffraction analysis (EBSD) mapping. (**d**) Table of characteristic temperature-dependent annealing behaviour. (**e**) Schematic of the deformed, barrelled crystal with areas of highest strain shaded in grey. The locations of A$_{ps+ss}$ closer to the centre and A$_{ps}$ peripheral to it are shown. (**f, g**) EBSD maps showing the two starting microstructures for the simulations. Maps are coloured in greyscale based on the angular deviation taken from the Euler angle orientation at the cross. The greyscale colouring (g) represents 0–10° and 0–8° angular deviation for (f) and (g), respectively. Subgrain boundaries are shown in an interpolated colour scale (g), with respect to misorientation blue to red showing 0–9°. (f) Starting microstructure A$_{ps}$ with white arrows showing primary LABS (no secondary LABs are present). (g) Starting microstructure A$_{ps+ss}$ with white arrows showing primary LABs and grey arrows showing secondary LABs.

Due to the starting asymmetry of the crystal prior to deformation, one set of slip planes was activated more easily than the other, thus we divided them into primary $(011)[0\bar{1}1]$ and $(0\bar{1}1)[011]$ and secondary $(101)[\bar{1}01]$ and $(\bar{1}01)[101]$, each causing the formation of slip-system-specific crystal structure distortions and well-defined subgrains with a misorientation of less than $10°$ (Fig. 1a, b). Secondary boundaries were only observed in the central zone of the crystal (e.g. area A_{ps+ss} in Fig. 1e), which is likely to be due to bulging of the outer surface allowing deformation to be accommodated. In the central regions, the additional stress can only be accommodated by the activation of the secondary, perpendicular slip planes.

(2)   During initial stages of recovery substructural behaviour can fluctuate suggesting that dislocations are not fixed in boundaries as initially thought. The same LABs can undergo both increases and decreases in misorientation (Fig. 1c). Behaviour is not directly related to the angle of LABs, but highly dependent on the character of the dislocations that make up the boundary.

(3)   Annealing behaviour is temperature dependent. Three temperature regimes could be identified (Fig. 1d). At temperatures lower than $c.$ 300 °C, the LABs related to the primary slip system increased in misorientation, while those related to the secondary slip system decreased. When $c.$ 300 °C $(T_h = 0.53)$ was reached, all LABs showed a decrease in misorientation, while above 300 °C they increased in misorientation. An error of ± 15 °C is estimated due to discrepancy between the sample temperature and the furnace temperature.

General conceptual model

The conceptual model that forms the basis for the presented numerical model is designed to encapsulate dislocation movement behaviour during post-deformational annealing and is based on the model presented in Borthwick & Piazolo (2010). In the most general sense, during post-deformational annealing, the stored energy of a grain is reduced by rearrangement of the dislocation-related structures. The stored energy of the system is characterized by the distribution and type of dislocations. Here we distinguish between so-called primary dislocations related to the dominant slip system and secondary dislocations related to the secondary, less dominant slip system. Primary dislocations form primary subgrain boundaries while secondary

dislocations form secondary boundaries. Commonly, the microstructure will be dominated by the primary boundaries and primary dislocation densities will be significantly higher than for secondary dislocations. At low temperatures (for halite, $T < 300$ °C), dislocation climb is not activated allowing only limited increase in misorientation angles of primary boundaries when primary dislocations are added. Secondary boundaries may show a decrease in misorientation due to annihilation and availability of secondary dislocations. At intermediate temperatures (halite, $T \sim 300$ °C) there is a decrease in boundary misorientation, because dislocation separation, which is related to the interacting stress fields of dislocations sited in an array (Hull & Bacon 2001), can no longer increase for LABs related to both primary and secondary slip systems. At high temperatures (halite, $T > 300$ °C) climb becomes a significant component of microstructural development. At the same time dislocations become more mobile since dislocation movement by climb is diffusion controlled and therefore temperature dependent. Separation can also increase as LAB-wall dislocations may climb to accommodate extra dislocations within the LAB wall. Furthermore, the range of influence of dislocations is expected to increase with increasing temperature (Kocks 1985; Hull & Bacon 2001).

Numerical implementation

Simulations were carried out using the open-source modelling platform Elle (Jessell *et al.* 2001; Bons *et al.* 2008; Piazolo *et al.* 2010; http://www.microstructure.info/elle). The data structure of this platform provides a framework for describing the arrangement of physical and chemical properties in a 2D microstructure. Previous studies using Elle include subgrain growth (Piazolo *et al.* 2004), isotropic and anisotropic grain growth (Bons *et al.* 2001; Jessell *et al.* 2003; Piazolo *et al.* 2004; Becker *et al.* 2008; Roessiger *et al.* 2011, 2012), dynamic recrystallization (Piazolo *et al.* 2002), strain localization (Jessell *et al.* 2005; Griera *et al.* 2011) and recovery and grain growth (Bergman & Piazolo 2011; Piazolo *et al.* 2012*a, b*).

The model described here used the so-called unconnected node, 'unode' grid structure which consists of a regular square grid of data points that each store physical and/or chemical properties at a specific point in space. The routine reduces the internal energy of the microstructure according to a set of rules which rotate the crystal lattice at each fixed datum point resulting in areas of more homogeneous lattice orientations, as well as production and/or disappearance of subgrain boundaries (see Fig. 2).

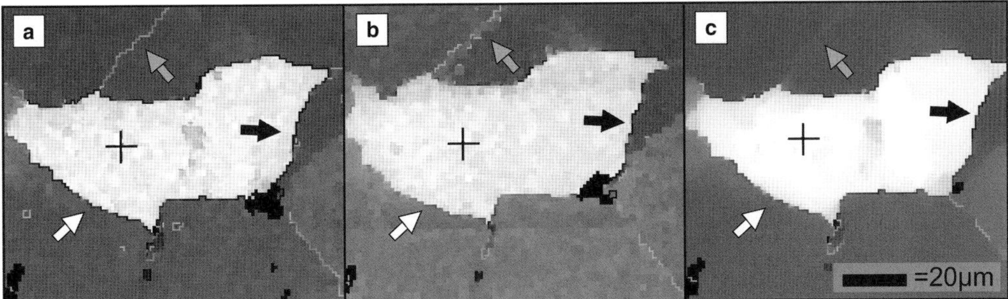

Fig. 2. Maps showing general behaviour of the numerical simulation compared to experimental results: (**a**) EBSD map before heating, (**b**) after heating for 1 h 30 min at 280 °C and (**c**) numerical simulation run for 50 steps with the before-heating map (a) as the starting point (size is 100×100 data points). The images show an angular deviation of 5° in greyscale taken from the Euler angle orientation at the black cross. Subgrain boundaries are shown in interpolated greyscale with silver $1-2°$, grey $2-3°$ and black $>3°$. White arrows show a boundary which decreases in misorientation in both the experiment (b) and the simulation (c). The black arrows show a boundary which does not decrease significantly in either (b) or (c). The grey arrow shows a boundary which partially dissipates after heating, as shown in (b). The simulation overestimates the amount of dissipation and in (c) the boundary is completely dissipated. A general smoothing of orientation variation is seen in both the experiment after heating (b) and the simulation (c).

A generalized flowchart of the simulation procedure is shown in Figure 3. The algorithm starts by choosing a random datum point and finding its neighbouring data points. To simplify the routine the boundary length of all data points is considered to be equal. The current stored energy (E_ρ) for the neighbourhood (with size N) is then

$$E_\rho = E\left(\left(\sum_{i=0}^{N} \theta_i\right)\Big/ N\right) \qquad (1)$$

where E_ρ is the stored energy in the neighbourhood N, E is the Read–Shockley energy (see equation A2) and θ is the misorientation between the central point and each individual neighbour i. Of the possible 24 symmetry operators for a cubic system, we select the misorientation/rotation-axis pair with the minimum misorientation angle (Wheeler *et al.* 2001).

A number of potential rotations around specific active slip-system-dependent crystallographic axes are used as trial rotations. Theoretically, the amount of rotation is a function of the temperature of annealing and the dislocation mobility, related to the type of dislocation and active slip systems (Hull & Bacon 2001). The stored energy (E_ρ) of the neighbourhood is compared with the current energy and the crystal orientation of the datum point is rotated towards the value which results in the maximum reduction in energy. If all trials result in an increase in the energy of the neighbourhood, the crystal orientation at the datum point is left unchanged. The rotation rate is varied to test different possible scenarios. The frequency of application

for each rotation is recorded for later analysis (see Appendix B). This procedure is repeated for each datum point in a random order. The reader is referred to the Appendix for a more detailed treatment of the numerical methodology.

The presented recovery routine is fully integrated into the Elle modelling platform and in the future it will be possible to use it in more complex models which require a series of different processes to be operating in conjunction. Examples are: dynamic recrystallization (Piazolo *et al.* 2002); strain localization (Jessell *et al.* 2005); crystal plasticity coupled with stored energy and surface-energy-driven grain-boundary migration (Griera *et al.* 2011; Piazolo *et al.* 2012*a, b*; Montagnat *et al.* 2013).

Investigating parameter space

Parameter tests: set-up

In order to explore how a given starting deformation microstructure changes at different temperatures, we performed numerical simulations investigating the effects of key parameters. The following three key parameters were chosen based on the conceptual model of dislocation behaviour and its calculated dependence on energies:

(a) *Type of deformation-induced dislocations.* We chose two microstructural datasets taken directly from the experiments on halite single crystals described above. They represent the end members of the observed variation in post-deformational microstructures (Borthwick & Piazolo 2010). Area A_{ps}, which is located

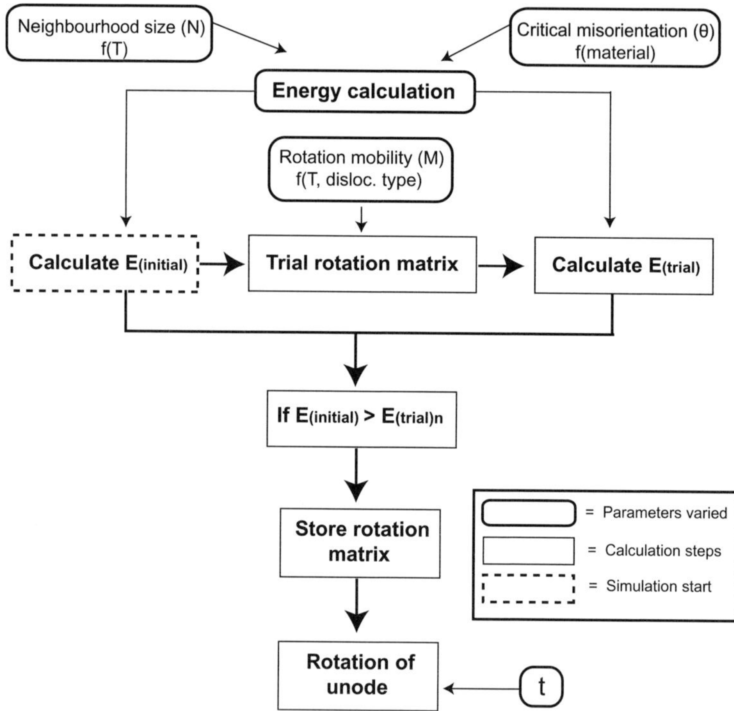

Fig. 3. Flowchart showing the set-up of the numerical simulation. Parameters surrounded by rounded boxes were changed and tested. The dashed box shows the starting point of the simulation. T is temperature; t is time.

away from the centre of the sample, has predominantly LABs that consist of primary dislocations with high misorientations (5–10°) between individual subgrains (Fig. 1e, f) while Area A_{ps+ss}, located at the centre of the sample, exhibits LABs formed by both primary dislocations and secondary dislocations with low misorientation (<5°) between individual subgrains (Fig. 1e, g).

(b) *Relative activity of different dislocation types.* Both theoretical considerations and experimental results suggest that boundary behaviour depends on the type of dislocations. These, in turn, determine the LAB characteristics comprising orientation in space and dominant rotation axes. In the numerical scheme, the relative activity of dislocation types is modelled by changing the relative rates of rotation around activated slip-system-dependent rotation axes. Because we apply the tests on microstructures from deformed halite we consider [100], [010] and [001] crystal directions as the dominant rotation axes related to the primary, secondary and auxiliary slip systems with corresponding dislocation types primary dislocations, secondary dislocations and auxiliary dislocations, respectively

(Guillope & Poirier 1979; Borthwick & Piazolo 2010). A change in the numerically applied rotation rate is conceptually related to a change in dislocation mobility. Since dislocation mobility is thermally activated with an Arrhenius-type relationship (e.g. Hu 1963; Kocks 1976, 1985), the angular velocity or rotation rates should be temperature dependent. Based on the theoretical change in relative mobility of dislocations associated with specific slip systems (Nicolas & Poirier 1976; Guillope & Poirier 1979; Carter & Hansen 1983), we have chosen four rotation-rate set-ups (refer to Table 1 for acronym definitions and experiments) with a variation of relative rotation rates for the three main rotation axes relevant for NaCl. R_{all} exhibits no dominance of any specific dislocation type (i.e. rotation axis). In contrast, R_{ss} shows a clear dominance of secondary dislocations, R_{ps+ss} equal dominance of primary and secondary dislocations over auxiliary, while R_{ps} shows slight dominance of primary dislocations over secondary with auxiliary dislocations playing a minor role (Table 1). Small normalized values, measured in degrees per time step, are chosen to ensure numerical

Table 1. *Parameters varied in parameter tests and used acronyms*

Parameters varied	Variant 1	Variant 2	Variant 3	Variant 4
Rotation rates R_i: [010][100][001]	R_{all} 1,1,1 All same	R_{ss} 0.25, 0.5, 1 [010] dominant	R_{ps+ss} 0.35, 0.35, 1 [010][100] same	R_{ps} 0.35, 0.25, 0.75 [100] dominant
Number of neighbours N	Small: 4–7	Medium: 8	Large: 24	

stability and reasonable calculation times, while relative rates are chosen according to known and experimentally observed difference in relative activation of dislocation types (Guillope & Poirier 1979). It should be noted that lower rotation rates are more likely to result in minimizing stored energy, as greater rates result in over-rotation and a higher-energy configuration. The specific rotation chosen is based on relative energy changes; thus it is independent of grid resolution, but for stability of the program the rotation rates need to be adjusted for different grid resolutions. Theoretically, R_{ss} is expected to be most relevant for low-, R_{ps+ss} for intermediate-, and R_{ps} for high-temperature conditions (Table 1). At low temperatures secondary dislocations should be significantly more mobile than the primary. At intermediate conditions both primary and secondary dislocations are mobile, while at higher temperatures the primary dislocations become even more mobile due to their possible movement via temperature-activated climb, while for secondary dislocations this is not the case. R_{all} is used as a reference only.

(c) *Number of neighbours (N)*. Theoretically, dislocation interaction and movement is strongly temperature dependent, (i.e. dislocation movement) (Hull & Bacon 2001). This is supported by experimental results (see above). Accordingly, in the model we test the different behaviour in response to the number of neighbouring data points (*N*) used in energy calculations. We tested three different scenarios where the neighbourhood size was small (randomly assigned number of first neighbours – on average 4), medium (all first neighbours – 8 neighbours) or large (all first and second neighbours – 24 neighbours) (see Table 1). Tests were conducted on both initial microstructures.

Parameter tests: results

In order to evaluate the results of the parameter tests, we studied the changes that occurred in misorientation distributions of the tested datasets. This method has previously proved valuable in identifying deformation mechanisms (e.g. Prior *et al.* 1999; Trimby *et al.* 2000; Halfpenny *et al.* 2006; Svahnberg & Piazolo 2010) and evaluation of experimentally observed temperature-dependent behaviour (Borthwick & Piazolo 2010).

Relative activity of different dislocation types. Results for Area A_{ps}, which is characterized by a dominance of primary dislocations, show a number of key features. All rotation-rate tests develop a distinctly bimodal distribution of misorientation angles (Fig. 4a–e). They demonstrate a shift of the initial *c.* 7.5° peak towards lower angles by *c.* 2°. R_{all} does not result in a change as significant as R_{ss}, R_{ps+ss} and R_{ps} rotation set-ups. R_{ss} favours development of the low- angle peak, while R_{ps+ss} shows minimum change and R_{ps} favours the high-angle peaks (Fig. 4; Table 1). Rotation around [001] is initially dominant for all four tests and with an increase in number of steps (a proxy for time), there is a decrease in the absolute number of successful trials (Fig. 4, and Appendix B). Statistics on trial rotations show that [010] was most significant in R_{ss}, [100] in R_{ps+ss}, [001] in R_{ps} and that no rotation occurred most frequently for R_{all}. Rotation axes pole figures show that all rotation-rate tests result in a tightening around the [100] axis (Fig. 4f–j).

For Area A_{ps+ss}, all rotation-rate tests result in an increase in the lower-angle peak, with R_{ss}, R_{ps+ss} and R_{ps} showing a large increase of >25% (Fig. 4k–o). Here, the high-angle peak (10.5°) is slightly more developed than in R_{all}. Trial rotation frequencies show similar results to Area A_{ps} (Fig. 4, and Appendix B). Rotation axes pole figures show that all rotation-rate tests result in disappearance of a low-angle cluster around [100] and a weakening of a cluster close to [111] (Fig. 4q–t). For all rotation-rate tests the rotation-axis cluster close to [001] is better defined. For R_{ps} the cluster around [010] is clearest.

Number of neighbours N. The difference in the number of neighbours N which are taken into account for the energy calculation, in other words the difference in the neighbourhood size, had a most pronounced effect on the microstructural behaviour in experiments using Area A_{ps} rather than Area A_{ps+ss} (Fig. 5). However, the misorientation

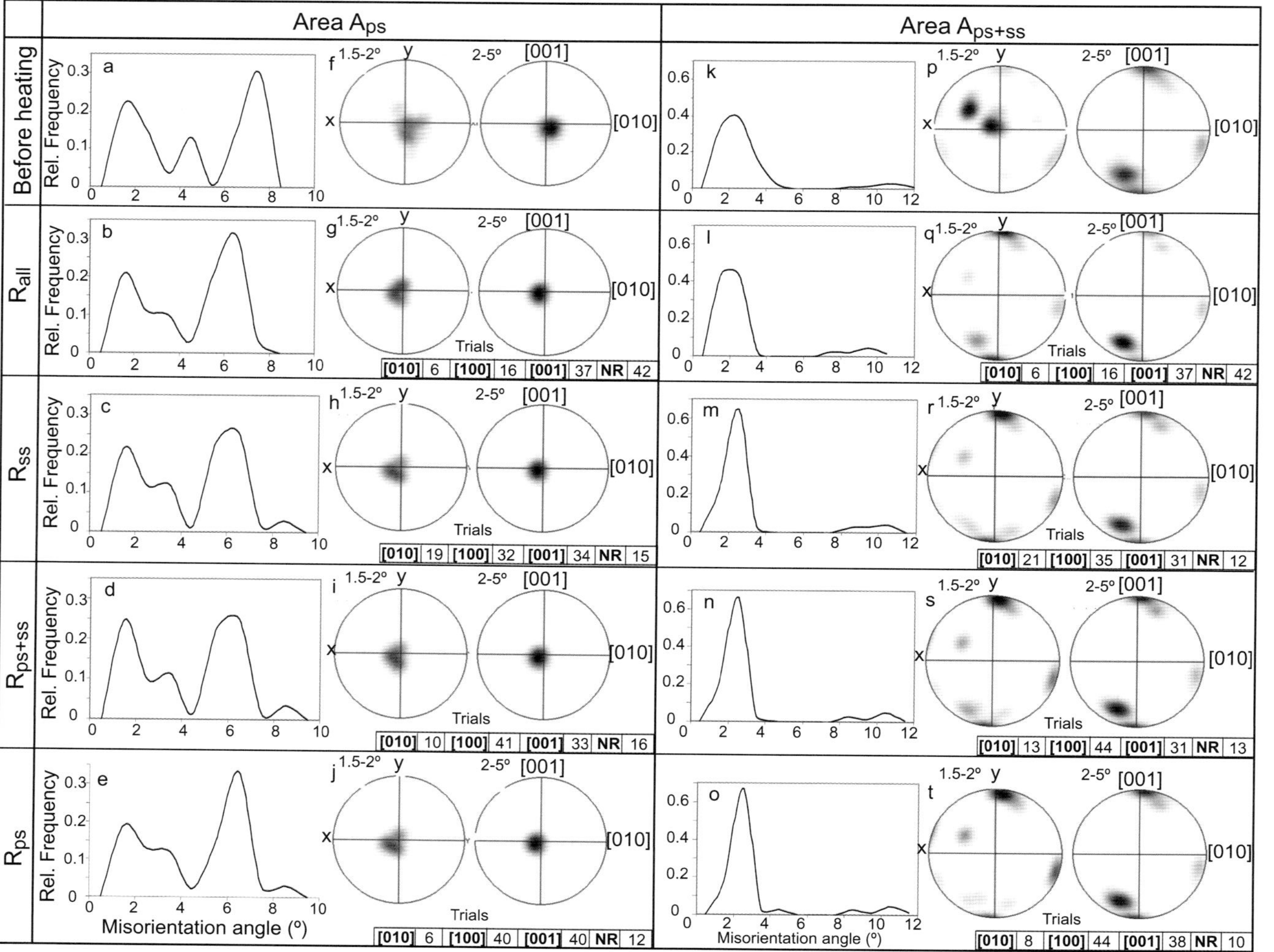
Area A_ps
Area A_ps+ss
Before heating
R_all
R_ss
R_ps+ss
R_ps
Rel. Frequency
Misorientation angle (°)
1.5-2°   y   2-5°   [001]
x   [010]
Trials
[010] 6 [100] 16 [001] 37 NR 42
[010] 19 [100] 32 [001] 34 NR 15
[010] 10 [100] 41 [001] 33 NR 16
[010] 6 [100] 40 [001] 40 NR 12
[010] 6 [100] 16 [001] 37 NR 42
[010] 21 [100] 35 [001] 31 NR 12
[010] 13 [100] 44 [001] 31 NR 13
[010] 8 [100] 44 [001] 38 NR 10

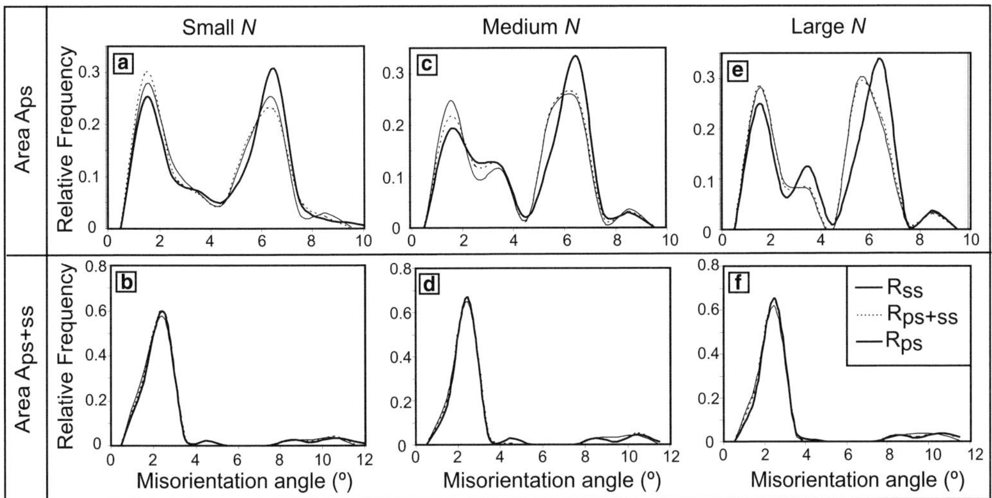

Fig. 5. Subgrain-boundary histograms displaying the relative frequency of different misorientation angles. Rows are labelled with the starting microstructure used. Columns are labelled with the varying neighbourhood size N. Simulations were run to 50 steps, with map size 100×100 data points, and a critical misorientation ϕ_m of $15°$. The legend in (f) applies to all figures.

frequency peaks for both areas show a gradual tightening with increased N. In addition, dips in Area A_{ps} become better defined as N increases. If we look specifically at the behaviour of different rotation rates in Area A_{ps}, we see some changes (Fig. 5a, c, e). For small N, there is an increase in lower angles with a corresponding decrease in the high-angle peak which occurs for R_{ss} and continues with R_{ps+ss}. R_{ps} shows the reverse behaviour, with an increased high-angle peak and decreased low-angle peak (Fig. 5a).

Medium N shows a decrease in the dominance of lower angles in R_{ps+ss}, while the high-angle peak stays the same as that for R_{ss}. R_{ps} exhibits a much larger increase in the high-angle peak, and more significant decrease in the lower-angle one. We see that misorientation angles $>8°$ are better developed for all rotation rates (Fig. 5c).

For large N, experiments show similar behaviour for R_{ss} and R_{ps+ss}, with increases in both low- and high-angle peaks, and a decrease at the dip between (Fig. 5e). There is also a shift of the peak towards lower angles. R_{ps} shows a decrease in the low-angle peak and an increase in the high-angle peak. Misorientation angles $>8°$ show the same development for all three rotation-rate tests.

Differences between results conducted with different N are less pronounced in Area A_{ps+ss} (Fig. 5b, c, f). We do see a minor shift towards higher-angle boundaries as N increases. R_{ps} results in a minor increase in intensity of higher angles as N increases. A near-unimodal distribution in angles $>8°$ is seen for R_{ss} with medium and large N, while for R_{ps+ss} and R_{ps} a slightly bimodal distribution is seen.

Parameter tests: discussion

The parameter variation tests demonstrate a number of key features. Varying the relative rotation rate of

---

Fig 4. Simulation to 50 steps using different rotation rates for the rotation axes. (**a–e**), (**k–o**) Subgrain-boundary histograms displaying the relative frequency of different misorientation angles. (a–e) are taken from the starting microstructure Area A_{ps} and (k–o) from Area A_{ps+ss}. (a), (k) The distribution for the before-heating microstructure in Area A_{ps} and Area A_{ps+ss} respectively. Histograms are shown for different rotation-rate set-ups: (b), (l) R_{all}; (c), (m) R_{ss}; (d), (n) R_{ps+ss}; (e), (o) R_{ps}. For significance of R_{all}, R_{ss} and R_{ps+ss}, see Table 1. Changes in the relative intensity with different rates can be seen. (**f–j**), (**p–t**) Lower-hemisphere, equal-area pole figures showing contoured intensities of rotation-axes orientation for the different mobilities. The rotation rates correspond to the rows defined by the histograms and labels. The number of points used to contour the pole figures are written above them. All mobility tests are run for 50 steps using medium n, a critical misorientation ϕ_m of $15°$ and a map size 100×100 data points. Beneath the pole figures for each corresponding rotation-rate test are the results of how many times each rotation axis was chosen as a percentage (percentages do not add up to 100% due to rounding). NR stands for no rotation and demonstrates that in these cases the datum point did not rotate.

the rotation axes has a significant effect, particularly when there are higher-angle boundaries involved (e.g. experimental Area A_{ps}). This suggests that both the initial dominance and the temperature-dependent relative mobility of different types of dislocations have a significant effect on post-deformational recovery behaviour. The two different starting microstructures, taken from the same crystal but with different dominant slip systems activated, show markedly different behaviour (Fig. 4, and Appendix B). Area A_{ps+ss} showed a dominance in the activation of the [010] rotation axis, with the [010] rotation axis chosen between 5% and 10% more frequently than for Area A_{ps} under the same conditions. Rotation-rate tests can be directly linked to different temperature-dependent regimes from the experiments as is detailed later in this paper.

An increase in the neighbourhood size, which theoretically is linked to increased temperature, allows long-range dislocation interaction and higher dislocation mobility facilitated by climb. Numerical results show that larger neighbourhood sizes resulted in better-defined peaks independent of rotation rates and initial microstructure. In summary, rocks that underwent post-deformational recovery at high temperatures are expected to show the following features: (a) distinct misorientation peaks, and (b) subgrains with little lattice bending due to long-range dislocation interaction, activation of climb and increased dislocation mobility. In a general sense, such features have been shown statistically to develop in geological materials (e.g. Heilbronner & Tullis 2002, fig. 10). In addition, after a period of pronounced microstructural adjustments, little further changes occur. At which angle these misorientation peaks occur depends on the initial post-deformational microstructure, and by correlation, the relative activity of dominant slip systems during deformation.

Numerical modelling of reported *in situ* experiments

In order to (a) validate the numerical model and the theoretical considerations on which the model is based and (b) offer an advanced interpretation of experimental results, we conducted numerical simulations of the *in situ* experiments, using the same experimental conditions. Here, experiments described in Borthwick & Piazolo (2010) and Borthwick *et al.* (2012) are compared with numerical modelling results.

Numerical set-up: parameter values used

The temperature-dependent parameters were changed in a stepwise manner in order to correspond to stepwise increase in temperature in the *in situ* experiments. Rotation rates were chosen based on theoretically and experimentally predicted dominant rotation axes at specific temperatures (Table 2). The parameter neighbourhood size N was increased with each temperature stage to simulate increasing long-range dislocation influences and climb (for parameter values see Table 2).

Simulations were conducted on the initial microstructural datasets for Area A_{ps} and Area A_{ps+ss} (cf. Fig. 1f, g). In order to replicate the increasing experimental temperature (cf. Fig. 1c), we began with the post-deformation, before-heating microstructure to which we then applied $<300\ °C$ parameters, followed by *c*. 300 °C and then $>300\ °C$ (Table 2). The number of steps for each temperature phase was determined by setting the same relative ratios of experimental heating duration to that of the *in situ* experiment. The resulting correlation of 1 min of experimental time to 2.5 steps in the simulation was applied to all other runs and also to Area A_{ps+ss}.

Numerical modelling: results and discussion

Subgrain-boundary histograms. Subgrain-boundary histograms for the experimental data and the simulation show distinct similarities (Fig. 6a, b). The simulation reproduces the general increasing and decreasing misorientation behaviour from the experiment reasonably well (Fig 6a, b); it also reproduces the increase in lower-angle boundaries (*c*. 2°) and decrease in higher-angle boundaries

Table 2. *Parameters used for numerical modelling of in situ experiment*

Parameter	Regime I	Regime II	Regime III
Temperature (°C)	<300	*c*. 300	>300
Rotation rate R_i	R_{ss}	R_{ps+ss}	R_{ps}
Number of neighbours N	small	small/med	large
Area A_{ps+ss} – number of steps	32	32	30
Area A_{ps} – number of steps	40	16	24

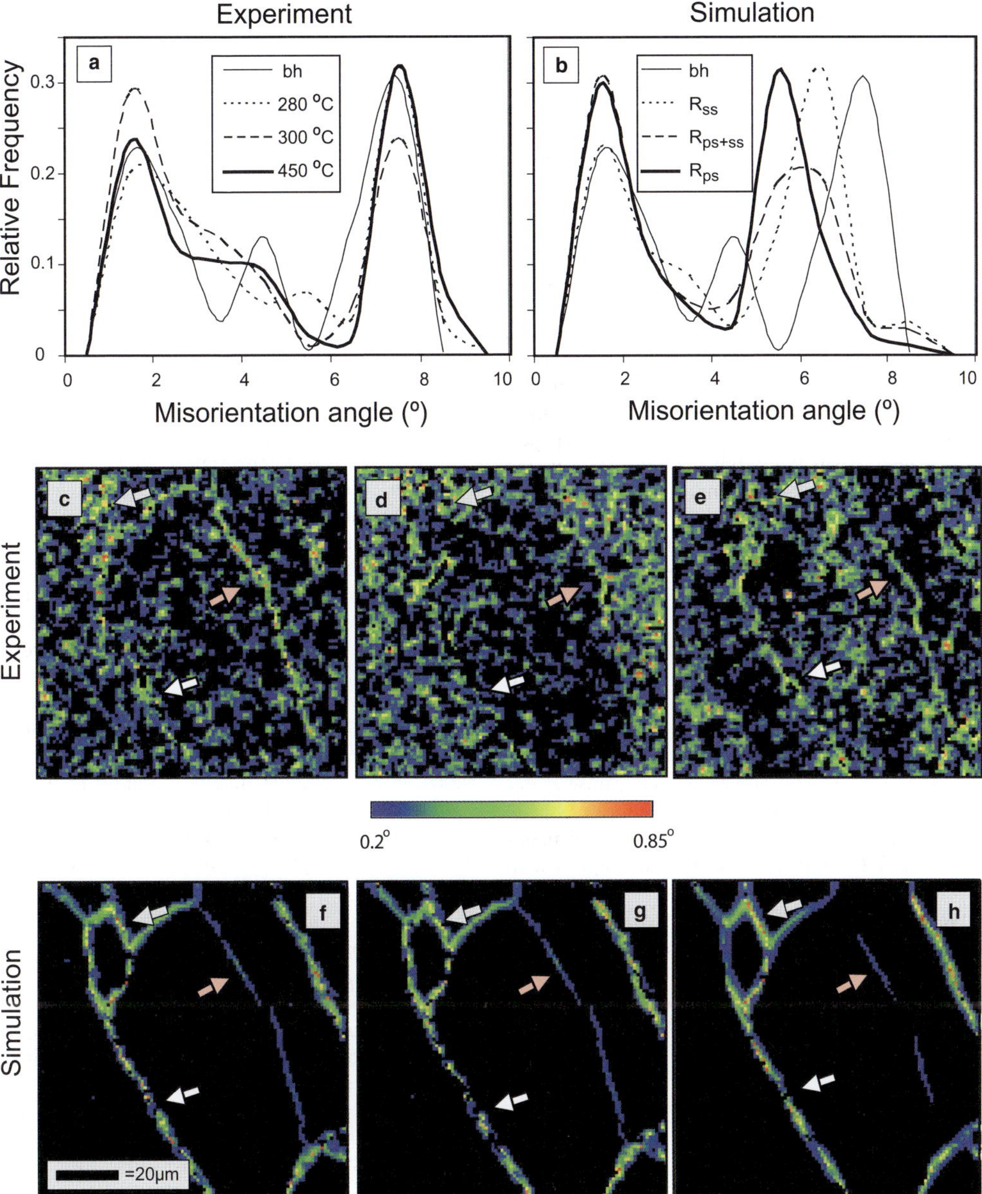

Fig. 6. Comparison of the simulation with the experimental data from A_{ps}. (**a**), (**b**) The subgrain-boundary histogram for the experiment and simulation, respectively, both exhibiting the fluctuating behaviour of the LABs. (**c–h**) Local misorientation maps using a filter size of 3×3 pixels and a subgrain angle of $1°$. Colour bar shows that the local misorientation varies between $0.2°$ and $0.85°$ (blue to red). Areas that are black are those with less than $0.2°$ of local misorientation. (c–e) Experimental data – noisier than the simulation data since there is some very low-angle damage from the heating process. Because of this we use a cut-off orientation of 0.2, while 0.1 is used for the simulation data. (f–h) Simulation maps. Interesting features from comparing the local misorientation maps include: white arrows showing a boundary that decreases in intensity at (d) and (g) *c*. 300 °C/R_{ps+ss} and increases again at (e) and (h) >300 °C/R_{ps}. Red arrows show a boundary which decreases in intensity as higher temperatures (R_{ps}) are reached. Grey arrows show an area which decreases in local misorientation intensity (d) and (g) *c*. 300 °C/R_{ps+ss} and increases again at (e) and (h) >300 °C/R_{ps}. In the simulation this area is clearly a boundary, while this is not so obvious in the experimental maps. Simulations that were run using a critical misorientation of $12°$ after testing (not shown here) showed that this best reproduced experimental results. bh, before heating.

(6–8°) at *c*. 300 °C. An increase in the misorientation of higher-angle boundaries at >300 °C is also reproduced. Even though the simulation histograms replicate many aspects of the experiment, the simulation does not retain higher values of LABs and the high-angle peak is shifted by *c*. 2°. This may be related to the lack of a physical law that distinguishes between LAB and grain boundaries on the basis of energy. According to Gottstein & Shvindlerman (2009), high-angle grain boundaries cannot be described as dislocation arrays alone, but have a grain-boundary energy specific to that mineral.

Local misorientation behaviour. Local misorientation maps show the local misorientation around a specific datum point which is a proxy for dislocation density (e.g. Takayama *et al.* 2005; Jorge-Badiola *et al.* 2007; Li *et al.* 2008; Wheeler *et al.* 2009). These maps confirm that similar behaviour is occurring in the experiments and the simulations (Fig. 6c–h). Experimental maps have significant noise, due to minor damage induced by the beam, but we can still see some key behaviour. During annealing at temperatures of *c*. 300 °C, there was a general decrease in LAB misorientation that is shown by reduction in local misorientation intensity. In Area A_{ps}, the white arrows (Fig. 6c–h) show a LAB that decreases in misorientation at *c*. 300 °C (Fig. 6d) and then increases again at temperature >300 °C (Fig. 6e). Similar behaviour is seen in the simulation, with the intensity decreasing with R_{ps+ss} (*c*. 300 °C) and increasing with R_{ps} (>300 °C; white arrows Fig. 6g, h). The red arrows show a LAB which decreases in intensity with each increase in annealing temperature (Fig. 6c–e), with lower intensity (red spots) appearing in the boundary. In Figure 6d, parts of the microstructure still have high local misorientation intensity and the LAB appears more disjointed (red arrows). At >300 °C, the LAB is clearer, but the intensity is lower. In the simulations, the LAB dissipates if R_{ps} is used (Fig. 6h), but we do not see exactly the same behaviour as at *c*. 300 °C in the experiment (Fig. 6d) as with R_{ps+ss} (Fig. 6g). In the experimental data, the grey arrows indicate an area in which the local misorientation intensity decreases at *c*. 300 °C (Fig. 6c, d) and then increases again at >300 °C (Fig. 6e). In the simulation this area is more clearly represented as an LAB and shows a corresponding increase in intensity and organization by R_{ps} (>300 °C; Fig. 6h).

Tests for temperature-dependent mobility of dislocations have been numerically implemented using differing rotation rates. The results of comparison with experimental data show that increasing temperature does cause significant changes in the relative mobility of the main dislocation types. Thus, the mode of microstructural evolution depends on the temperature of recovery, where it is significant if recovery temperature is below or above the temperature when climb is activated for one or several of the main slip systems.

Rotation-axis development. The post-deformation, pre-heating rotation axes were dominantly around [001] for higher-angle boundaries (2–5°), while for lower angles (1.5–2°), the rotation axis was closer to [100] (Fig. 7). During *in situ* experiments (Fig. 7e–g) the cluster of [100] misorientation axes had disappeared at *c*. 300 °C. This is clearly reproduced in the simulation, with all traces of the orientation gone once R_{ps+ss} is activated (Fig. 7c). The experimentally observed spread of low-angle orientations at >300 °C is, however, not well reproduced in the simulation. We suggest that the misorientation-axis distribution was not well defined for the deformed microstructure and low-temperature annealing, as misorientations were still dominantly low. At low misorientation angles, rotation axes are not well defined especially as significant surface damage occurred during heating experiments (Prior 1999). For higher angles, the development of rotation axes is similar in both experiments and numerical simulations, showing a rotation away from the [001]. This occurs more gradually in the simulation than in the experiments (Fig. 7b–d), with similar rotation occurring once high temperatures are reached and R_{ps} is used (Fig. 7d). The rotation axis sub-parallel to [010] does not disappear in the simulations as it does in the experiment by *c*. 300 °C (Fig. 7f), but remains in the equivalent stages (Fig. 7c). This may indicate that, in contrast to numerical simulations, [010] is not at all activated or retained at temperatures above 300 °C.

Summary and discussion: a refined conceptual model for recovery at and below the temperature of deformation

We present a refined model for recovery of halite at temperatures at or below that of deformation, with reference to presented numerical simulations and their comparison to physical experimental data (Fig. 8).

A starting dislocation budget is envisaged where primary dislocations have been mostly arranged into higher-misorientation LABs, while the more sparse secondary dislocations mostly remain free in the subgrain interior, with only a few arranged into LABs. Recovery behaviour is highly temperature dependent, as temperature controls dislocation mobility and range of dislocation interaction.

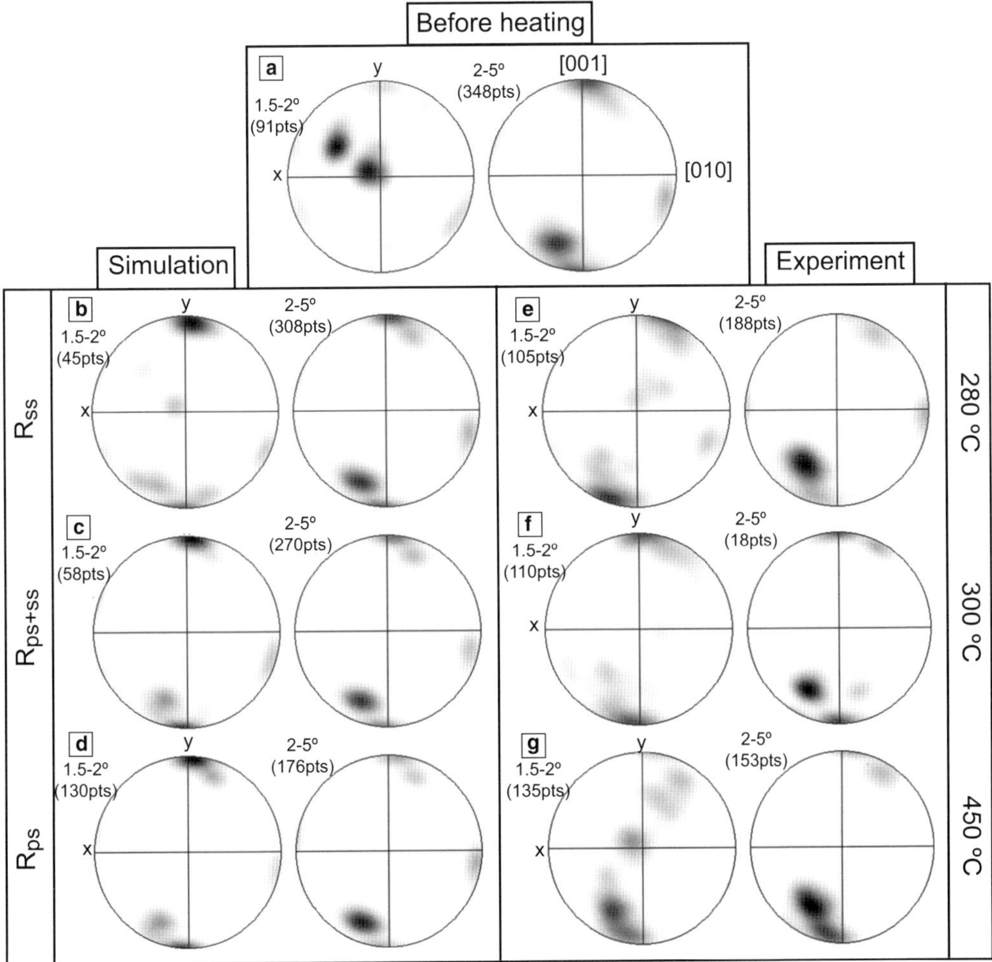

Fig. 7. Pole figures (**a–g**) showing the distribution of rotation axes in A_{ps+ss} at different temperatures in the experiment and stages in the simulation. Pole figures are lower hemisphere, equal area with contouring. (a) Pole figures before heating, which exhibit clusters at [100] and [111] for the lower angles and clustering sub-parallel to [001] for higher angles. The figure is arranged so that the pole figures at different temperatures are in the same row as the corresponding rotation-rate test.

Three distinct regimes can be distinguished on this basis. The absolute value for the temperature at which annealing regimes I, II and III occur are anticipated to be strongly material dependent. However, the temperature at which dislocation climb is possible for one of the main dislocation types governs the change from regime I (below the temperature to activate significant climb), to regime II (at the temperature to activate significant climb), to regime III (above the temperature to activate significant climb and at climb temperature for second and/or third main dislocation type). Controls on microstructural evolution can be described for these regimes:

(1) *Regime I* (Fig. 8a). Dislocation movement occurs predominantly by glide. Primary dislocations glide into LABs and increase their misorientation, while the large number of free secondary dislocations results in a decrease in secondary LABs. In order to model this with the simulation, the secondary rotation axis was made dominant by varying rotation mobilities (R_{ss}). Dislocations also annihilate in the subgrain interior and, when there are none available of opposite sign, begin to line up into low-energy arrays (i.e. geometrically necessary dislocations). Long-range dislocation effects are not as significant in regime I and this could be reproduced in the numerical model by using an

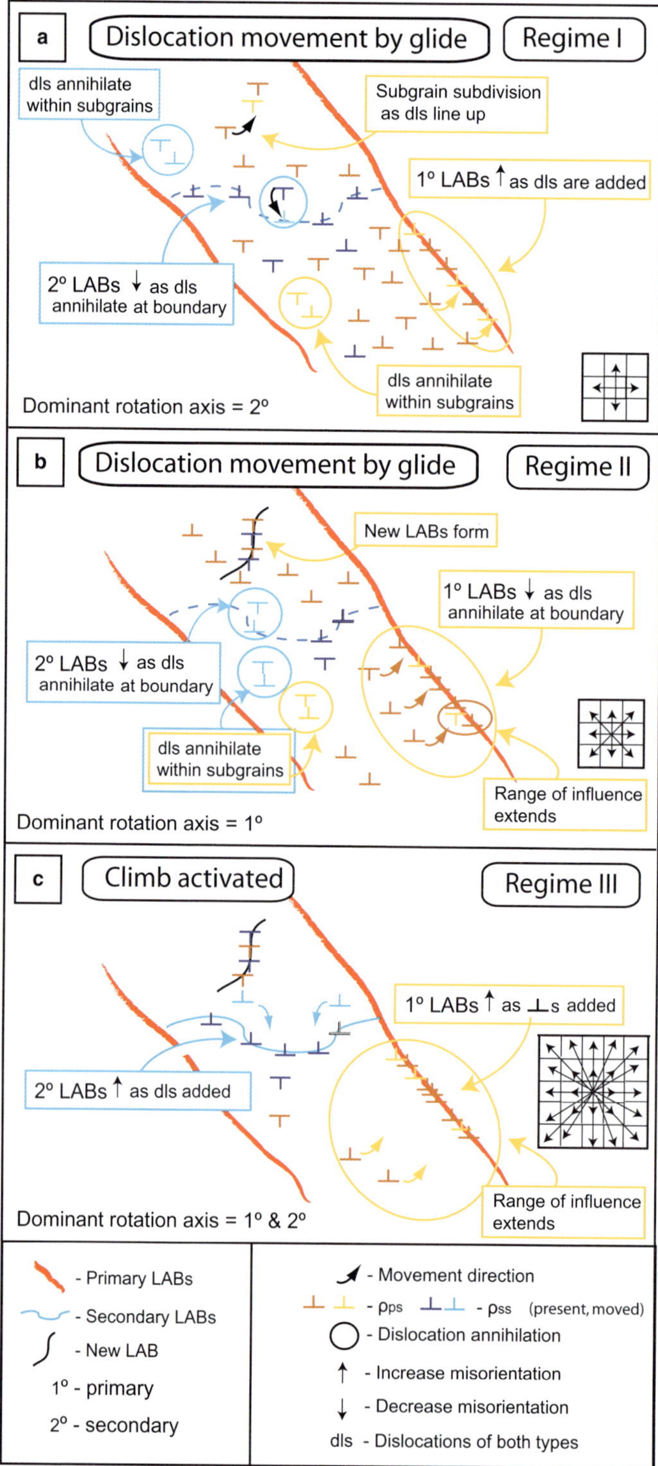

Fig 8. Summary of conceptual model for dislocation behaviour during post-deformational annealing (ρ_{ps} are primary dislocations and ρ_{ss} are secondary dislocations).

energy calculation which assumes a short-range neighbourhood influence.

(2) *Regime II* (Fig. 8b). Dislocation movement still occurs predominantly by glide. Secondary boundaries continue to decrease in misorientation, but at a reduced rate as the number of secondary dislocations decreases. To model the activity of both dislocation types in the simulation, primary and secondary rotation axes were given the same dominance (R_{ps+ss}). Primary LABs decrease in misorientation also by annihilation in the boundary vicinity, since dislocation separation can only increase by climb between lattice planes. Dislocation annihilation and low-energy-array building continues and new boundaries begin to form. The long-range influence of dislocations increases, which was modelled by increasing the neighbourhood range for the energy calculation.

(3) *Regime III* (Fig. 8c). The temperature to activate significant climb is now reached. LABs increase in misorientation as dislocation separation within boundaries can increase. Long-range influence of dislocations extends significantly as modelled by further increasing the neighbourhood range for the energy calculation.

In short, regimes I, II, and III correspond to changes in the characteristics of dislocation movement from glide, glide and climb and finally climb dominated from low, medium to high temperature, respectively. In addition, we show that dislocations in low-angle subgrain boundaries remain mobile and can rearrange significantly. This can lead to subgrain-boundary angle increase and decrease depending on the temperature, type and number of dislocations present.

Conclusion

By identifying different types of LABs related to different slip systems (primary and secondary) and applying this knowledge to model design, we were able to reproduce experimental results of post-deformational annealing. In accordance with this, relative dislocation mobilities, numerically expressed as rotation rates, change with temperature and may result in fluctuating misorientation behaviour. This suggests that dislocations can still be relatively independent up to misorientation angles higher than previously suggested (Humphreys & Hatherly 2004). We also infer that the mutual interaction range of dislocations has a significant effect on the annealing behaviour. Increasing the neighbourhood size which mimics the extension of dislocation influence with increasing temperature was essential to reproduce the experimental results. This confirms that dislocations have an important effect on one another during the annealing process.

All experiments show that, at some temperatures, microstructural changes are rapid and significant in initial stages of deformation, followed by much less pronounced changes. However, any increase in temperature, even below the temperature of deformation, has again a significant effect on the microstructural evolution.

The presented recovery model is particularly relevant for geological materials as it focuses on recovery at temperatures below the deformation temperature. This is a common situation for rocks in the crust which experience significant annealing during residence at elevated temperatures once deformation at high temperatures has ceased (e.g. Molli *et al.* 2000; Park *et al.* 2001; Bergman & Piazolo 2011). This model is significant for materials with a primarily activated slip system and a secondary slip system which is activated only later in the deformation process. As this is the case for many geological materials it provides a useful analogue for post-deformational annealing at temperatures below deformation.

Appendix A: Details on the physical basis for the conceptual model for recovery

The presence of dislocations is associated with stored energy due to the distortion of the crystal structure. For a certain angle of rotation of a bent structure it is possible to calculate the number or density (ρ) of geometrically necessary dislocations to achieve this rotation (Ashby 1970). This is proportional to the rotation angle (ϕ) divided by the distance (δx) over which the rotation occurs, and inversely proportional to the magnitude of lattice distortion, which is given by the Burgers vector (b):

$$\rho_i = \frac{\phi_i}{\delta x_i b_i} \tag{A1}$$

where the subscript i, denotes the dislocation type under consideration with an axis of rotation that lies on the slip plane. Any arbitrary angle of rotation can thus be achieved by the summation of the rotation of individual dislocation types. Whereas the number of dislocations is proportional to the rotation angle, the energy (γ) associated with the rotation is not. Here we employ the Read–Shockley boundary energy relationship, which assumes a simple tilt-boundary geometry (Read & Shockley 1950):

$$\gamma_{(\phi)} = \frac{\gamma_m \phi}{\phi_m (1 - \ln(\phi/\phi_m))} \tag{A2}$$

Here ϕ_m is the critical or maximum rotation angle for which the equations hold and at which the boundary energy is γ_m.

It has been suggested that the rotation rate of a grain/subgrain is proportional to the torque (Q) generated by the change of surface energy associated with the reduction in misorientation (Li 1962; Erb & Gleiter 1979; Randle 1993). Analogous to grain boundaries where the velocity is proportional to a driving force (e.g. Urai *et al.* 1986), we assume a linear relationship between the angular velocity ω and a driving torque Q:

$$\omega = MQ \qquad (A3)$$

where M is the rotational mobility (Moldovan *et al.* 2002). Considering a 2D microstructure, the torque acting on a group of data points of neighbourhood size N with datum point i in the centre is given by:

$$Q = \sum_{j=0}^{N} l d\gamma_{ij}/d\theta_{ij} \qquad (A4)$$

where l denotes the boundary length between two data points (i.e. the distance from one datum point to the reference datum point), γ_{ij} is the surface energy and θ_{ij} is the misorientation angle across the subgrain boundary between the datum point i and the neighbouring datum point j. Equation (A4) assumes the torque is calculated with respect to an axis through the mass centre of data points i. This theoretical treatment assumes that subgrain rotation is accommodated by either co-operative motion/rearrangement of boundary dislocations (Li 1962) and/or by boundary and lattice diffusion (Moldovan *et al.* 2001, 2002). The rotation between two adjacent data points with orientations g_i and $g_{j.}$ can be expressed by the tensor $\mathbf{M}$:

$$\mathbf{M} = \mathbf{R}_i \times \mathbf{R}_j^{-1} \qquad (A5)$$

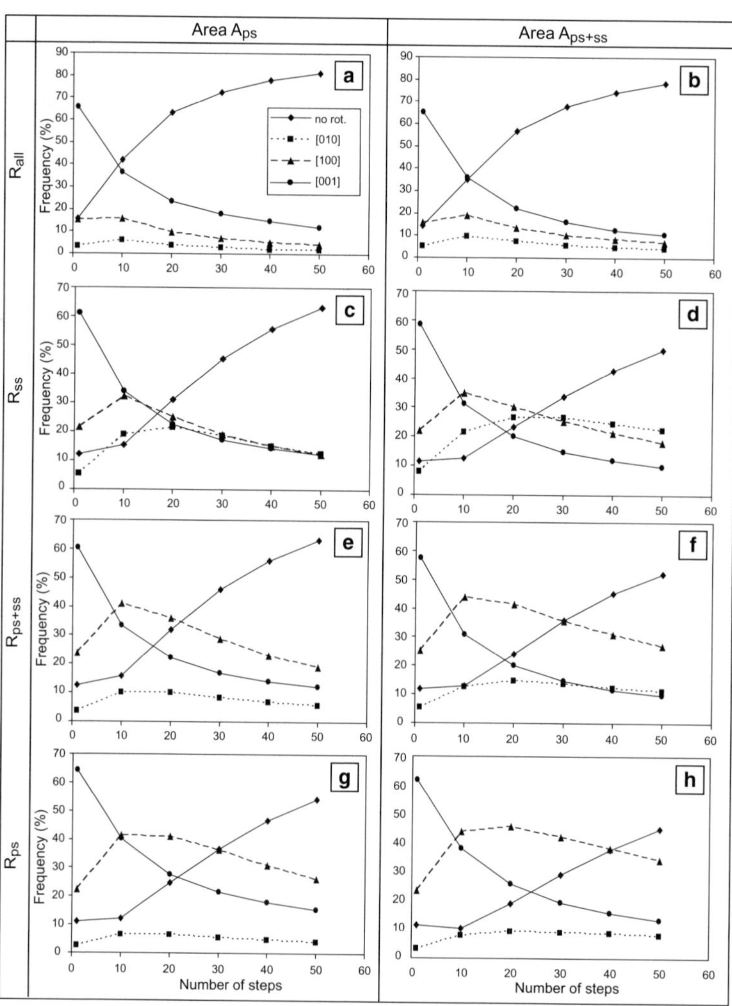

Fig. B1. The recorded frequency of chosen trial rotation resulted in the largest energy reduction. The data are cumulative with increasing number of steps. This figure was produced from the results of the simulations run in Figure 4 and thus had the same numerical set-up including parameter values and number of numerical steps.

where the rotation matrices $\mathbf{R}_i$ and $\mathbf{R}_j$ correspond to orientations g_i and g_j. Thus, the misorientation angle between both data points can be obtained from the tensor $\mathbf{M}$ as (e.g. Weiss & Wenk 1985):

$$\theta = \arccos[(tr(\mathbf{M}) - 1)/2] \qquad (A6)$$

For materials with crystal symmetry higher than triclinic, equivalent rotations need to be considered in order to obtain the minimum misorientation θ between two data points. For cubic materials, such as halite, the 24 symmetry operators were applied and the rotation tensor were calculated using

$$\mathbf{M} = \mathbf{O}^{crys}\mathbf{R}_i \times (\mathbf{O}^{crys}\mathbf{R}_j)^{-1} \qquad (A7)$$

where $\mathbf{O}^{crys}$ denotes the symmetry operators (e.g. Kocks *et al.* 1998). The rotation tensor that yields a minimum θ was utilized (Bunge 1985).

Appendix B: The recorded frequency of chosen trial rotation

Figure B1 shows the recorded frequency of chosen trial rotation that resulted in the largest energy reduction.

We would like to thank T. J. M. van der Linden, C. Peach and G. Pennock for their part in deforming and providing the samples. J. Särnesjö is acknowledged for his assistance in developing the model. M. W. Jessell is acknowledged for providing very helpful advice on the modelling process. The authors would like to thank the reviewers, J. C. White and L. Morales, for their helpful comments which significantly improved the article. V. E. Borthwick and S. Piazolo acknowledge the financial support of the European Science Foundation under the EUROCORES Programme, EuroMinSci, MinSubStrDyn, No. ERAS-CT-2003-980409 of the European Commission, DG Research, FP6. P. D. Bons acknowledges financial support from the German Research Foundation (DFG, grant BO-1776/7. This is contribution 341 from the ARC Centre of Excellence for Core to Crust Fluid Systems (http://www.ccfs.mq.edu.au) and 900 in the GEMOC Key Centre (http://www.gemoc.mq.edu.au).

References

ASHBY, M. F. 1970. The deformation of plastically non-homogeneous materials. *Philosophical Magazine*, **21**, 399–424, http://dx.doi.org/10.1080/1478643700 8238426

AVE LALLEMANT, H. G. & CARTER, N. L. 1970. Syntectonic recrystallization of olivine and models of flow in the upper mantle. *Geological Society of America Bulletin*, **81**, 2203–2220, http://dx.doi.org/10.1130/0016-7606(1970)81[2203:SROOAM]2.0.CO;2

BARNHOORN, A., BYSTRICKY, M., BURLINI, L. & KUNZE, K. 2005. Post-deformational annealing of calcite rocks. *Tectonophysics*, **403**, 167–191, http://dx.doi.org/10.1016/j.tecto.2005.04.008

BARRIE, C. D., BOYLE, A. P., COX, S. F. & PRIOR, D. J. 2008. Slip systems and critical resolved shear stress in pyrite: an electron backscatter diffraction (EBSD) investigation. *Mineralogical Magazine*, **72**, 1181–1199, http://dx.doi.org/10.1180/minmag.2008.072.6.1181

BECKER, J. K., BONS, P. D. & JESSELL, M. W. 2008. A new front-tracking method to model anisotropic grain and phase boundary motion in rocks. *Computers and Geosciences*, **34**, 201–212, http://dx.doi.org/10.1016/j.cageo.2007.03.013

BERGMAN, H. & PIAZOLO, S. 2011. The recognition of multiple magmatic events and pre-existing deformation zones in metamorphic rocks as illustrated by CL signatures and numerical modelling: examples from the Ballachulish contact aureole, Scotland. *International Journal of Earth Science*, **101**, 1127–1148, http://dx.doi.org/10.1007/s00531-011-0731-6

BESTMANN, M., PIAZOLO, S., SPIERS, C. J. & PRIOR, D. J. 2005. Microstructural evolution during initial stages of static recovery and recrystallization: new insights from in-situ heating experiments combined with electron backscatter diffraction analysis. *Journal of Structural Geology*, **27/3**, 447–457, http://dx.doi.org/10.1016/j.jsg.2004.10.006

BONS, P. D., JESSELL, M. W., EVANS, L., BARR, T. D. & STÜWE, K. 2001. Modelling of anisotropic grain growth in minerals. *In*: KOYI, H. A. & MANCKTELOW, N. S (eds) *Tectonic Modeling: A Volume in Honor of Hans Ramberg*. Geological Society of America Memoirs, **193**, 39–49.

BONS, P. D., KOEHN, D. & JESSELL, M. W. 2008. *Microdynamics Simulation*. Springer-Verlag, Berlin.

BORTHWICK, V. E. & PIAZOLO, S. 2010. Post-deformational annealing at the subgrain scale: temperature dependent behaviour revealed by in-situ heating experiments on deformed single crystal halite. *Journal of Structural Geology*, **32/7**, 982–996, http://dx.doi.org/10.1016/j.jsg.2010.06.006

BORTHWICK, V. E., SCHMIDT, S., PIAZOLO, S. & GUNDLACH, C. 2012. Quantification of mineral behaviour in 4 dimensions: grain boundary and substructure dynamics in salt. *Geochemistry, Geophysics and Geosystems*, **13**, http://dx.doi.org/10.1029/2012GC004057

BUNGE, H. J. 1985. Representation of preferred orientations. *In*: WEISS, L. E. & WENK, H. R. (eds) *Preferred Orientation in Deformed Metals: An Introduction to Modern Texture Analysis*. Academic Press, London, 73–104.

CARTER, N. L. & HANSEN, F. D. 1983. Creep of rock-salt. *Tectonophysics*, **92**, 275–333, http://dx.doi.org/10.1016/0040-1951(83)90200-7

CHOPRA, P. N. & PATERSON, M. S. 1981. The experimental deformation of dunite. *Tectonophysics*, **78**, 453–473, http://dx.doi.org/10.1016/0040-1951(81)90024-X

DINGLEY, D. J. 1984. Diffraction from sub-micron areas using electron backscattering in a scanning electron microscope. *Scanning Electron Microscopy*, **2**, 569–575.

DINGLEY, D. J. & RANDLE, V. 1992. Microtexture determination by electron backscatter diffraction. *Journal of Materials Science*, **27**, 4545–4566, http://dx.doi.org/10.1007/BF01165988

ERB, U. & GLEITER, H. 1979. The effect of temperature on the energy and structure of grain boundaries. *Scripta Metallurgica*, **13**, 61–64, http://dx.doi.org/10.1016/0036-9748(79)90390-9

FAIRBAIRN, H. W. 1949. *Structural Petrology of Deformed Rocks*. Addison-Wesley, Cambridge.

FROST, H. J., THOMPSON, C. V. & WALTON, D. T. 1990. Simulation of thin film grain structures: I grain growth stagnation. *Acta Metallurgica et Materialia*, **38**, 1455–1468, http://dx.doi.org/10.1016/0956-7151(90)90114-V

GOTTSTEIN, G. & SHVINDLERMAN, L. S. 2009. *Grain Boundary Migration in Metals: Thermodynamics, Kinetics, Applications*. 2nd edn. CRC Press, Florida.

GRIERA, A., BONS, P. D., JESSELL, M. W., LEBENSOHN, R., EVANS, L. & GOMEZ-RIVAS, E. 2011. Strain localization and porphyroclast rotation. *Geology*, **39**, 275–278, http://dx.doi.org/10.1130/G31549.1

GRUBER, J., MILLER, H. M., HOFFMANN, T. D., ROHRER, G. S. & ROLLETT, A. D. 2009. Misorientation texture development during grain growth. *Part I: Simulation and Experiment. Acta Materialia*, **57**, 6102–6112, *http://dx.doi.org/10.1016/j.actamat.2009.08.036*

GUILLOPE, M. & POIRIER, J. P. 1979. Dynamic recrystallization during creep of single-crystal halite: An experimental study. *Journal of Geophysical Research*, **84**, 5557–5567, http://dx.doi.org/10.1029/JB084iB10p05557

HALFPENNY, A., PRIOR, D. J. & WHEELER, J. 2006. Analysis of dynamic recrystallization in a quartzite mylonite. *Tectonophysics*, **427**, 3–14, http://dx.doi.org/10.1016/j.tecto.2006.05.016

HEILBRONNER, R. & TULLIS, J. 2002. The effect of static annealing on microstructures and crystallographically preferred orientations of quartzites experimentally deformed in axial compression and shear. *In*: DE MEER, S., DRURY, M. R., DE BRESSER, J. H. P. & PENNOCK, G. (eds) *Deformation Mechanisms, Rheology and Tectonics: Current Status and Future Perspectives*. Geological Society, London, Special Publications, **200**, 191–218, http://dx.doi.org/10.1144/GSL.SP.2001.200.01.12

HOBBS, B. E. 1968. Recrystallization of single crystals of quartz. *Tectonophysics*, **6**, 353–340, http://dx.doi.org/10.1016/0040-1951(68)90056-5

HOLM, E. A., MIODOWNIK, M. A. & ROLLETT, A. D. 2003. On abnormal subgrain growth and the origin of recrystallization nuclei. *Acta Materialia*, **51**, 2701–2716, http://dx.doi.org/10.1016/S1359-6454(03)00079-X

HOLM, E. A., MIODOWNIK, M. A. & HEALEY, K. J. 2004. A subgrain growth model for strain-free grain nucleation during recrystallization. *Materials Science Forum*, **467–470**, 611–616, http://dx.doi.org/10.4028/www.scientific.net/MSF.467-470.611

HONDOH, T. 2000. Nature and behaviour of dislocation in ice. *In*: HONDOH, T. (ed.) *Physics of Ice Core Records 1*. Hokkaido University Press, 3–24.

HU, H. 1963. Annealing of silicon-iron single crystals. *In*: HIMMERL, L. (ed.) *Recovery and Recrystallization of Metals*. Wiley, New York, 311–378.

HULL, D. & BACON, D. J. 2001. *Introduction to Dislocations*. 4th edn. Butterworth & Heinemann, Oxford.

HUMPHREYS, F. J. 2001. Review grain and subgrain characterisation by electron backscatter diffraction. *Journal of Materials Science*, **36**, 3833–3854, http://dx.doi.org/10.1023/A:1017973432592

HUMPHREYS, F. J. & HATHERLY, M. 2004. *Recrystallization and Related Annealing Phenomena*. 3rd edn. Elsevier, Oxford.

JESSELL, M., BONS, P., EVANS, L., BARR, T. & STÜWE, K. 2001. Elle: the numerical simulation of metamorphic and deformation microstructures. *Computers and Geosciences*, **27**, 17–30, http://dx.doi.org/10.1016/S0098-3004(00)00061-3

JESSELL, M. W., KOSTENKO, O. & JAMTVEIT, B. 2003. The preservation potential of microstructures during static grain growth. *Journal of Metamorphic Geology*, **21**, 481–491, http://dx.doi.org/10.1046/j.1525-1314.2003.00455.x

JESSELL, M. W., SIEBERT, E., BONS, P. D., EVANS, L. & PIAZOLO, S. 2005. A new type of numerical experiment on the spatial and temporal patterns of localization of deformation in a material with a coupling of grain-size and rheology. *Earth and Planetary Science Letters*, **239**, 309–326, http://dx.doi.org/10.1016/j.epsl.2005.03.030

JORGE-BADIOLA, D., IZA-MENDIA, A. & GUTIERREZ, I. 2007. Evaluation of intragranular misorientation parameters measured by EBSD in a hot worked austenitic stainless steel. *Journal of Microscopy*, **228**, 373–383, http://dx.doi.org/10.1111/j.1365-2818.2007.01850.x

KOCKS, U. F. 1976. Laws for work-hardening and low-temperature creep. *Journal of Engineering Materials and Technology*, **98**, 76–85, http://dx.doi.org/10.1115/1.3443340

KOCKS, U. F. 1985. Dislocation interactions: flow stress and strain hardening. *In: Proceedings of the conference to celebrate the fiftieth anniversary of the concept of dislocation in crystals: Dislocations and Properties of Real Materials*. The Institute of Metals, London, 125–143.

KOCKS, U. F., TOMÉ, C. N. & WENK, H.-R. 1998. *Texture and Anisotropy: Preferred Orientations in Polycrystals and their Effect on Materials Properties*. Cambridge University Press.

LAW, R. D., SCHMID, S. M. & WHEELER, J. 1990. Simple shear deformation and quartz crystallographic fabrics: a possible natural example from the Torridon area of NW Scotland. *Journal of Structural Geology*, **12**, 29–45, http://dx.doi.org/10.1016/0191-8141(90)90046-2

LE GALL, R., LIAO, G. & SAINDRENAN, G. 1999. In-situ SEM studies of grain boundary migration during recrystallization of cold-rolled nickel. *Scripta Materialia*, **41**, 427–432, http://dx.doi.org/10.1016/S1359-6462(99)00109-8

LI, J. C. M. 1962. Possibility of subgrain rotation during recrystallization. *Journal of Applied Physics*, **33**, 2958–2965, http://dx.doi.org/10.1063/1.1728543

LI, H., HSU, E., SZPUNAR, J., UTSUNOMIYA, H. & SAKAI, T. 2008. Deformation mechanism and texture and microstructure evolution during high-speed rolling of AZ31B Mg sheets. *Journal of Materials Science*, **43**, 7148–7156, http://dx.doi.org/10.1007/s10853-008-3021-3

McQueen, H. J. & Evangelista, E. 1988. Substructures in aluminium from dynamic and static recovery. *Czechoslovak Journal of Physics B*, **38**, 359–372, http://dx.doi.org/10.1007/BF01605405

Miodownik, M. A. 2002. A review of microstructural computer models used to simulate grain growth and recrystallisation in aluminium alloys. *Journal of Light Metals*, **2**, 125–135, http://dx.doi.org/10.1016/S1471-5317(02)00039-1

Moldovan, D., Wolf, D. & Phillpot, S. R. 2001. Theory of diffusion-accommodated grain rotation in columnar polycrystalline microstructures. *Acta Materialia*, **49**, 3521–3532, http://dx.doi.org/10.1016/S1359-6454(01)00240-3

Moldovan, D., Wolf, D., Phillpot, S. R. & Haslam, A. J. 2002. Role of grain rotation during grain growth in a columnar microstructure by mesoscale simulation. *Acta Materialia*, **50**, 3397–3414, http://dx.doi.org/10.1016/S1359-6454(02)00153-2

Molli, G. & Heilbronner, R. 1999. Microstructures associated with static and dynamic recrystallization of Carrara marble (Alpi Apuane, NW Tuscany, Italy). *Geologie en Mijnbouw*, **78**, 119–126, http://dx.doi.org/10.1023/A:1003826904858

Molli, G., Conti, P., Giorgetti, G., Meccheri, M. & Oesterling, N. 2000. Microfabric study on the deformational and thermal history of the Alpi Apuane marbles (Carrara marbles), Italy. *Journal of Structural Geology*, **22**, 1809–1825, http://dx.doi.org/10.1016/S0191-8141(00)00086-9

Montagnat, M., Castelnau, O. *et al.* 2013. Multiscale modeling of ice deformation behavior. *Journal of Structural Geology*, in press, http://dx.doi.org/10.1016/j.jsg.2013.05.002

Nicolas, A. & Poirier, J. P. 1976. *Crystalline Plasticity and Solid State Flow in Metamorphic Rocks*. Wiley, New York.

Olsen, T. S. & Kohlstedt, D. L. 1984. Analysis of dislocations in some naturally deformed plagioclase feldspars. *Physics and Chemistry of Minerals*, **11**, 153–160, http://dx.doi.org/10.1007/BF00387845

Park, Y., Ree, J. H. & Kim, S. 2001. Lattice preferred orientation in deformed-then-annealed material: observations from experimental and natural polycrystalline aggregates. *International Journal of Earth Sciences*, **90**, 127–135, http://dx.doi.org/10.1007/s005310000163

Piazolo, S., Bons, P. D., Jessell, M. W., Evans, L. & Passchier, C. W. 2002. Dominance of microstructural processes and their effect on microstructural development: insights from numerical modelling of dynamic recrystallization. *In*: De Meer, S., Drury, M. R., De Bresser, J. H. P. & Pennock, G. (eds) *Deformation Mechanisms, Rheology and Tectonics: Current Status and Future Perspectives*. Geological Society, London, Special Publications, **200**, 149–170, http://dx.doi.org/10.1144/GSL.SP.2001.200.01.10

Piazolo, S., Jessell, M. W., Prior, D. J. & Bons, P. D. 2004. The integration of experimental in-situ EBSD observations and numerical simulations: a novel technique of microstructural process analysis. *Journal of Microscopy*, **213/3**, 273–284, http://dx.doi.org/10.1111/j.0022-2720.2004.01304.x

Piazolo, S., Sursaeva, V. G. & Prior, D. J. 2005. The influence of triple junction kinetics on the evolution of polycrystalline materials during normal grain growth: new evidence from in-situ experiments using columnar Al foil. *Zeitschrift für Metallkunde*, **96**, 1152–1157.

Piazolo, S., Bestmann, M., Prior, D. J. & Spiers, C. J. 2006. Temperature dependent grain boundary migration in deformed-then-annealed material: observations from experimentally deformed synthetic rocksalt. *Tectonophysics*, **427**, 55–71, http://dx.doi.org/10.1016/j.tecto.2006.06.007

Piazolo, S., Montagnat, M. & Blackford, J. R. 2008. Sub-structure characterization of experimentally and naturally deformed ice using cryo-EBSD. *Journal of Microscopy*, **230**, 509–519, http://dx.doi.org/10.1111/j.1365-2818.2008.02014.x

Piazolo, S., Jessell, M. W., Bons, P. D., Evans, L. & Becker, J. K. 2010. Numerical simulations of microstructures using the *Elle* platform: a modern research and teaching tool. *Journal of the Geological Society of India*, **75**, 110–127, http://dx.doi.org/10.1007/s12594-010-0028-6

Piazolo, S., Austrheim, H. & Whitehouse, M. 2012*a*. Brittle–ductile microfabrics in naturally deformed zircon: deformation mechanisms and consequences for U–Pb dating. *American Mineralogist*, **97**, 1544–1563, http://dx.doi.org/10.2138/am.2012.3966

Piazolo, S., Borthwick, V. E., Griera, A., Montagnat, M., Jessell, M. W., Lebensohn, R. & Evans, L. 2012*b*. Substructure dynamics in crystalline materials: new insight from in-situ experiments, detailed EBSD analysis of experimental and natural samples and numerical modeling. *Materials Science Forum*, **715–716**, 502–507, http://dx.doi.org/10.4028/www.scientific.net/MSF.715-716.502

Prior, D. J. 1999. Problems in determining the misorientation axes, for small angular misorientations, using electron backscatter diffraction in the SEM. *Journal of Microscopy*, **195**, 217–225, http://dx.doi.org/10.1046/j.1365-2818.1999.00572.x

Prior, D. J., Boyle, A. P. *et al.* 1999. The application of electron backscatter diffraction and orientation contrast imaging in the SEM to textural problems in rocks. *American Mineralogist*, **84**, 1741–1759.

Raabe, D. & Becker, R. C. 2000. Coupling of a crystal plasticity finite-element model with a probabilistic cellular automaton for simulating primary static recrystallization in aluminium. *Modelling Simulation in Materials Science and Engineering*, **8**, 445–462, http://dx.doi.org/10.1088/0965-0393/8/4/304

Ranalli, G. 1995. *Rheology of the Earth*, 2nd edn. Chapman and Hall, London.

Randle, V. 1993. Microtexture investigation of the relationship between strain and anomalous grain growth. *Philosophical Magazine A*, **67**, 1301–1313, http://dx.doi.org/10.1080/01418619308225356

Read, W. T. & Shockley, W. 1950. Dislocation models of crystal grain boundaries. *Physical Review*, **78**, 275–289.

Reddy, S. M., Timms, N. E., Pantleon, W. & Trimby, P. 2007. Quantitative characterization of plastic deformation of zircon and geological implications.

Contributions to Mineral Petrology, **153**, 625–645, http://dx.doi.org/10.1007/s00410-006-0174-4

REE, J. H. & PARK, Y. 1997. Static recovery and recrystallisation microstructures in sheared octachloropropane. *Journal of Structural Geology*, **19/12**, 1521–1526, http://dx.doi.org/10.1016/S0191-8141(97)00067-9

ROESSIGER, J., BONS, P. D. *ET AL.* 2011. Competition between grain growth and grain-size reduction in polar ice. *Journal of Glaciology*, **57**, 942–948, http://dx.doi.org/10.3189/002214311798043690

ROESSIGER, J., BONS, P. D. & FARIA, S. H. 2012. Influence of bubbles on grain growth in ice. *Journal of Structural Geology*, in press, http://dx.doi.org/10.1016/j.jsg.2012.11.003

SELLARS, C. M. 1978. Recrystallization of metals during hot deformation. *Philosophical Transactions of the Royal Society of London A*, **288**, 147–158, http://dx.doi.org/10.1098/rsta.1978.0010

SEWARD, G. G. E., PRIOR, D. J., WHEELER, J., CELOTTO, S., HALLIDAY, D. J. M., PADEN, R. S. & TYE, M. R. 2002. High-temperature electron backscatter diffraction and scanning electron microscopy imaging techniques: in-situ investigations of dynamic processes. *Scanning*, **24**, 232–240, http://dx.doi.org/10.1002/sca.4950240503

SREEKALA, S. & HAATAJA, M. 2007. Recrystallization kinetics: a coupled coarse-grained dislocation density and phase-field approach. *Physical Review B*, **76**, http://dx.doi.org/10.1103/PhysRevB.76.094109

STOECKERT, B. & DUYSTER, B. 1999. Discontinuous grain growth in recrystallized vein quartz – implications for grain boundary structure, grain boundary mobility, crystallographic preferred orientation, and stress history. *Journal of Structural Geology*, **21**, 1477–1490, http://dx.doi.org/10.1016/S0191-8141(99)00084-X

SVAHNBERG, H. & PIAZOLO, S. 2010. The initiation of strain localisation in plagioclase-rich rocks: insights from detailed microstructural analyses. *Journal of Structural Geology*, **32**, 1404–1416, http://dx.doi.org/10.1016/j.jsg.2010.06.011

TAKAYAMA, Y., JERZY, A., SZPUNAR, J. A. & HAJIME KATO, H. 2005. Analysis of intragranular misorientation related to deformation in an Al–Mg–Mn alloy. *Materials Science Forum*, **495–497**, 1049–1054, http://dx.doi.org/10.4028/www.scientific.net/MSF.495-497.1049

TRIMBY, P. W., DRURY, M. R. & SPIERS, C. J. 2000. Misorientations across etched boundaries in deformed rocksalt: a study using electron backscatter diffraction. *Journal of Structural Geology*, **22**, 81–89, http://dx.doi.org/10.1016/S0191-8141(99)00126-1

TURNER, F. J. & WEISS, L. E. 1963. *Structural Analysis of Metamorphic Tectonites*. McGraw-Hill, New York.

URAI, J. L., MEANS, W. D. & LISTER, G. S. 1986. Dynamic recrystallization of minerals. *In*: HOBBS, B. E. & HEARD, H. C. (eds) *Mineral and Rock Deformation: Laboratory Studies (the Paterson Volume)*. Geophysical Monograph Series of the American Geophysical Union, **36**, 161–200, http://dx.doi.org/10.1029/GM036p0161

VANDERMEER, R. A. & JUUL JENSEN, D. 1998. The migration of high angle grain boundaries during recrystallization. *Interface Science*, **6**, 95–104, http://dx.doi.org/10.1023/A:1008668604733

VENABLES, J. A. & HARLAND, C. J. 1973. Electron backscattering patterns – a new technique for obtaining crystallographic information in the scanning electron microscope. *Philosophical Magazine*, **27**, 1193–1200, http://dx.doi.org/10.1080/14786437308225827

WAKAHAMA, G. 1964. On the plastic deformation of ice. V. Plastic deformation of polycrystalline ice. *Low Temperature Science Series A*, **22**, 1–24.

WEISS, L. E. & WENK, H. R. 1985. *Preferred Orientation in Deformed Metals and Rocks: An Introduction to Modern Texture Analysis*. Academic Press, London.

WEYGAND, D., BRÉCHET, Y. & LÉPINOUX, J. 2001. Mechanics and kinetics if recrystallisation: a two dimensional vertex dynamics simulation. *Interface Science*, **9**, 311–317, http://dx.doi.org/10.1023/A:1015175231826

WHEELER, J., PRIOR, D. J., JIANG, Z., SPIESS, R. & TRIMBY, P. W. 2001. The petrological significance of misorientations between grains. *Contributions to Mineral Petrology*, **141**, 109–124, http://dx.doi.org/10.1007/s004100000225

WHEELER, J., MARIANI, E., PIAZOLO, S., PRIOR, D. J., TRIMBY, P. & DRURY, M. R. 2009. The Weighted Burgers Vector: a new quantity for constraining dislocation densities and types using electron backscatter diffraction on 2D sections through crystalline materials. *Journal of Microscopy*, **233**, 482–494, http://dx.doi.org/10.1111/j.1365-2818.2009.03136.x

WILSON, C. 1982. Texture and grain growth during the annealing of ice. *Textures and Microstructures*, **5**, 19–31, http://dx.doi.org/10.1155/TSM.5.19

ZHANG, Y. B., GODFREY, A., LIU, Q., LIU, W. & JUUL JENSEN, D. 2009. Analysis of the growth of individual grains during recrystallization in pure nickel. *Acta Materialia*, **57**, 2631–2639, http://dx.doi.org/10.1016/j.actamat.2009.01.039

Index

Page numbers in *italic* denote Figures. Page numbers in **bold** denote Tables.

A000022893680